MEDICINAL CHEMISTRY WITH PHARMACEUTICAL PRODUCT DEVELOPMENT

AAP Research Notes on Chemistry

MEDICINAL CHEMISTRY WITH PHARMACEUTICAL PRODUCT DEVELOPMENT

Edited by

Debarshi Kar Mahapatra, PhD
Sanjay Kumar Bharti, PhD

Apple Academic Press Inc.
3333 Mistwell Crescent
Oakville, ON L6L 0A2 Canada

Apple Academic Press Inc.
1265 Goldenrod Circle NE
Palm Bay, Florida 32905 USA

© 2019 by Apple Academic Press, Inc.
Exclusive worldwide distribution by CRC Press, a member of Taylor & Francis Group
No claim to original U.S. Government works
International Standard Book Number-13: 978-1-77188-710-6 (Hardcover)
International Standard Book Number-13: 978-1-77463-409-7 (Paperback)
International Standard Book Number-13: 978-0-42948-784-2 (eBook)

Library and Archives Canada Cataloguing in Publication

Medicinal chemistry with pharmaceutical product development / edited by
Debarshi Kar Mahapatra, PhD, Sanjay Kumar Bharti, PhD.

(AAP research notes on chemistry)
Includes bibliographical references and index.
Issued in print and electronic formats.
ISBN 978-1-77188-710-6 (hardcover).--ISBN 978-0-429-48784-2 (PDF)

1. Drug development. 2. Pharmaceutical chemistry. I. Mahapatra, Debarshi Kar, editor
II. Bharti, Sanjay Kumar, editor III. Series: AAP research notes on chemical engineering

| RM301.25.M43 2018 | 615.1'9 | C2018-906021-2 | C2018-906022-0 |

Library of Congress Cataloging-in-Publication Data

Names: Mahapatra, Debarshi Kar, editor. | Bharti, Sanjay Kumar, editor.

Title: Medicinal chemistry with pharmaceutical product development / editors,
Debarshi Kar Mahapatra, Sanjay Kumar Bharti.

Description: Toronto ; New Jersey : Apple Academic Press, 2019. | Series: AAP research notes on
chemistry | Includes bibliographical references and index.

Identifiers: LCCN 2018049885 (print) | LCCN 2018050211 (ebook) |
ISBN 9780429487842 (ebook) | ISBN 9781771887106 (hardcover)

Subjects: | MESH: Chemistry, Pharmaceutical

Classification: LCC RM301.25 (ebook) | LCC RM301.25 (print) | NLM QV 744 |
DDC 615.1/9--dc23

LC record available at https://lccn.loc.gov/2018049885

Apple Academic Press also publishes its books in a variety of electronic formats. Some content that appears in print may not be available in electronic format. For information about Apple Academic Press products, visit our website at **www.appleacademicpress.com** and the CRC Press website at **www.crcpress.com**

AAP RESEARCH NOTES ON CHEMISTRY

This series reports on research developments and advances in the ever-changing and evolving field of chemistry for academic institutes and industrial sectors interested in advanced research books.

Richard A. Pethrick, PhD, DSc
Research Professor and Professor Emeritus, Department of Pure and Applied Chemistry, University of Strathclyde, Glasgow, Scotland, UK

Charles Wilkie, PhD
Professor, Polymer and Organic Chemistry, Marquette University, Milwaukee, Wisconsin, USA

Georges Geuskens, PhD
Professor Emeritus, Department of Chemistry and Polymers, Universite de Libre de Brussel, Belgium

BOOKS IN THE AAP RESEARCH NOTES ON CHEMISTRY SERIES

Chemistry and Chemical Biology: Methodologies and Applications
Editors: Roman Joswik, PhD, and Andrei A. Dalinkevich, DSc
Reviewers and Advisory Board Members: A. K. Haghi, PhD, and
Gennady E. Zaikov, DSc

Functional Materials: Properties, Performance, and Evaluation
Editor: Ewa Kłodzińska, PhD
Reviewers and Advisory Board Members: A. K. Haghi, PhD, and
Gennady E. Zaikov, DSc

High Performance Elastomer Materials: An Engineering Approach
Editors: Editors: Dariusz M. Bielinski, DSc, Ryszard Kozlowski, PhD,
and Gennady E. Zaikov, DSc

**Chemical Analysis: Modern Materials Evaluation and
Testing Methods**
Editors: Ana Cristina Faria Ribeiro, PhD, Cecilia I. A. V. Santos, PhD,
and Gennady E. Zaikov, DSc

Medicinal Chemistry with Pharmaceutical Product Development
Editors: Debarshi Kar Mahapatra, PhD, and Sanjay Kumar Bharti, PhD

CONTENTS

ABOUT THE EDITORS

Debarshi Kar Mahapatra, PhD
*Assistant Professor, Department of Pharmaceutical Chemistry,
Dadasaheb Balpande College of Pharmacy, Rashtrasant Tukadoji
Maharaj Nagpur University, Nagpur, Maharashtra, India*

Debarshi Kar Mahapatra, PhD, is currently Assistant Professor, Department of Pharmaceutical Chemistry, Dadasaheb Balpande College of Pharmacy, Rashtrasant Tukadoji Maharaj Nagpur University, Nagpur, Maharashtra, India. He was formerly Assistant Professor in the Department of Pharmaceutical Chemistry, Kamla Nehru College of Pharmacy, RTM Nagpur University, Nagpur, India. He taught medicinal and computational chemistry at both the undergraduate and postgraduate levels and has mentored students in their various research projects. His area of interest includes computer-assisted rational designing and synthesis of low molecular weight ligands against druggable targets, drug delivery systems, and optimization of unconventional formulations. He has published research, book chapters, reviews, and case studies in various reputed journals and has presented his work at several international platforms, for which he received several awards by a number of bodies. He has also authored the book titled *Drug Design*. Presently, he is serving as reviewer and editorial board member for several journals of international repute. He is a member of a number of professional and scientific societies, such as the International Society for Infectious Diseases (ISID), the International Science Congress Association (ISCA), and ISEI.

Sanjay Kumar Bharti, PhD
*Assistant Professor, Institute of Pharmaceutical Sciences, Guru Ghasidas
Vishwavidyalaya (A Central University), Bilaspur, India*

Sanjay Kumar Bharti, PhD, is an Assistant Professor at the Institute of Pharmaceutical Sciences, Guru Ghasidas Vishwavidyalaya (A Central University), Bilaspur, India. He has working experience in several organizations, such as Win-Medicare Pvt. Ltd, Meerut, India (as a chemist) and the National Institute of Pharmaceutical Education and Research (NIPER), Hajipur (as a lecturer). He has published several research papers, one book, several book chapters, and review articles in various reputed journals. His research interest includes synthesis of Schiff's base, heterocyclic compounds, and metallopharmaceuticals for therapeutics. He is an active member of various scientific and pharmaceutical organizations, such as IPA, IPGA, ISCA, etc. Dr. Bharti has completed a BPharm from IT-BHU, Varanasi, India, in 2003; an MPharm from RGPV, Bhopal, India, in 2004; and a PhD from IIT-BHU, Varanasi, India, in 2011.

CONTRIBUTORS

Vivek Asati
Department of Pharmaceutical Chemistry, NRI Institute of Pharmaceutical Sciences, Bhopal 462021, Madhya Pradesh, India

Sanjay Kumar Bharti
Institute of Pharmaceutical Sciences, Guru Ghasidas Vishwavidyalaya (A Central University), Bilaspur–495009, Chhattisgarh, India, E-mail: skbharti.ggu@gmail.com

Manik Das
Department of Pharmacy, Tripura University (A Central University), Suryamaninagar–799022, Tripura, India

Ashwini Deshpande
SVKM's NMIMS, School of Pharmacy & Technology Management, Shirpur, Dhule, Maharashtra, India, E-mail: Ashwini.deshpande@nmims.edu

Mayuresh S. Garud
Shobhaben Pratapbhai Patel School of Pharmacy and Technology Management, SVKM's NMIMS, Vile Parle (West), Mumbai – 400056, India

R. S. Gaud
Shobhaben Pratapbhai Patel School of Pharmacy and Technology Management, SVKM's NMIMS, Vile Parle (West), Mumbai – 400056, India

Sayan Dutta Gupta
Department of Pharmaceutical Chemistry, Gokaraju Rangaraju College of Pharmacy, Hyderabad, Telangana, India, E-mail: sayandg@rediffmail.com

Abhay Ittadwar
Department of Pharmaceutics, Gurunanak College of Pharmacy, Rashtrasant Tukadoji Maharaj Nagpur University, Nagpur, Maharashtra, India

Urmila Jarouliya
School of Studies in Biotechnology, Jiwaji University, Gwalior (M.P.), 474011, India

Gunjan Jeswani
Department of Pharmaceutics, Faculty of Pharmaceutical Sciences, Shri Shankaracharya Group of Institutions, SSTC, Bhilai, Chhattisgarh, India

Raj K. Keservani
School of Pharmaceutical Sciences, Rajiv Gandhi Proudyogiki Vishwavidyalaya, Bhopal (M.P.), 462036, India, 462036, India, Tel. +9178978-03904, E-mail: rajksops@gmail.com

Vaishali Kilor
Department of Pharmaceutics, Gurunanak College of Pharmacy, Rashtrasant Tukadoji Maharaj Nagpur University, Nagpur, Maharashtra, India

Yogesh A. Kulkarni
Shobhaben Pratapbhai Patel School of Pharmacy and Technology Management,
SVKM's NMIMS, Vile Parle (West), Mumbai 400056, India,
Tel. +91-22-42332000, Fax: +91-22-26185422, E-mail: yogeshkulkarni101@yahoo.com

Debarshi Kar Mahapatra
Department of Pharmaceutical Chemistry, Dadasaheb Balpande College of Pharmacy,
Nagpur 440037, Maharashtra, India

Kuntal Manna
Department of Pharmacy, Tripura University (A Central University), Suryamaninagar–799022,
Tripura, India

Parveen Parasar
Division of Surgery, Department of Gynecology, Children's Hospital Boston,
Harvard Medical School, Boston, Massachusetts, 02115, USA,
Tel. +(1)617-637-3144, E-mail: suprovet@gmail.com

Tulshidas S. Patil
SVKM's NMIMS, School of Pharmacy & Technology Management, Shirpur, Dist. Dhule,
Maharashtra, India, E-mail: tulshidaspatil01@gmail.com

Swarnali Das Paul
Department of Pharmaceutics, Faculty of Pharmaceutical Sciences,
Shri Shankaracharya Group of Institutions, SSTC, Bhilai, Chhattisgarh, India,
Tel. +91-9977258200, E-mail: swarnali34@gmail.com, swarnali4u@rediffmail.com

Vivek Singh
Department of Veterinary Physiology and Biochemistry, Junagadh Agricultural University,
Junagadh, Gujarat, 362001, India, Tel. +(91)-8306857587, E-mail: suprovet@gmail.com

Alok Ubgade
Department of Pharmaceutics, Gurunanak College of Pharmacy,
Rashtrasant Tukadoji Maharaj Nagpur University, Nagpur, Maharashtra, India

Shobha Ubgade
Department of Pharmaceutics, Gurunanak College of Pharmacy,
Rashtrasant Tukadoji Maharaj Nagpur University, Nagpur, Maharashtra, India,
Tel. +91-97634-03953, E-mail: shobha_yadav1402@yahoo.co.in

ABBREVIATIONS

AADC	amino acid decarboxylase
AAV	adeno-associated virus
ACS	acute coronary syndromes
ACT	activated clotting time
AD	Alzheimer's disease
ADA	American Diabetes Association
ADP	adenosine diphosphate
AFM	atomic force microscopy
AFP	alpha fetoprotein
ALC	aminoclay-lipid hybrid composite
ALK	anaplastic lymphoma kinase
AMAD	azide methyl anthraquinone derivative
AMTA	anti-mitochondrial antibodies
ANA	anti-nuclear antibodies
AP	alkaline phosphatase
API	active pharmaceutical ingredient
APTT	activated partial thromboplastin time
AR	adenosine receptor
AR	amphiregulin
ASMA	anti-smooth muscle antibodies
ASR	analyte specific reagent
AT	antithrombin
ATIII	antithrombin III
AUC	area under curve
BCS	biopharmaceutics classification system
BDNF	brain-derived neurotrophic factor
BHA	butylated hydroxyl anisole
BHT	butylated hydroxyl toluene
BME	β-mercaptoethanol

CAD	coronary artery disease
CAGR	compound annual growth rate
CFA	circulating filariasis antigen
CMC	critical micelle concentration
CMIA	chemiluminescent microparticle immunoassay
CPP	critical packing parameter
CREB	cAMP response element binding protein
CS	chitosan
CSII	continuous subcutaneous insulin infusion
DA	dopamine
DAR	dopamine receptor
DAT	dopamine transporter
DCP	dicetyl phosphate
DE	definitive endoderm
DLB	dementia with Lewy bodies
DM	diabetes mellitus
DSC	differential scanning calorimetry
DVT	deep vein thrombosis
ECT	ecarin clotting time
EDTA	ethylene diamine tetra-acetic acid
EFM	electrical field gradient microscopy
EGF	epidermal growth factor
EGFR	epidermal growth factor receptor
EPR	enhanced permeation retention
ERK	extracellular signal-regulated kinase
FDA	food and drug administration
FFPE	formalin-fixed paraffin-embedded
FFPET	formalin-fixed paraffin-embedded tumor tissue
FGF	fibroblast growth factor
FHIC	Fisetin hydroxypropyl β-cyclodextrin
FISH	fluorescence in situ hybridization
FST	Fisetin
FTIR	Fourier transform infrared spectroscopy

GA	geldanamycin
GABA	gamma-aminobutyric acid
GI	gastrointestinal
GIP	glucose insulinotrophic peptide
GMS	glycerylmonostearate
GP	glycoprotein
GPCR	g-protein-coupled receptor
GT	gut tube
GVHD	graft versus-host disease
GWAS	genome wide analysis study
HGF	hepatocyte growth factor
HIC	hydrophobic interaction
HIT	heparin-induced thrombocytopenia
HIV	human immunodeficiency virus
HMS	hexagonal mesoporous silica
HMWK	high-molecular-weight kininogen
HPH	high-pressure homogenization
HPMC	hydroxypropyl methyl cellulose
HRI	heme-regulated eIF2a kinase
HRPII	histidine-rich protein II
HSPC	hydrogenated soyphosphatidylcholine
HTS	high throughput screening
ICM	inner cell mass
IFD	induced fit docking
IHC	immunohistochemical
IL	interleukins
IR	insulin receptor
IRAK	interleukin-1 receptor-associated kinase
ISL	isoliquiritigenin
IU	intended use
IVD	*in vitro* diagnostic
LA	lupus anticoagulants
LAR	leukocyte antigen-related tyrosine phosphatase

LD	laser diffractometry
LFCS	lipid formulation classification system
LMWH	low molecular weight heparin
MAC	mycobacterium avium complex
MAP	mitogen-activated protein
MAPK	mitogen-activated protein kinase
MC	mesoporous carbon
MD	molecular dynamic
MDMA	methylenedioxymethamphetamine
ME	microemulsion
MPS	mononuclear phagocyte system
MR	mannose receptor
MRP	multidrug resistance associated protein
MRT	mean residence time
MSN	mesoporous silica nanoparticles
MW	molecular weight
NCC	N-carboxymethyl chitosan
NCS	nanotoxicological classification system
NDA	new drug application
NMDAR	n-methyl-d-aspartate receptor
NMR	nuclear magnetic resonance
NNI	National Nanotechnology Initiative
NR	nuclear receptor
NSCLC	non-small cell lung cancer
OGTT	oral glucose tolerance test
OLM	Olmesartan medoxomil
ONR	orphan nuclear receptor
PAMP	pathogen-associated molecular patterns
PC	phosphatidyl choline
PCI	percutaneous coronary intervention
PCR	polymerase chain reaction
PCS	photon correlation spectroscopy
PD	Parkinson's disease

PDB	protein data bank
PDE	phosphodiesterases
PDGF	platelet-derived growth factor
PDGFR	platelet-derived growth factor receptor
PDI	Polydespersity index
PE	pulmonary embolism
PEG	polyethylene glycols
PG	phosphatidyl glycerol
PI	phosphatidyl inositol
PI	polydispersity index
PKA	protein kinase A
PKC	protein kinase C
PLC	phospholipase C
PLH	precipitation-lyophilization-homogenization
PP	pancreatic progenitors
PPAR	peroxisome proliferator activated receptor
PPI	polypropylene imine
PPPMM	pluronic-phosphatidylcholinepolysorbate 80 mixed micelles
PS	phosphatidyl serine
PT	prothrombin time
PTB	phosphotyrosine-binding
PTCA	percutaneous transluminal coronary angioplasty
PVD	peripheral vascular disease
RD	radicicol
RE	response element
RESAS	rapid expansion from supercritical to aqueous solutions
RESS	rapid expansion from supercritical solutions
RPB	rotating packed bed
RPR	rapid plasma reagin
SA	stearylamine
SAP	sensor-augmented pump
SAS	supercritical anti-solvent method
SCF	supercritical fluid

SEM	scanning electron microscopy
SFN	supercritical fluid nucleation
SLE	systemic lupus erythematosus
SLS	sodium laurylsulfate
SPI	soybean protein isolate
STAT	signal transducers and activators of transcription
STM	scanning tunneling microscope
TB	tuberculosis
TEM	transmission electron microscopy
TF	tissue factor
TFA	trans-ferulic acid
TGA	therapeutic goods administration
TGF	transforming growth factor
TPR	tetratroicopeptide repeat
TS	threshold suspends
UFH	unfractionated heparin
VEGF	vascular endothelial growth factor
VEGFR	vascular endothelial growth factor receptor
VKA	vitamin K antagonist
VTE	venous thrombo embolisms
WPI	whey protein isolate
XRD	X-ray diffraction
XRPD	X-ray powder diffraction

PREFACE

In recent years, drug resistance emerged as a global health challenge for people around the world. Antibiotic-resistant microorganisms, ineffectiveness of traditional chemotherapeutics, tolerance and/or toxicity associated with current chemotherapeutics, etc., necessitate the discovery and development of novel and more effective and selective chemotherapeutics. Medicinal chemistry is considered as the mother of all subjects related to pharmaceuticals.

This volume is prepared with the aim to keep readers abreast of knowledge related to current theoretical and practical aspects of pharmaceuticals for the discovery and development of novel therapeutics for current health problems. The book, *Medicinal Chemistry with Pharmaceutical Products Development,* is comprised of chapters that are quite up-to-date, complete, written in easy-to-understand language, and concise. The book includes the therapeutic regulations of the USP (*United States Pharmacopeia*) along with all the latest therapeutic guidelines put forward by WHO (World Hunger Organization) and the U.S. Food and Drug Administration.

The book is primarily focused on novel therapeutics and strategies for the development of pharmaceutical products, keeping drug molecule as the central component. It aims to explain the necessary features essential for pharmacological activity. Adopting a user-friendly format of studying (introduction, classification, mechanism of action, tools and techniques, synthesis, applications, etc.) makes this book unique compared to other already published books. The book takes an interdisciplinary approach by including a unique combination of pharmacy, chemistry, and medicine along with clinical aspects. It is designed to fulfill the theoretical and practical demands of students and professionals.

The book provides a broad exposure to the essentials of pharmaceutical chemistry. We hope that after reading the contents of this book, readers will definitely find the difference and may be able to develop novel schemes by themselves. Specifically, the protein as drug targets, PTP-1B inhibitors for the treatment of diabetes, Hsp90 inhibitors as anticancer agents, natural products as anti-glycating agents, nanomedicine, nanocarriers, diagnostic devices, stem cell research and therapy have been focused on in detail. The figures, illustrations, flowcharts, and diagrams embedded in the text are helpful for explaining the content in a few words. The high level of illustrations and important facts, figures, flowcharts, and diagrams relevant to the content in the simplest and understandable form is designed to be beneficial to students, faculty, and industry professionals.

—Debarshi Kar Mahapatra, PhD
Sanjay Kumar Bharti, PhD

FOREWORD

It is my pleasure to write the foreword to this book, *Medicinal Chemistry with Pharmaceutical Product Development*, edited by Debarshi Kar Mahapatra and Sanjay K. Bharti. This book focuses on novel therapeutics and strategies for the treatment and/or management for various human ailments. I am delighted to read the contents of this book, which encompasses various principles and applications associated with medicinal chemistry and pharmaceutical product development. As the title implies, this book integrates insights of drug discovery and development through various pharmacological aspects of products, druggable targets of drug candidates, and approaches including nanomedicines. The unprecedented pace observed in the field of drug design and development in the recent years has emphasized the relevance of pharmaceutical technology. With the development of understanding in scientific thoughts and practices, this subject is no longer restricted to any boundaries. This book is prepared with an aim to stay abreast of the latest advancements in the various fields of pharmaceutical product development. The book is comprised of nine well-written chapters by various reputed authors across the globe. The usefulness of this book relies on the contents written by expert authors in their respective fields and the incorporation of contemporary relevant literature. This book contains several examples, illustrations, diagrams, and figures that are presented in such a manner that even an average reader would be able to understand the contents. I am confident that the book will be useful for students, researchers, scientists, and teachers working in cutting-edge research in pharmaceutical technology.

—G. S. Singh, PhD, FISCA
Professor
Chemistry Department
University of Botswana
Gaborone, Botswana

CHAPTER 1

PROTEIN FUNCTION AS CELL SURFACE AND NUCLEAR RECEPTOR IN HUMAN DISEASES

URMILA JAROULIYA[1] and RAJ K. KESERVANI[2]

[1]School of Studies in Biotechnology, Jiwaji University, Gwalior (M.P.), 474011, India

[2]School of Pharmaceutical Sciences, Rajiv Gandhi Proudyogiki Vishwavidyalaya, Bhopal (M.P.), 462036, India, Mobile: +91-7897803904, E-mail: rajksops@gmail.com

ABSTRACT

Protein is the major component of all cells in the body and plays numbers of functions in the biological world, from catalyzing chemical reactions to build the structures of all living things. Cell receptors are made up of protein and they play a major role in signal transduction. Out of the many cell receptors, the well-known G-protein-coupled receptors (GPCRs) represent the most important targets in modern pharmacology because of the different functions they mediate, especially within the brain and peripheral nervous system. Other GPCRs like β1-adrenergic receptor (β1AR) that plays a paramount role in chronic heart failure, 5-HT$_4$ receptor and acetylcholine receptor (AChR) has role in the treatment of Alzheimer's disease (AD), N-methyl-D-aspartate receptor (NMDAR) is a stimuli for neuroautoimmune disorder, dopamine receptor (DAR) improves the stimulants in the treatment of Parkinson's disease (PD), cannabinoid (CB1 and CB2) receptor in the brain, and are involved in a variety of physiological processes including appetite, pain-sensation, mood, and memory. Adenosine receptor (AR) on erythrocytes reduces the HbS oxygen affinity and promotes its polymerization and red blood cell sickling. Chemokine receptors are cytokine receptor that interacts with the leukocyte cells on the surface. Another category of receptor involves receptor tyrosine kinases (RTKs) are the high-affinity cell surface receptors for many polypeptide growth factors, cytokines, and hormones, that phosphorylates the cascades of the various signaling pathways. The transforming growth factor-β (TGF-β) system signals via protein kinase receptors and regulates the biological processes, including morphogenesis, embryonic development, adult stem cells differentiation, immune regulation, wound healing, or inflammation. Signaling through platelet-derived growth factor (PDGF) receptors contribute to multiple tumors associated processes. The ErbB/HER protein-tyrosine kinases, which include the epidermal growth factor receptor, are associated with the development of neurodegenerative diseases, such as multiple sclerosis and Alzheimer's disease. Insulin receptor (IR) plays a key role in glucose homeostasis. Among other receptors in plants, one of them is receptor-like kinases (RLKs) which control a wide range of processes, including development, disease resistance, hormone perception, and self-incompatibility in plants. Toll-like receptors (TLRs) are non-catalytic receptors expressed in sentinel cells and play a key role in the innate immune response to invading pathogens. Apart from cell surface receptors, there is a nuclear receptor (NR) that regulates transcription in response to small lipophilic compounds and plays a role in every aspect of development, physiology, and disease in humans. Nuclear receptor (NR) includes estrogen (ERNR), progesterone (PRNR), androgen (ARNR), Vitamin D (VDRNR), and thyroid hormone (TRNR) receptors. In this chapter, we focus mainly on a signaling pathway of various receptors and their significance in various human ailments.

1.1   INTRODUCTION

The receptors are the protein molecule, usually found in the cells' plasma membrane; facing extracellular (cell surface receptors), cytoplasmic (cytoplasmic receptors), or in the nucleus (nuclear receptors). The chemical signals such as hormones, cytokines, growth factors, enzyme or any ligand bind to a receptor that is triggering changes in the function of the cellular activity by means of various signaling pathways, this process is called signal transduction. There are several receptor components present in the cell and can be classified into the following categories:

- G protein-coupled receptors: also known as seven transmembrane G-protein receptor that includes the receptors for several hormones and slow transmitters
- Receptor tyrosine kinase and enzyme-linked receptor: plasma membrane receptors that are also enzymes. When one of these receptors is activated by its extracellular ligand, catalyzes the production of an intracellular second messenger ex: the insulin receptor
- Nuclear receptors: When binding to their specific ligand (such as the hormone), alter the rate at which specific genes are transcribed and translated into cellular proteins. Steroid and thyroid hormone receptors are examples of such receptors.
- Ionotropic receptors: these are gated-ion channels in the plasma membrane that open and close in response to the binding of chemical ligands such as acetylcholine (nicotinic) and gamma-aminobutyric acid (GABA) and activation of these receptors, results in changes in ion movement across the membrane.

In this chapter, we focus on G protein-coupled receptor and the receptors involved in human diseases such as β_1-adrenergic receptor, acetylcholine receptor, dopamine receptor, growth receptor, plant receptor-like kinase, toll-like receptors, and the nuclear receptors.

1.2   G-PROTEIN-COUPLED RECEPTORS (GPCRS)

It is a cell surface receptor (membrane receptors or transmembrane receptors) that is made up of specialized integral membrane proteins that takes part in communication between the cell and the extracellular environment of the cell. It is a receptor with seven transmembrane helical segments and distributed across nearly all of the body's organs and tissues. In the cell, GPCRs plays as a key role in signal transducers that makes GPCRs key regulatory elements in a broad range of normal and pathological processes [1]. In addition to the cell surface, GPCRs are present in the endoplasmic reticulum, Golgi apparatus, nuclear membrane and even inside the nucleus itself [2–4]. The variety of endogenous ligands (stimuli) activate GPCRs including biogenic amines, neuropeptides, amino acids, ions, hormones, chemokines, lipid-derived mediators, proteases peptides, proteins [5], protons (H^+) and ions (Ca^{2+}) [6]. GPCRs lacks internal enzymatic activity and are coupled to heterotrimeric guanosine nucleotide–binding protein (G protein), which consist of Gα, Gβ and Gγ subunits. The binding of ligand(s) stabilizes the occupied GPCR in an active signaling conformation during which the heterotrimeric G-proteins dissociate in GTP-bound Gα and Gβγ subunits (Figure 1.1). These regulate the activity of several enzymes such as adenylate cyclase, phospholipase C (PLC) isoforms and kinases, resulting in the generation of intracellular second messengers that control cellular functions, which are responsible for triggering different signaling responses [7].

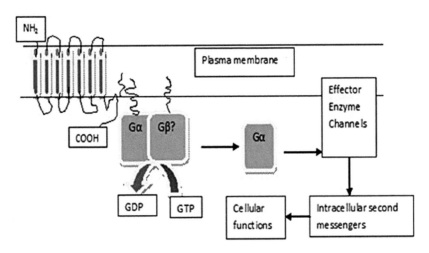

FIGURE 1.1   (See color insert.) G-protein-coupled receptor (GPCR).

Phosphorylation and desensitization of the GPCRs bound signaling molecule were executed by G protein-coupled receptor kinases (GRKs) that are truly recognized as the kinases. GPCR bound signaling molecules are phosphorylated by GRKs that leads to the translocation and binding of β-arrestins to the receptors, which further inhibits G protein activation by blocking receptor-G protein coupling [8, 9]. This process also promotes signaling molecules bound GPCR internalization [10–12] Thus, the GRK catalyzed phosphorylation and binding of β-arrestin to the receptors are believed to be the common mechanism of GPCR desensitization [13, 14].

GPCR desensitization is important for maintaining homeostasis in the cell, as the impairment of the desensitization process could cause various diseases such as heart failure [15, 16], asthma [17], Parkinson's disease [18], and autoimmune disease [19]. Thus, GRKs play an essential role in maintaining cells and tissues in normal states. Here we, discuss the GPCRs malfunctioning in concern to various human diseases such as cardiac disease, Alzheimer's disease, autoimmune, and Parkinson's disease.

1.2.1 ADRENERGIC RECEPTORS AND HEART FAILURE

It is a GPCR receptor. There are four forms of adrenergic receptors they are α_1, α_2, β_1, and β_2 defined by minute differences in their affinities and responses to a group of agonists and antagonists. Here we focus on the β_1-adrenergic receptors (β_1AR) that appear to be a common target of several agonistic autoantibody-diseases that lead to chronic heart failure. The β1AR is the most-abundant in the human heart, approaching 75% of the total number of receptors [20]. The β_1AR are coupled primarily to the heterotrimeric G protein (Gs), to stimulate adenylyl cyclase activity. This association generates intracellular cyclic adenosine monophosphate (cAMP) and protein kinase A (PKA) activation, which regulate cardiac contractility and heart rate [21]. The activity of cAMP/PKA is kept in check by the activity of cAMP phosphodiesterases (PDE) and protein phosphatases [22–25], and the activation of these enzymes leads to the breakdown of cAMP and inactivation of PKA. During stress hormones released such as epinephrine (catecholamine) or the adrenaline, which binds to β_1AR and induce signaling by the activity of cAMP/PKA pathways that increases the cardiac activity and results in heart failure so the GRKs activity in the heart appears to play a critical role, especially in heart failure. Heart failure following chronic β_1AR stimulation is typically associated with a reduction of β_1AR-mediated regulatory amplitude due to an increase in baseline cAMP and a decrease in maximal chronotropic catecholamine response [26]. In general, the current management of heart failure includes the use of angiotensin-converting enzyme inhibitors or angiotensin receptor blockers, combined with β_1-adrenergic receptor (β_1AR) blockers, diuretics, aldosterone antagonists, digitalis, and nitrates. Already there exist a variety of specific therapy options for β_1AR autoantibody-mediated heart disease. These include non-selective IgG and IgG3, which are largely established therapy options [27].

1.2.2 NICOTINIC ACETYLCHOLINE RECEPTOR (NACHRS) AND ALZHEIMER'S DISEASE

It is ligand-gated ion channels that are widely distributed in the human brain where they have a modulatory function associated with numerous transmitter systems. The receptor channel opens in response to the neurotransmitter acetylcholine (and to nicotine). This receptor is found in the postsynaptic membrane of neurons at certain synapses and in the muscle fibers (myocytes) at neuromuscular junctions. Acetylcholine released by an excited neuron diffuses a few micrometers across the synaptic cleft or the neuromuscular junction to the postsynaptic neuron or myocyte, where it interacts with the acetylcholine receptor and triggers electrical excitation (depolarization) of the receiving cell. Reductions in nAChR density have been identified in a number of neurodegenerative disorders, including Alzheimer's disease (AD), dementia with Lewy bodies (DLB), and Parkinson's disease (PD) [28–30].

$$CH_3$$
$$|$$
$$CH_3\text{-}N\text{-}CH_2CH_2O\text{-}C\text{-}CH_3$$
$$|$$
$$CH_3$$

Acetylcholine (Ach)

Alzheimer's disease (AD) is one of the neurodegenerative diseases presenting with dementia, and there are no definitive treatments or prophylactic agents. The presence of two types of abnormal deposits, senile plaques, and neurofibrillary tangles, and extensive neuronal loss characterize the pathology of AD [31]. AD is clinically characterized by memory loss and cognitive functions, including deterioration of language as well as defects in visual and motor coordination, and eventual death [32]. Cholinergic abnormalities have been also observed in AD brains [33]. It has been reported that the protein level of acetylcholine receptors is reduced in AD

[34], and that dysfunction of cholinergic signal transmission could be responsible for the symptoms of AD. β-amyloid peptide (Aβ) its solubility and quantity in a pool is significantly involved in Alzheimer's disease, as it is the main component of the *amyloid* plaques found in the brains of Alzheimer patients. Accumulation of these Aβ is toxic to cultured neurons in the brains of patients having AD [35]. Biochemical studies of brains of patients with AD reveal deficiencies in nAChRs with an elevation in butyrylcholinesterase activity, reduction in ACh, and changes in activity of cholinergic synthetic; e.g., choline acetyltransferase (ChAT) which, in turn, inactivates the (AChE) enzymes [35-36]. Butyrylcholinesterase and AChE terminate the ACh signaling by hydrolyzing the transmitter, thereby inactivating it. These findings have led to the cholinergic hypothesis of AD and the development therapies targeting cholinergic molecular components by pharmaceutical companies, so far mainly targeting the hydrolytic breakdown of ACh by AChE [37]. Currently available treatment of AD includes three acetylcholinesterase (AChE) inhibitors: rivastigmine, galantamine, and donepezil [38, 39]. Therapeutic agents in clinical development have limited benefits to the patient, they may either offer symptomatic relief to patients or provide pure disease modifications. One of the well-known agonist receptor is 5-HT$_4$ (5-hydroxytryptamine) is a better option for the treatment of AD patients [40]. In the biochemical studies, it was observed that the activation of 5-HT$_4$ receptor under preclinical improve neurotransmission and increases the release of acetylcholine which results in the memory formation.

5-HT$_4$ (5-hydroxytryptamine) receptor is a G protein-coupled receptor (GPCR) which belongs to serotonin receptor family and is coupled to G-protein containing Gαs subunit [41]. Upon activation by an agonist (are structural analogs that bind to a receptor and mimic the effects of its natural ligand), the receptor leads to the formation of intracellular cyclic AMP (cAMP) which, in turn, activates Protein kinase A. A cascade of signaling events results in the phosphorylation of cAMP response element binding protein (CREB) which in the form of dimer binds to its response element leading to the expression of a number of genes involved in cell survival. CREB mediated memory formation is likely mediated through the expression of brain-derived neurotrophic factor (BDNF) and other trophic and pro-cognitive factors. Activation of the 5-HT$_4$ receptor leads to the release of various neurotransmitters mediated through calcium ion influx, as a result of blockage of potassium ion channels, with the release of BDNF that may help in memory formation. It was also reported that the activation of the receptor elevates the release of soluble amyloid precursor protein alpha (sAPPα), which along with BDNF induced neurogenesis. Thus, in the brain by elevating the release of acetylcholine, which would, in turn, increase the cholinergic transmission may improve memory formation in the AD patients those who are suffering from the memory loss. Major efforts to develop disease-modifying drugs for AD treatment focused on identifying potent and selective acetylcholinesterase (AChE) inhibitors.

1.2.3   DOPAMINE RECEPTOR AND PARKINSON'S DISEASE

Parkinson's disease (PD) is a common neurodegenerative disorder. It is progressive in nature and causes a movement disorder characterized by bradykinesia, resting tremor, rigidity and postural instability along with non-motor symptoms that mainly include autonomic dysfunction and cognitive impairment [42]. PD occurs when nerve cells, or neurons, in an area of the brain that controls movement become impaired and/or die. Normally, these neurons produce an important brain chemical known as dopamine, but when the neurons die or become impaired, they produce less dopamine. Dopamine (DA) may play an important role in learning about sequences of environmentally important stimuli. Also, it is a proto-typical slow neurotransmitter that plays significant roles in a variety of not only motor functions, but also cognitive, motivational, and neuroen-docrine [43]. Parkinson's disease can occur through both genetic mutation (familial) and exposure to environmental and neurotoxins (sporadic). The DA receptor shares the common features of GPCRs such as seven hydro-phobic transmembrane helices.

Significant changes in the amino acid sequence between different subfamilies, posttranslational modifications such as glycosylation and phosphorylation and conserved amino acid residues that are involved in interaction of G-protein and in binding ligand in the transmembrane regions [44]. Based on its physiological and biochemical role DA recep-tors have been classified into two subfamilies, termed as D_1 and D_2. In the normal state, release of the neurotransmitter dopamine in the presynaptic neuron results in signaling in the postsynaptic neuron through D_1 and D_2 type dopamine receptors.

D_1 receptor transduce a signal through activation of cAMP/PKA signaling pathway. One of the pKA substrate is DA and cAMP-regulated phosphoprotein 32-kDa (DARPP-32). It is a multiple neurotransmitters act as a merger involved in the cell signaling response. Activated DARPP-32 inhibits protein phosphatase 1 (PP1) and activates mitogen-activated protein (MAP) kinases cascades which involves the activation of extracellular signal-regulated kinase (ERK) and MAP/ERK kinase (MEK). MAP kinase signaling pathway plays an important role in the regulation of DA-associated behavior. Thus, the activation of D1 receptor/PKA/DARPP-32 signaling cascade leads to the inactivation of PP1 and allows for the activation of MEK and its downstream kinase ERK [45].

Besides this D2 type receptor block this signaling by inhibiting adenylate cyclase (a second messenger), resulting in a decrease in PKA activity. D2 subfamily regulates the ion channels or by trigger the release of intracellular calcium ions [46]. D2 receptors regulate $G\beta\gamma$ subunits signaling. This subunit complex activates phospholipase C (PLC) that is specific for the plasma membrane lipid phosphatidylinositol 4, 5-bisphosphate. This hormone-sensitive enzyme catalyzes the formation of two potent second messengers: diacylglycerol and inositol 1,4,5-trisphosphate (IP_3). Inositol trisphosphate (IP_3) a water-soluble compound, diffuses from the plasma membrane to the endoplasmic reticulum, where it binds to specific IP_3 receptors (rhodamine) and causes Ca^{2+} channels to open within the ER. Ca^{2+} is thus released into the cytosol and its elevated level activates the protein kinase C (PKC). Phosphorylation of targeted cellular proteins by PKC produces cellular responses to specific proteins, changing their catalytic activities for target proteins. D2 receptor-mediated $G\beta\gamma$ subunits signaling regulate not only calcium channels, but also potassium channels. D2 receptors also regulate the desensitization of the receptor by binding to the β-arrestin (βarr), the protein also called arrestin 2 and binding of β-arrestin effectively prevents interaction between the receptor and the G protein. The scaffolding proteins β-arrestin 1 and β-arrestin 2 have been traditionally associated with the termination of GPCR signaling and with receptor internalization. The binding of β-arrestin also facilitates the removal of receptors from the plasma membrane by endocytosis into small intracellular vesicles. These receptors in the endocytic vesicles are then dephosphorylated again returned to the plasma membrane for resensitizing and complete the signaling pathway. In the study by Beaulieu et al.

[47] *β-arrestin* 2 is a signaling intermediate implicated in the cAMP-independent regulation of Akt and glycogen synthase kinase 3 (GSK-3) by DA. In the other study, they demonstrated that D2 DA receptor-mediated Akt/GSK3-signaling is disrupted in mice lacking *β*-arrestin 2. Thus, it resulted that DA influences neuronal activity, synaptic activity, and behavior. So, the reduction in dopamine causes neurodegenerative disorder, e.g., Parkinson's disease (PD). In a report given by America's biopharmaceutical research companies [48] therapies for PD involves the various diagnostic methods: (i) A gene therapy in this adeno-associated virus (AAV) vector that delivers the gene for aromatic L-amino acid decarboxylase (AADC) to cells in a part of the brain that controls the movement; (ii) An intraduodenal gel formation in a combination with levodopa (a chemical precursor of dopamine) and carbidopa (prevents the levodopa degradation); and (iii) A molecular imaging agent that uses SPECT (single photon emission computed tomography) in the diagnosis of Parkinson's disease. The imaging agent binds to the dopamine transporter (DAT) protein found on the surface of dopamine-producing neurons and this agent is designed to measure movement of DATs and their numbers in the region of the brain. Parkinson's patients have a less number of dopamine-producing neurons and a significantly lower number of DATs. Although the actual cause or causes of Parkinson's disease is unknown, neurologist assumed that the individual suffering from Parkinson's have some portion/area in the brain non-functional. The major target of drug therapies have tended to focus on replacing dopamine or addressing specific symptoms associated with the disease.

1.3   RECEPTOR TYROSINE KINASES/TRANSMEMBRANE KINASES

Receptors on the cell surface act as messengers for transmitting information from the extracellular environment to inside the cell. The cell-cell communication is important for the survival of organisms from multicellular to unicellular organisms. Different types of cell surface receptors that mediate different biological functions have been discovered and characterized. One large group of growth factors act by binding to and activating surface receptors with intrinsic tyrosine kinase activity is a receptor tyrosine kinase (RTKs) or transmembrane kinases. These receptor tyrosine kinases (RTKs) or transmembrane kinases have been extensively studied.

The cellular receptor with tyrosine kinase enzymatic activity phosphory-lates the tyrosine residues of proteins. The ligands for these membrane-bound receptors are growth factors, cytokines and hormones. Upon activation by ligand binding, they can transduce signals and regulates the various cellular functions such as cell proliferation, apoptosis, motility, angiogenesis, or cell differentiation. All Protein kinases are enzymes that play a key regulatory role in nearly every aspect of cell biology. They regulate apoptosis, cell cycle progression, cytoskeletal rearrangement, differentiation, development, the immune response, nervous system func-tion, and transcription. Dysregulation of protein kinases occurs in a variety of diseases, including cancer, diabetes, and autoimmune, cardiovascular, inflammatory, and nervous disorders. There are about more than 50 known tyrosine kinase receptors in humans, classified into 20 families [49]. Some of them are Insulin receptor (IR), the epidermal growth factor receptor (EGFR), the vascular endothelial growth factor receptor (VEGFR), or the platelet-derived growth factor receptor (PDGFR) family. All tyro-sine kinase receptors consist of an extracellular ligand binding domain; a membrane-spanning domain with a cytoplasmic domain [50]. The extra-cellular domain transmits the signal via the cytoplasmic domain to intra-cellular target proteins. The cytoplasmic domain contains, in addition to the catalytic protein tyrosine kinase, distinct regulatory sequences with tyrosine, serine, and threonine phosphorylation sites. Then, these phos-phorylated residues are recognized by cytoplasmic proteins containing Src homology-2 (SH2) or phosphotyrosine-binding (PTB) domains, trig-gering different signaling cascades. Cytoplasmic proteins with SH2 or PTB domains can be effectors proteins with enzymatic activity, or adaptor proteins that mediate the activation of enzymes lacking these recogni-tion sites. Some examples of signaling molecules are: phosphoinositide 3-kinase (PI3K), phospholipase C (PLC), growth factor receptor-binding protein (Grb), or the kinase Src, The main signaling pathways activated by RTK are: PI3K/Akt, Ras/Raf/ERK and signal transduction and acti-vator of transcription (JAK-STAT) pathways [51]. Other than the family of receptor tyrosine kinases (RTKs), receptor ser/thr kinases (STKR) mediate many signaling events at the cell surface [52, 53] (Figure 1.2).

When any external ligand binds to the receptor on the cell surface, they activate the enzyme phosphoinositide-3-Kinase (PI-3K) because of RTK phosphorylation. PI-3K phosphorylates and converts phosphatidylinositol

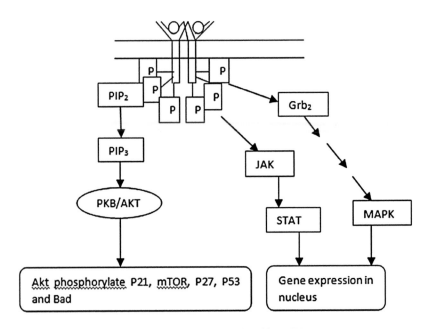

FIGURE 1.2   Signal transduction pathways initiated by RTK.

4,5-bisphosphate (PIP_2) to phosphatidylinositol 3,4,5-triphosphate (PIP_3) and generates diacylglycerol (DAG) respectively, which mediates the activation of the serine/threonine kinase Akt (also known as protein kinase B). When bound to PIP_3, PDK1 another protein kinase that phosphorylates and activates protein kinase B (PKB). This, in turn, phosphorylates Ser or Thr residues in its target proteins. Upon phosphorylation, Akt is able to phosphorylate the substrates involved in the various cellular activities such as cell cycle regulation, apoptosis and protein synthesis and glucose metabolism.

The Ras/Raf/ERK pathway is the main signaling pathway initiated by receptor tyrosine kinase. This signaling pathway is triggered upon binding of the signaling molecule to receptor and that undergoes autophosphorylation on its carboxyl-terminal Tyr residues, SH2 domain of Grb2 binds to P–Tyr which is located in receptor cytoplasmic tails. This binding produces a conformational change in SoS (Son of sevenless), which recruits and activates the GTP hydrolase (GTPase) Ras. Ras functions to relay information to the mitogen-activated protein kinase

(MAPK) signaling pathway. Subsequently, Ras activates the serine/ threonine kinase Raf, which activates MEK phosphorylation that activates extracellular signal-regulated kinase (ERK), which moves to the nucleus and phosphorylates nuclear transcription factor by activating them.

The Janus kinase (JAK)/signal transducers and activators of transcription (STAT) pathway are initiated upon binding of ligands (cytokines, interferons and growth factors) to the receptor. Binding of ligands activates the cell surface receptor; the receptor dimerizes and can now bind the soluble protein kinase JAK (Janus kinase). This binding activates JAK, which phosphorylates several Tyr residues in the cytoplasmic domain of the receptor. A family of transcription factors, collectively called STAT, is also targets of the JAK kinase activity. The SH2 domain in STAT binds P–Tyr residues in the receptor, positioning it for this phosphorylation by JAK. When STAT is phosphorylated in response to ligand, it forms dimers, exposing a signal for its translocation to the nucleus and induce expression of regulatory proteins. Thus, the JAK/ STAT pathway communicates between extracellular signals and transcriptional responses within the nucleus. STAT may also be directly phosphorylated by RTK such as EGFR and PDGFR and by non-receptor tyrosine kinases. These signaling pathways are frequently disturbed in many diseases.

Many growth factors, such as epidermal growth factor (EGF), platelet-derived growth factor (PDGF) and fibroblast-derived growth factor (FGF) and insulin receptor (IR) functions through tyrosine kinase activity. Activation of tyrosine kinase receptors takes place by dimerization of the receptors followed by autophosphorylation, which mainly occurs when one receptor molecule phosphorylates the other molecule in the dimer. The autophosphorylation occurs on two different classes of tyrosine residues. The autophosphorylation takes place on a conserved tyrosine residue within the kinase domain. The phosphorylation of the tyrosine residue at this end-neighboring site leads to an increase in the kinase activity and precedes phosphorylation of other sites in the receptor or substrates in cases of receptors for insulin and hepatocyte growth factor (HGF) [54, 55]. Some autophosphorylation sites are normally localized outside the kinase domains and create docking sites for downstream signal transduction containing SH2 domains.

1.3.1   EPIDERMAL GROWTH FACTOR RECEPTOR (EGFR)

EGFR is a transmembrane receptor (glycoprotein) located on the surface of epithelial cells, expressed in normal human tissues. It is a member of the tyrosine kinase family of growth factor receptors and encoded by ErbB gene in humans after the v-ErbB oncogene of avian erythroblastosis virus. EGFR belongs to the family of ErbB receptors together with ErbB-2, ErbB-3, and ErbB-4. Cohen and co-workers discovered that the receptor for epidermal growth factor is a protein-tyrosine kinase. These studies showed that a single integral membrane protein recognizes the growth factor and possesses protein-tyrosine kinase activity [56].

The well-known ligands for Erb receptors are epidermal growth factor (EGF), amphiregulin (AR), transforming growth factor alpha (TGFα), betacellulin (BTC), epiregulin (ERG), and heparin-binding epidermal growth factor (HB-EGF). These ErbB receptors are recognized by different structurally related growth factors. Over-expression or abnormal functioning causes widespread tumors, including breast, colorectal, pancreatic, ovarian, and non-small cell lung cancers. The ErbB receptors are targets for anticancer drugs. Two strategies for blocking the action of these proteins include antibodies directed against the ectodomain and drugs that inhibit protein tyrosine kinase activity. An ErbB2/HER2 ectodomain directed antibody (trastuzumab, or Herceptin) has also been approved for the treatment of various cancers. Current research promises to produce additional agents based upon these approaches [57].

1.3.2   PLATELET-DERIVED GROWTH FACTOR RECEPTOR (PDGFR)

This factor was purified by Heldin et al. in the mid-1970s; several other researcher groups demonstrated the existence of a major serum growth factor for fibroblasts, smooth muscle cells and glial cells, all are derived from platelets. This factor was further known as a platelet-derived growth factor (PDGF). PDGF receptor for PDGFs is a receptor tyrosine kinase and is made up of by PDGFRα and PDGFRβ. PDGFs are highly expressed in the heart, skeletal muscle, pancreas and moderate levels in other organs. PDGF family stimulates the proliferation, survival, and motility of connective tissue cells and certain other cell types. PDGF isoforms have important roles during the embryonal development, particularly to promote the development of various mesenchymal cell types in different organs [58]. In the adult, PDGF stimulates normal wound healing [59] and regulates the interstitial fluid pressure (IFP) of tissues [60].

PDGF has many other properties such as chemoattractant, involved in bone formation, erythropoiesis, angiogenesis, and in the development of the kidney, brain, cardiovascular and respiratory systems [61]. Heparin and collagens, the molecules of extracellular matrix plays a major role in the binding of PDGF to collagen, and the PDGF-heparin-collagen complex promotes proliferation of fibroblasts, cell migration and vascularization [62].

Over or under expression of PDGF has been linked to tumorigenesis, as well as to the development of other diseases involving excessive cell proliferation, such as atherosclerosis and various fibrotic conditions.

The research is going on to identify and target the protein that is known to be essential in the tumor growth and that is the valid target for the anticancer therapy targets. Various combination therapies such as with hydroxyurea and imatinib for glioblastomas [63–65]. Sunitinib (Sutent) and sorafenib (Nexavar) for renal cell cancer, both belong to a class of multikinase inhibitors with the ability to block both VEGF and PDGF receptors. Continued clinical studies with multikinase inhibitors will reveal if different tumor types will display large variations with regard to sensitivity to combined VEGF and PDGF receptor inhibitors. For the detailed study of activated PDGFR there is a strong need for more specific antibodies that are suitable for immunohistochemistry, and based on the expected impact on clinical decision a careful validation of their specificity will be required. Based on the PDGF/PDGFR inhibitor studies on biomarkers research should provide data to support the realistic approach of antitumor and anti-angiogenic therapy. Finally, it should be noted that the available methods for determination of PDGF receptor expression and activation status in tumor tissue are still a limiting factor for rational development of PDGF receptor-based cancer therapies.

1.3.3   VASCULAR ENDOTHELIAL GROWTH FACTOR RECEPTOR (VEGFR)

Vascular endothelial growth factor (VEGF) was also identified as vascular permeability factor, a protein secreted by tumor cells that promote the accumulation of ascites fluid. The VEGF family of related proteins includes VEGF-A (induce vascular permeability), VEGF-B (normal heart function and a potential anti-cancer target), VEGF-C (essential for embryonic lymphangiogenesis), VEGF-D (important for lymphangiogenesis and a potential anti-cancer and antimetastasis target), and placenta growth factor, each of these proteins contains a signal sequence that is cleaved during biosynthesis. These ligands interact with VEGF receptor tyrosine kinases including VEGFR-1, VEGFR-2, and VEGFR-3. Each VEGF isoform binds to a particular subset of these receptors, giving rise to the formation of receptor homo and heterodimers that activate discrete signaling pathways. In addition, VEGF family ligands can interact with neuropilin-1 and neuropilin-2, which are non-kinase receptors known for their role in semaphorin signaling. Signal specificity of VEGF receptors is further modulated upon recruitment of coreceptors, such as neuropilins, heparan sulfate, integrins, or cadherins. VEGF is essential in embryogenesis. VEGF also functions as a neurotrophic, neuroprotective, and hematopoietic growth factor [66]. VEGF regulate blood and lymphatic vessel development and homeostasis but also have profound effects on neural cells. VEGF are predominantly produced by endothelial, hematopoietic and stromal cells in response to hypoxia and upon stimulation with growth factors such as transforming growth factors, interleukins or platelet-derived growth factor.

It was well understood in different cell types that the signal transduction pathways in response to VEGF are activated through receptor tyrosine kinase activity [67]. VEGF receptors (VEGFR1 and VEGFR2) are present on vascular endothelial cells and their signaling is mediated by receptor dimerization leading to autophosphorylation of the cytosolic domains of the receptors. Phosphorylated VEGF receptors serve as docking sites for adapter molecules or signaling enzymes such as PI3K. VEGF has been shown to activate PI3K generating phosphatidylinositol (3,4,5)-trisphosphates. Activation of PI3K signaling pathway regulates cell survival, cell differentiation, and cell growth.

Dysfunction or loss of function of the VEGF pathway has been identified in a large number of disease processes ranging from cancer to autoimmunity, retinopathy, and many more, which has led to the common perception that inhibition of the pathway would result in rapid and sustained clinical response. The development of highly specific inhibitors of both the VEGF ligand (bevacizumab, VEGF-Trap, ranibizumab) as well as the VEGF receptor (cediranib, pazopanib, sorafenib, sunitinib, vandetanib, axitinib, telatinib, semaxanib, motesanib, vatalanib, zactima) relates to the central role that this pathway plays in disease [68–74]. Blocking VEGF function has been used to treat cancer and ocular angiogenesis. VEGF could potentially be used to promote angiogenesis in clinical settings such as ischemic cardiovascular disease. VEGF gene therapy is a promising alternative treatment method for patients with severe cardiovascular diseases. VEGF inhibitor therapy either single agent or with the combination with traditional chemotherapy and/or radiation have entered the new path in human clinical trials for a wide range of diseases/therapy and/or which can be compared with the standard therapy [75]. Thus, VEGF function can be a pathologic or a beneficial agent depending upon the clinical conditions.

1.3.4   OTHER RECEPTORS

1.3.4.1   MACROPHAGE MANNOSE RECEPTOR

The mannose receptor (MR) is a transmembrane glycoprotein and is expressed predominantly by most tissue macrophages, dendritic cells (DCs) and selected lymphatic or liver endothelial cells. The MR serves as a homeostatic receptor by binding and scavenging unwanted high mannose N-linked glycoproteins as well as pituitary hormones from the circulation. Since many pathogenic microbes are coated with mannose-containing structures, the macrophage MR interacts with these pathogens in a form of the host molecular mimicry [76]. The MR recognizes a variety of microorganisms, including bacteria, fungi, virus, and parasites. Notably, *Mycobacterium tuberculosis*, *Streptococcus pneumonia*, *Yersinia pestis*, *Candida albicans*, *Pneumocystis carinii*, *Cryptococcus neoformans*, HIV, influenza virus, dengue virus, and *Leishmania* species bind to the MR and are engulfed by macrophages. Many MR-binding organisms are intracellular pathogens. Since the MR is an abundantly expressed endocytic, recycling receptor, targeting this receptor is a viable and attractive strategy for the delivery of carbohydrate-containing imaging/diagnostic agents as well as the intracellular delivery of therapeutics for many infectious diseases.

1.3.4.2   TRANSFORMING GROWTH FACTOR-β (TGF-β)

Transforming growth factor-β (TGF-β) family, including TGF-β, activin, nodal, bone morphogenetic proteins (BMPs), and others, play vital roles in development, tissue homeostasis and some disease development [77]. This growth factor signal via protein (serine/threonine) kinase receptor, and Smad mediators to regulate a large number of biological processes, including morphogenesis, embryonic development, adult stem cell differentiation, immune regulation, wound healing or inflammation. TGF-β acts via specific receptors, activating multiple intracellular pathways resulting in phosphorylation of receptor-regulated Smad2/3 proteins that associate with the common mediator, Smad4. Such complex translocates to the nucleus, binds to DNA and regulates transcription of many genes. Furthermore, TGF-β-activated kinase-1 (TAK1) is a component of TGF-β signaling and activates mitogen-activated protein kinase (MAPK) cascades. Alterations of specific components of the TGF-β signaling pathway may contribute to a broad range of pathologies such as cancer, cardiovascular pathology, fibrosis, or congenital diseases. The knowledge about the mechanisms involved in TGF-β signal transduction has allowed a better understanding of the disease pathogenicity as well as the identification of several molecular targets with great potential in therapeutic interventions.

1.3.4.3   TUMOR NECROSIS FACTOR

Tumor necrosis factor (TNF) plays a pivotal role in various immune and inflammatory processes, including cellular activation, survival and proliferation, as well as cell death by necrosis and apoptosis. As a regulatory cytokine, TNF coordinates communication between immune cells, and controls many of their functions [78]. TNF is best known for its role in leading immune defenses to protect a localized area from invasion or injury, but it is also involved in controlling whether target cells live or die, increased understanding of the complex roles of TNF, its receptor activity and signaling pathways is therefore crucial for the identification of new molecular targets and the development of safer and more effective medications. TNF is produced primarily by cells of hematopoietic origin, including myeloid lineage such as monocytes and macrophages.

Dysregulation of inflammatory pathways driven by cytokines such as TNF is believed to be a common underlying mechanism leading to immune-mediated inflammatory diseases, mainly autoimmune diseases such as rheumatoid arthritis, Crohn's disease, multiple sclerosis, lupus, type-I diabetes and Sjogren's syndrome. TNF can bind to two structurally distinct membrane receptors TNFR1 and TNFR2 (TNF receptor family members can be categorized according to the presence or absence of a death domain, DD) on target cells to activate two separate intracellular signaling pathways to gene transcription.

1.3.4.4 CANNABINOID RECEPTOR

Cannabinoids are a class of diverse chemical compounds that act on canna-binoid receptors on cells those suppresses the neurotransmitter release in the brain. Cannabinoids exert their effects by interacting with cannabinoid receptors present on the surface of cells in different parts of the central nervous system. The cannabinoid receptor family currently includes two pharmacologically distinct receptors: the CB_1 cannabinoid receptor, predominantly found in the brain and other nervous tissues, and the CB_2 cannabinoid receptor, mainly associated with immune tissues but also expressed at a lower density in the brain. Both cannabinoid receptors regu-late a variety of central and peripheral physiological functions, including neuronal development, neuromodulatory processes, energy metabolism as well as cardiovascular, respiratory and reproductive functions. In addi-tion, these receptors also modulate proliferation, motility, adhesion and apoptosis of cells. As members of the GPCR superfamily both the CB_1 and CB_2 cannabinoid receptors were initially reported to exert these reported biological effects by activating heterotrimeric G proteins. Both the CB_1 and CB_2 cannabinoid receptors regulate the phosphorylation and acti-vation of different members of the family of mitogen-activated protein kinases (MAPKs), including extracellular signal-regulated kinase-1 and -2 (ERK1/2), p38 MAPK and c-Jun N-terminal kinase (JNK). The CB_1 cannabinoid receptor may also induce elevations in intracellular Ca^{2+} through G protein-dependent activation of phospholipase C-β (PLC-β) [79]. Activation of cannabinoid receptors primarily leads to the inhibi-tion of adenylyl cyclase and reductions in cyclic AMP accumulation in most tissues and models. In some diseases, such as neuropathic pain and multiple sclerosis, increases in cannabinoid receptor expression are thought to reduce symptoms and/or inhibit the progression of disease and thus serve a protective role [80]. In other diseases, alterations in receptor expression are not adjusting properly to the environment, examples being CB_1R up-regulation in liver fibrosis and down-regulation in colorectal cancer [81]. Regulation of cannabinoid receptor expression is of interest from a therapeutic point of view.

1.3.4.5   RECEPTOR-LIKE KINASES

Signal transduction through cell-surface receptors is a common feature among living organisms. In plants, several different types of cell-surface receptors perceive diverse signals and stimuli from the environment (both abiotic and biotic) [82]. Among other receptors in plants, one of them is Receptor-like kinases (RLKs) is a class of transmembrane kinases similar in basic structure of receptor tyrosine kinases (RTKs) present in plants. RLKs are a single helical transmembrane segment that connects a ligand binding receptor domain to the outside of the membrane with a kinase protein (Ser/Thr kinase) on the cytoplasmic tail. These RLKs receptors mainly participate in the defense mechanism triggered by infection with a bacterial pathogen [83]. Some major functions of RLK members include developmental processes, such as the regulation of meristem proliferation, organ specification, reproduction and hormone signal transduction. Some of the receptor classes such as the nucleotide-binding site-leucine-rich repeat (NBS-LRR) receptors and histidine kinase receptors, can mediate responses to organic chemicals such as the hormones ethylene and cyto-kinin. Plants have threats from various pathogenic microbes and resist attacking pathogens through both constitutive and inducible defenses [84]. Pathogen entry into host tissue is a critical, first step in causing plant infection. The major protein of the bacterial flagellum is 'flagellin' when any pathogen attacks the plants; the signal to turn on the genes needed for defense against infection is a peptide (flg22) that is released by the breakdown of flagellin protein. Binding of flg22 to the FLS2 receptor of *Arabidopsis* induces receptor dimerization which, in turn, autophosphory-lates the Ser and Thr residues and the downstream effect is activation of a MAPK pathway, this pathway activates a specific transcription factor (Jun, Fos, NFκB) that triggers the synthesis of the proteins that protect against the bacterial infection. Another receptor kinase in plants is lectin receptor kinases which have also a role in innate immunity signaling [85]. These lectin receptor kinases are classified into three types: G, C, and L [86]. G-type lectin receptor kinases are known as S-domain RLKs and are involved in self-incompatibility in flowering plants. C-type (calcium-dependent) lectin can be found in a large number of mammalian proteins that mediate innate immune responses and play a major role in pathogen recognition [87], but are rare in plants. L-type lectin receptor kinases (LecRKs) are characterized by an extracellular legume lectin-like domain, a transmembrane domain and an intracellular kinase domain [88]; they were suggested to play a role in abiotic stress signal transduction. In the coming years, researchers need to explore the effect of protein structure on both upstream (ligands and their cofactors) and downstream (effectors and their signaling cascade) targets to develop a clearer picture of plant signal transduction pathways. Information obtained from such studies could lead to novel methods for managing plant disease resistance and signals are increased transcription of specific genes.

1.3.4.6   TOLL-LIKE RECEPTORS

It is well-known that innate immunity is very essential to human survival. Researches from decade in animal models have discovered the powerful toll-like receptors (TLRs) that play a central role in the immune defense system. The first mammalian homolog of Toll was identified in 1997 and after that 10 different members of TLR was identified and they are responsible for recognizing molecular patterns associated with pathogens (PAMP, pathogen-associated molecular patterns), and expressed by a broad spectrum of infectious agents. During any infection The primary response to pathogens in the innate immunity system is mediated by pattern recognition receptors (PRR) which recognize pathogen-associated molecular patterns (PAMP) present in a wide array of microorganisms [89] The most important PRR (pattern recognition receptor) are Toll-like receptors (TLR) which selectively recognize a large number of varied and complex PAMP, characteristic molecules of microorganisms such as lipopolysaccharides, flagellin, mannose, or nucleic acids from virus and bacteria. Once the PRR, and in particular the TLR, recognize these microorganism-specific molecules, an innate immune response is triggered which activates the production of inflammatory mediators such as a large number of inter-leukins (IL), interferons (IFN) and tumor necrosis factor alpha (TNF-α) [90]. Within the group of TLRs, two types have been identified: surface-expressed TLRs, which are predominantly active against bacterial cell wall compounds; and intracellular receptors, which preferentially recognize virus-associated pattern molecules. In addition, surface-expressed recep-tors trigger phagocytotic and maturation signals, while the intracellular TLRs lead to the induction of antiviral genes [91]. TLRs are type I trans-membrane proteins that consist of a cytoplasmic Toll/IL-1 (TIR) domain and an extracellular domain having leucine-rich repeats (LRR). This TIR domain has the ability to bind and activate distinct molecules, among them MyD88 (myeloid differentiation factor 88), the Toll/IL1-R domain-containing adaptor protein (TIRAP), Toll/IL-1R domain-containing adaptor inducing IFN-beta (TRIF), the TRIF-related adaptor molecule (TRAM), Interleukin-1 receptor-associated kinase (IRAK), tumor necrosis factor (TNF), and TNF receptor-associated factor 6 (TRAF6); all necessary to activate different pathways such as mitogen-activated protein kinases (MAPK), signal transducers and activators of transcription (STAT) and the nuclear factor-kappa B (NF-κB) pathway and interferon regulatory factor 3 (IRF3), which, in turn, induce various immune and inflammatory genes [92, 93]. A greater understanding of the TLRs and their roles in immunity holds potential for the development of therapeutics for bacterial and viral infections, allergies and cancer, and also to limit the damage caused by autoimmune disorders. Moreover, the role of TLRs in tissue repair and regeneration provides a further avenue for drug targeting [94]. It is recog-nized that TLRs bind to specific ligands, distribute on different cell types, and play key roles in the pathophysiology of various disorders involving both the innate and adaptive immunity [95] (Table 1.1).

1.3.4.7 NUCLEAR RECEPTORS

Nuclear receptors (NRs) are proteins that share considerable amino acid sequence similarity in two highly conserved domains – the DNA binding (DBD) and the ligand binding domains (LBD). These domains are respon-sible for binding specific DNA sequences and small lipophilic ligands [96]. NRs Comprises a large, ancient, superfamily of eukaryotic tran-scription factors that govern a wide range of metabolic, homeostatic, and developmental pathways, and which have been implicated in disease states

TABLE 1.1 Toll-Like Receptors: Their Binding Ligands, Localization and Their Role in Regulation of Cytokine in Humans

Receptors	Ligands	Localization	Regulation of cytokines
TLR 2	Lipoproteins (bacteria), lipoarabinomanan (Mycobacterium tuberculosis), porins zimosan (fungi), Pam 3 cys	Cellular surface	IL-6 and TNF-α, IL-1β and IL-10
TLR 3	Bacterial DNA and viral double-chain RNA	Intracellular compartment	IFN-δ, IL-1β, IL-6, IL-10 and TNF-α
TLR 4	Taxol, HSPs, fibronectin, heparan sulfate, F-protein	Cellular surface	IFN-δ and IL-1β
TLR 5	Flagellin	Cellular surface	IL-6, IFN-δ and IL-1β
TLR 6	Macrophage-activating lipopeptide-2	Cellular surface	IFN-δ and IL-1β
TLR 7/8	Single chain RNA, Loxoribine, Imidazoquinolin	Intracellular compartment	IL-6, IFN-δ, IL-1β, IL-6, IL-10 and TNF-α
TLR 9	Bacterial DNA and viral RNA	Intracellular compartment	IFN-δ, IL-1β, IL-6, IL-10 and TNF-α
TLR 10	Associated with TLR1 and TLR6	Cellular surface	IFN-δ, IL-1β, IL-6, IL-10 and TNF-α

including cancer, inflammation, and diabetes. The ability of NRs to activate or repress gene transcription is modulated through direct binding of small lipophilic ligands which induce conformational changes in their related receptor. These changes are structural in nature and lead to the recruitment of coactivator or corepressor complexes, ultimately regulating the expression of target genes to whose response elements NRs are bound (Figure 1.3).

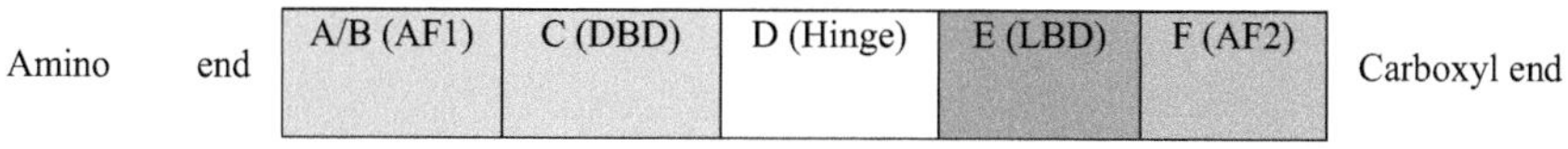

Amino end	A/B (AF1)	C (DBD)	D (Hinge)	E (LBD)	F (AF2)	Carboxyl end

FIGURE 1.3 (See color insert.) Nuclear receptor and its structure.

AF-1: Activation factor 1, it is a ligand-independent factor and responsible for gene activation

DBD: DNA binding domain is composed of two highly conserved zinc fingers and responsible for targeting the receptor to highly specific DNA sequences comprising a hormone response element (RE), it also includes the hinge region.

LBD: Ligand binding domain, capable of binding to small lipophilic molecules such as steroids, retinoids, and vitamins which regulate the activity of these receptors.

AF-2: Activation factor 2, it is a legend dependent and responsible for the gene activation function.

Numerous studies on nuclear receptor which includes the steroid, orphan and other have divulged that the nuclear receptors are involved in many metabolic and inflammatory diseases, such as diabetes, dyslipidemia, cirrhosis, and fibrosis [97]. Some of them are described here.

1.3.4.8   ORPHAN NUCLEAR RECEPTOR (ONR)

Orphan nuclear receptors are proteins that bind and are activated by unknown signaling molecules (called ligands, neurotransmitters, or hormones). However, they share structural components with identified receptors (steroid and thyroid) whose signaling mechanism is already known. These receptors play essential roles in development, cellular homeostasis, and disease, including cancer and diabetes [98].

1.3.4.9   THYROID RECEPTOR

Nuclear receptors, especially thyroid hormone receptors (TRs), are also involved in pathogenesis and in the development of several types of human diseases and malignancies. It is known that mutations of TRs genes are the main cause of resistance to thyroid hormone syndrome [99]. TRs are encoded by two genes, *THRA* and *THRB*, located on chromosomes 17 and 3. Further, genetic alterations and/or aberrant expression of the TRs are reported to be associated with human malignancies such as breast, liver, thyroid, pituitary and renal cancers [100]. Some other nuclear receptor and their subfamilies are given in Table 1.2.

It is clear from the scientific research and observed data that over and under expression of nuclear receptors involved in the contribution of various diseases. As ligands (agonists or antagonists) binding play a crucial role in regulating nuclear receptor activity, meanwhile the discovery of novel ligands for nuclear receptors represents an interesting and promising therapeutic approach. Some of the well-known drugs include Phenobarbital, Tamoxifen, Dexamethasone, Rifampicin, RU-486, etc. Therefore, further investigations are needed in order to use these receptors as therapeutic targets or as biological markers to decide on appropriate forms of treatment.

TABLE 1.2   Subfamilies of the Nuclear Receptor

Name of the receptor	Ligand
Steroid hormone receptors	
Oestrogen receptor	Oestrogens
Progesterone receptor	Progesterones
Androgen receptor	Androgens
Glucocorticoid receptor	Glucocorticoids
Mineralocorticoid receptor	Mineralocorticoids
Vitamin D receptor	Vitamin D
Retinoic acid receptor	Retinoic acid
Orphan receptors	
Photoreceptor-specific nuclear receptor	None of the ligand is known
RAR-related orphan receptor	
Oestrogen-related receptor	
Nuclear receptor subfamily	
Germ cell nuclear factor 1	
Hepatocyte nuclear factor 4	
COUP transcription factor (I–III)	
Other nuclear receptors	
Retinoid X receptor	Retinoic acid
Peroxisome proliferator activated receptor (PPAR)	Fatty acids
Pregnane X receptor	Xenobiotics
Steroidogenic factor 1	Phosphatidylinositols
Liver receptor homolog 1	Phosphatidylinositols

1.4   CONCLUSION

As discussed above, several cell surface and nuclear receptor that are constituted of proteins have a major role in cell signaling for the transcription of the function genes involved in the various cellular activity. Overexpression or misfunctioning of these receptors causes various diseases in human beings. For the therapeutic purpose, these receptors and their target ligand is the major area of scientific study.

KEYWORDS

- Alzheimer's disease
- epidermal growth factor receptor (EGFR)
- G protein-coupled receptor (GPCR)
- Nuclear receptors (NR)
- Parkinson's disease
- platelet-derived growth factor receptor (PDGFR)
- protein
- receptor-like kinase (RLK)
- receptor tyrosine kinase (RTK)
- toll-like receptors (TL)
- vascular endothelial growth factor receptor (VEGFR)

REFERENCES

1. Salon, J. A., Lodowski, D. T., & Palczewski, K., (2011). The significance of G protein-coupled receptor crystallography for drug discovery. *Pharmacol Rev., 63*, 901-937.
2. Calebiro, D., Nikolaev, V. O., Persani, L., & Lohse, M. J., (2010). Signaling by internalized G-protein-coupled receptors. *Trends Pharmacol. Sci., 31*(5), 221–228.
3. Cheng, S. B., Graeber, C. T., Quinn, J. A., & Filardo, E. J., (2011). Retrograde transport of the transmembrane estrogen receptor, G protein-coupled-receptor-30 (GPR30/GPER) from the plasma membrane towards the nucleus. *Steroids, 76(9)*, 892–896.
4. Re, M., Pampillo, M., Savard, M., Dubuc, C., Mc Ardle, C. A., Millar, R. P., et al., (2010). The human gonadotropin releasing hormone type I receptor is a functional intracellular GPCR expressed on the nuclear membrane. *PLo S One, 5*(7), e11489.
5. Hazell, G. G., Hindmarch, C. C., Pope, G. R., Roper, J. A., Lightman, S. L., Murphy, D., et al., (2012). G protein-coupled receptors in the hypothalamic paraventricular and supraoptic nuclei–serpentine gateways to neuroendocrine homeostasis. *Front. Neuroendocrinol., 33*(1), 45–66.
6. Lagerstrom, M. C., & Schioth, H. B., (2008). Structural diversity of G protein-coupled receptors and significance for drug discovery. *Nat. Rev. Drug Discov., 7*(4), 339–357.
7. Cattaneo, F., Guerra, G., Parisi, M., De Marinis, M., Tafuri, D., Cinelli, M., & Ammendola, R., (2014). Cell-surface receptors transactivation mediated by g protein-coupled receptors. *Int J Mol Sci, 15*, 19700-19728.

8. Bouvier, M., Hausdorff, W. P., De Blasi, A., O'Dowd, B. F., Kobilka, B. K., & Caron, M. G., (1988). Removal of phosphorylation sites from the beta 2-adrenergic receptor delays onset of agonist-promoted desensitization. *Nature, 333,* 370–373.

9. Lohse, M. J., Benovic, J. L., Codina, J., Caron, M. G., & Lefkowitz, R. J., (1990). Beta-arrestin: a protein that regulates beta-adrenergic receptor function. *Science, 248,* 1547–1550.

10. Ferguson, S. S., Downey, 3rd, W. E., Colapietro, A. M., Barak, L. S., Menard, L., & Caron, M. G., (1996). Role of beta-arrestin in mediating agonist-promoted G protein-coupled receptor internalization. *Science., 271,* 363–366.

11. Ferguson, S. S., Menard, L., Barak, L. S., Koch, W. J., Colapietro, A. M., & Caron, M. G., (1995). Role of phosphorylation in agonist-promoted beta 2-adrenergic receptor sequestration. Rescue of a sequestration-defective mutant receptor by beta ARK1. *J. Biol. Chem., 270,* 24782–24789.

12. Tsuga, H., Kameyama, K., Haga, T., Kurose, H., & Nagao, T., (1994). Sequestration of muscarinic acetylcholine receptor m2 subtypes. Facilitation by G protein-coupled receptor kinase (GRK2) and attenuation by a dominant negative mutant of GRK2. *J. Biol. Chem., 269,* 32522–32527.

13. Reiter, E., & Lefkowitz, R. J., (2006). GRKs and beta-arrestins: roles in receptor silencing, trafficking and signaling. *Trends Endocrinol. Metab., 17,* 159–165.

14. Ferguson, S. S., (2001). Evolving concepts in G protein-coupled receptor endocytosis: the role in receptor desensitization and signaling. *Pharmacol. Rev., 53,* 1–24.

15. Tachibana, H., Naga Prasad, S. V., Lefkowitz, R. J., Koch, W. J., & Rockman, H. A., (2005). Level of beta-adrenergic receptor kinase 1 inhibition determines degree of cardiac dysfunction after chronic pressure overload-induced heart failure. *Circulation., 111,* 591–597.

16. Lymperopoulos, A., Rengo, G., Funakoshi, H., Eckhart, A. D., & Koch, W. J., (2007). Adrenal GRK2 upregulation mediates sympathetic overdrive in heart failure. *Nat. Med., 13,* 315–323.

17. Wang, W. C., Mihlbachler, K. A., Brunnett, A. C., & Liggett, S. B., (2009). Targeted transgenesis reveals discrete attenuator functions of GRK and PKA in airway beta-2-adrenergic receptor physiologic signaling. *Proc. Natl. Acad. Sci., USA, 106,* 15007–15012.

18. Gainetdinov, R. R., Bohn, L. M., Sotnikova, T. D., Cyr, M., A. Laakso, A. D., Macrae, et al. (2003). Dopaminergic supersensitivity in G protein-coupled receptor kinase 6-deficient mice. *Neuron., 38,* 291–303.

19. Balabanian, K., Lagane, B., Pablos, J. L., Laurent, L., Planchenault, T., Verola, O., et al. (2005). WHIM syndromes with different genetic anomalies are accounted for by impaired CXCR4 desensitization to CXCL12. *Blood., 105,* 2449–2457.

20. Brodde, O. E., (1993). Beta-adrenoceptors in cardiac disease. *Pharmacol. Ther., 60,* 405–443.

21. Bristow, M. R., Herschberger, R. E., Port, J. D., & Rasmussen, R., (1989). β1 and β2 adrenergic receptor mediated adenylyl cyclase stimulation in non-failing and failing human ventricular myocardium. *Mol. Pharmacol., 35,* 395–399.

22. Cohen, P. T., (2002). Protein phosphatase 1-targeted in many directions. *J. Cell. Sci., 115,* 241–256.

23. Ceulemans, H., & Bollen. M., (2004). Functional diversity of protein phosphatase-1, a cellular economizer and reset button. *Physiol. Rev., 84,* 1–39.

24. Herzig, S., & Neumann, J., (2000). Effects of serine/threonine protein phosphatases on ion channels in excitable membranes. *Physiol. Rev., 80,* 173–210.

25. Marks, A. R., Marx, S. O., & Reiken, S., (2002). Regulation of ryanodine receptors via macromolecular complexes: a novel role for leucine/isoleucine zippers. *Trends Cardiovasc. Med. 12,* 166–170.

26. Gollub, S. B., Elkayam, U., Young, J. B., Miller, L. W., & Haffey, K. A., (1991). Efficacy and safety of a short-term (6-h) intravenous infusion of dopexamine in patients with severe congestive heart failure: A randomized, double-blind, parallel, placebo-controlled multicenter study. *J. Am. Coll. Cardiol., 18,* 383–390.

27. Beatrice, B., Dirk, R., Roland, J., & Fritz, B., (2014). Diagnostic and therapeutic aspects of β1-adrenergic receptor autoantibodies in human heart disease. *Autoimmun. Rev., 13,* 954–962.

28. Rinne, J. O., Myllykyla, T., Lonnberg, P., & Marjamaki, P., (1991). A postmortem study of brain nicotinic receptors in Parkinson's and Alzheimer's disease. *Brain Res., 547,* 167–170.

29. Perry, E. K., Morris, C. M., Court, J. A., Fairbairn, A. F., & Mc Keith, I. G., (1995). Alteration in nicotine binding sites in Parkinson's disease, Lewy body dementia and Alzheimer's disease: possible index of early neuropathology. *Neurosci., 64,* 385–395.

30. Court, J. A., Martin-Ruiz, C., Graham, A., & Perry, E., (2000a). Nicotinic receptors in human brain: topography and pathology. *J. Chem. Neuroanat., 20,* 281–298.

31. Kihara, T., & Shimohama, S., (2004) Alzheimer's disease and acetylcholine receptors. *Acta Neurobiol Exp. 64*(1), 99–106.

32. Cummings, J. L., (2004). Alzheimer's disease. *N. Engl. J. Med., 351,* 56–67.

33. Shimohama, S., Taniguchi, T., Fujiwara, M., & Kameyama, M., (1986). Changes in nicotinic and muscarinic cholinergic receptors in Alzheimer-type dementia. *J. Neurochem., 46,* 288–293.

34. Nordberg, A., (2001). Nicotinic receptor abnormalities of Alzheimer's disease: therapeutic implications. *Biol. Psychiatry, 49,* 200–210.

35. Bartus, R. T., Dean, R., Beer, B., & Lippa, A. S. (1982). The cholinergic hypothesis of geriatric memory dysfunction. *Science. 217*(4558), 408–414.

36. Francis, P. T., Palmer, A. M., Snape, M., & Wilcock, G. K. (1999). The cholinergic hypothesis of Alzheimer's disease: a review of progress. *J Neurol Neurosurg Psychiat. 66*(2), 137–147.

37. Yao, M., Nguyen, T. V., & Pike, C. J., (2005). β-amyloid-induced neuronal apoptosis involves *c*-Jun N-terminal kinase-dependent down regulation of Bcl-w. *J. Neurosci., 25,* 1149–1158.

38. Arneric, S. P., Holladay, M., & Williams, M., (2007). Neuronal nicotinic receptors: A perspective on two decades of drug discovery research. *Biochem. Pharmacol., 74,* 1092–1101.

39. Aguglia, E., Onor, M. L., Saina, M., & Maso, E., (2004). An open-label, comparative study of rivastigmine, donepezil and galantamine in a real-world setting. *Curr. Med. Res.Opin., 20,* 1747–1752.

40. Ritchie, C. W., Ames, D., Clayton, T., & Lai, R., (2004). Metaanalysis of randomized trials of the efficacy and safety of donepezil, galantamine, and rivastigmine for the treatment of Alzheimer disease. Am. J. Geriatr. Psychiatry., 12, 358–369.

41. Gerald, C., Adham, N., Kao, H. T., Olsen, M. A., Laz, T. M., Schechter, L. E., et al. (1995) The 5-HT4 receptor: molecular cloning and pharmacological characterization of two splice variants. The EMBO J, 14(12), 2806–2815.

42. Ahmad, I., & Nirogi. R., (2011). 5-HT4 Receptor agonists for the treatment of Alzheimer's disease. Neurosci. Med., 2, 87–92.

43. Lees, A. J., Hardy, J., & Revesz, T., (2009). Parkinson's disease. Lancet, 373, 2055–2066.

44. Carlsson, A., (2001). "A paradigmshift in brain research." Science., 294, 1021–1024.

45. Hisahara, S., & Shimohama, S. (2011). Dopamine receptors and Parkinson's disease. Int J Med Chem. Article ID 403039, 16 pages.

46. Niznik, H. B., & Van., & Tol, H. H. M., (1992). Dopamine receptor genes: new tools for molecular psychiatry. J. Psychiatry Neurosci., 17, 158–180.

47. Missale, C., Russel Nash, S., Robinson, S. W., Jaber, M., & Caron, M. G., (1998). Dopamine receptors: from structure to function. Physiol. Rev., 78, 189–225.

48. Beaulieu, J. M., Gainetdinov, R. R., & Caron, M. G., (2007). The Akt-GSK-3 signaling cascade in the actions of dopamine. Trends Pharmacol. Sci., 28, 166–172.

49. Medicines in Development Parkinson's Disease (2014). PhRMA Research Reports. Available: https://www.phrma.org/report/medicines-in-development-for-parkinson-s-disease-2014-report (Accessed: 10 January 2018)

50. Robinson, D. R., Wu, Y. M., & Lin., S. F., (2000). The protein tyrosine kinase family of the human genome. Oncogene., 19, 5548–5557.

51. Carrasco-García, E., Saceda, M., & Martínez-Lacaci, I. (2014). Role of receptor tyrosine kinases and their ligands in glioblastoma. Cells. 3(2), 199–235.

52. Ullrich, A., & Schlessinger, J., (1990). Signal transduction by receptors with tyrosine kinase activity. Cell., 61(2), 203–212.

53. Van der, G. P., Hunter, T., &Lindberg, R. A., (1994). Receptor protein-tyrosine kinases and their signal transduction pathways. Annu. Rev. Cell. Biol., 10, 251–337.

54. Massagué, J. (1998). TGF-beta signal transduction. Annu. Rev. Biochem., 67, 753–91.

55. Naldini, L., Vigna, E., Ferracini, R., Longati, P., Gandino, L., Prat, M., & Comoglio, P. M., (1991). The tyrosine kinase encoded by the MET proto-oncogene is activated by autophosphorylation. Mol. Cell. Biol., 11(4), 1793–803.

56. White, M. F., Shoelson, S. E., Keutmann, H., & Kahn, C. R. (1988). A cascade of tyrosine autophosphorylation in the beta-subunit activates the phosphotransferase of the insulin receptor. J. Biol. Chem., 263(6), 2969–2980.

57. Cohen, S., Ushiro, H., Stoscheck, C., & Chinkers, M., (1982). A native 170, 000 epidermal growth factor receptor-kinase complex from shed plasma membrane vesicles. J. Biol. Chem., 257, 1523–1531.

58. Roskoski Jr, R. (2004). The ErbB/HER receptor protein-tyrosine kinases and cancer. Biochem Biophys Res Commun, 319, 1-11.

59. Betsholtz, C. (2004). Insight into the physiological functions of PDGF through genetic studies in mice. Cytokine Growth. Factor Rev., 15, 215–228.

60. Robson, M. C., Phillips, L. G., Thomason, A., Robson, L. E., & Pierce, G. F., (1992). Platelet-derived growth factor BB for the treatment of chronic pressure ulcers. *Lancet., 339*, 23–25.

61. Rodt, S. A., Ahle'n, K., Berg, A., Rubin, K., & Reed, R. K., (1996). A novel physiological function for plateletderived growth factor-BB in rat dermis. *J. Physiol., 495*, 193–200.

62. Li, W. L., Yamada, Y., Ueno, M., Nishikawa, S., Nishikawa, S. I., & Takakura, N., (2006). Platelet-derived growth factor receptor alpha is essential for establishing a microenvironment that supports definitive erythropoiesis. *J. Biochem. 140*, 267–273.

63. Sun, B., Chen, B., Zhao, Y., Sun, W., Chen, K., & Zhang, J., (2009). Crosslinking heparin to collagen scaffolds for the delivery of human platelet-derived growth factor. *J. Biomed. Mater. Res. Part B: Appl. Biomater., 91B*, 366–372.

64. Dresemann, G. (2005). Imatinib and hydroxyurea in pretreated progressive glioblastoma multiforme: A patient series. *Ann. Oncol., 16*, 1702–1708.

65. Reardon, D. A., Egorin, M. J., Quinn, J. A., Sr. Rich, J. N., Gururangan, I., Vredenburgh, J. J., et al., (2005). Phase II study of imatinib mesylate plus hydroxyurea in adults with recurrent glioblastoma multiforme. *J. Clin. Oncol., 23*, 9359–9368.

66. London, N. R., & Gurgel, R. K., (2014). The role of vascular endothelial growth factor and vascular stability in diseases of the ear. *Laryngoscope, 124*(8), E340–E346.

67. Wang, S., Li, X., Parra, M., Verdin, E., Bassel-Duby, R., & Olson, E. N., (2008). Control of endothelial cell proliferation and migration by VEGF signaling to histone deacetylase 7. *Proc. Natl. Acad. Sci., USA, 105*, 7738–7743.

68. Faivre, S., Demetri, G., Sargent, W., & Raymond, E., (2007). Molecular basis for sunitinib efficacy and future clinical development. *Nat. Rev. Drug Discov., 6*, 734–745.

69. Tabernero, J. (2007). The role of VEGF and EGFR inhibition: Implications for combining anti-VEGF and anti-EGFR agents. *Mol. Cancer Res., 5*, 203–220.

70. Choueiri, T. K., (2008). Axitinib, a novel anti-angiogenic drug with promising activity in various solid tumors. *Curr. Opin. Investig. Drugs., 9*, 658–671.

71. Dadgostar, H., & Waheed, N., (2008). The evolving role of vascular endothelial growth factor inhibitors in the treatment of neovascular age-related macular degeneration. *Eye (Lond)., 22*, 761–767.

72. Sloan, B., & Scheinfeld, N. S., (2008). Pazopanib, a VEGF receptor tyrosine kinase inhibitor for cancer therapy. *Curr. Opin. Investig. Drugs., 9*, 1324–1335.

73. Lindsay, C. R., Mac Pherson, I. R., & Cassidy, J., (2009). Current status of cediranib: The rapid development of a novel antiangiogenic therapy. *Future Oncol., 5*, 421–432.

74. Porta, C., Paglino, C., Imarisio, I., & Ferraris, E., (2009). Sorafenib tosylate in advanced kidney cancer: Past, present and future. *Anticancer Drugs., 20*, 409–415.

75. Kessler, T., Bayer, M., Schwoppe, C., Liersch, R., Mesters, R. M., & Berdel, W. E., (2010). Compounds in clinical Phase III and beyond. *Recent Results Cancer Res., 180*, 137–163.

76. Azad, A.K., Rajaram, M. V. S., & Schlesinger, L. S., (2014). Exploitation of the macrophage mannose receptor (CD206) in infectious disease diagnostics and therapeutics. *J. Cytol. Molecul. Biol., 1*, 5.

77. Huang, F., & Ye-Guang, C., (2012). Regulation of TGF-β receptor activity. *Cell Biosci., 2*, 9.

78. Croft, M., (2009). The role of TNF superfamily members in T-cell function and diseases. *Nature Rev. Immunol., 9*, 271–285.

79. Howlett, A. C., (2005). Cannabinoid receptor signaling. Handb. *Exp. Pharmacol.,* 53–79.

80. Pertwee, R. G., (2009). Emerging strategies for exploiting cannabinoid receptor agonists as medicines. *British J Pharmacol, 156*(3), 397–411.

81. Wang, D., Wang, H., Ning, W., Backlund, M. G., Dey, S. K., & Du Bois, R. N., (2008). Loss of cannabinoid receptor 1 accelerates intestinal tumor growth. *Cancer Res., 68*, 6468–6476.

82. Miya, A., Albert, P., Shinya, T., Desaki, Y., Ichimura, K., Shirasu, K., et al., (2007). CERK1, a Lys M receptor kinase, is essential for chitin elicitor signaling in Arabidopsis. *Proc. Natl. Acad. Sci. USA, 104*, 19613–19618.

83. Bent, A. F., & Mackey, D., (2007). Elicitors, effectors, and R genes: the new paradigm and a lifetime supply of questions. *Annu. Rev. Phytopathol., 45,* 399–436.

84. Jones, J. D., & Dang, J. L., (2006). The plant immune system. *Nature., 16,* 323–329.

85. Singh, P., & Zimmerli, L. Z., (2013). Lectin receptor kinases in plant innate immunity. *Front Plant Sci,* 4, 124.

86. Vaid, N., Pandey, P. K., & Tuteja, N., (2012). Genome-wide analysis of lectin receptor-like kinase family from *Arabidopsis* and rice. *Plant Mol. Biol., 80*, 365–388.

87. Cambi, A., Koopman, M., & Figdor, C. G. (2005). How C-type lectins detect pathogens. *Cell Microbiol. 7*(4), 481–488.

88. Bouwmeester, K., & Govers, F., (2009). Arabidopsis L-type lectin receptor kinases: Phylogeny, classification, and expression profiles. *J. Exp. Bot., 60,* 4383–4396.

89. Duez, C., Gosset, P., & Tonnel, A., (2006). Dendritic cells and toll-like receptors in allergy and asthma [revisión]. *Eur. J. Dermatol., 16,* 12–6.

90. Gon, Y., (2008). Toll-Like receptors and airway inflammation [revision]. *Allergol Int., 57,* 33–37.

91. Crispín, J. C., Kyttaris, V. C., Terhorst, C., & Tsokos, G. C., (2010). T cells as therapeutic targets in SLE. *Nat Rev Rheumatol. 6*(6), 317.

92. Fritz, J. H., & Girardon, D. E., (2005). How toll-like receptors and Nod-like receptors contribute to innate immunity in mammals. *J. Endotoxin. Res., 11*, 390–394.

93. Takeda, K., Kaisho, T., & Akira, S., (2006).Toll-like receptors. *Annu. Rev. Immunol., 27,* 352–357.

94. Coyne, L., (2008). Target analysis: Toll-like receptors. *Pharma. Projects., 29,* 1–4.

95. O'Neill, L. A., Bryant, C. E., & Doyle, S. L., (2009). Therapeutic Targeting of Toll-Like Receptors for Infectious and Inflammatory Diseases and Cancer. *Pharmacol. Rev., 61*, 177–197.

96. Sladek, F. M. (2010). What are nuclear receptor ligands? *Mol. Cell. Endocrinol.,* doi:10.1016/j.mce.2010.06.018.

97. Cermenati, G., Brioschi, E., Abbiati, F., Melcangi, R. C., Caruso, D., & Mitro, N., (2013). Liver X receptors, nervous system, and lipid metabolism. *J. Endocrinol. Investig., 36,* 435–443.

98. Safe, S., Jin, U. H., Hedrick, E., Reeder, A., & Lee, S. O., (2014). Minireview: role of orphan nuclear receptors in cancer and potential as drug targets. *Mol Endocrinol, 28,* 157-172.

99. Rosen, M. D., & Privalsky, M. L. (2009) Thyroid hormone receptor mutations found in renal clear cell carcinomas alter corepressor release and reveal helix 12 as key determinant of corepressor specificity. *Mol Endocrinol, 23*(8), 1183–1192.

100. Rebai, M., Kallel, I., & Rebai, A., (2012). Genetic features of thyroid hormone receptors. *J. Genet. 91,* 367–374.

CHAPTER 2

ISLET TRANSPLANTATION IN TYPE 1 DIABETES: STEM CELL RESEARCH AND THERAPY

PARVEEN PARASAR[1] and VIVEK SINGH[2]

[1]Division of Surgery, Department of Gynecology, Children's Hospital Boston, Harvard Medical School, Boston, Massachusetts, 02115, USA, Tel: +(1)617-637-3144, E-mail: suprovet@gmail.com

[2]Department of Veterinary Physiology and Biochemistry, Junagadh Agricultural University, Junagadh, Gujarat, 362001, India, Tel: +(91)+8306857587, E-mail: drvivekndri@gmail.com

ABSTRACT

Type 1 diabetes mellitus is an autoimmune disease, which results in loss of pancreatic β cells. A myriad of eminent discoveries in stem cell biology have provided insights into the therapeutic potential of stem cells. A significant advance in cell therapy for diabetes has been the extraction of pancreatic islet cells and transplantation into recipients as per the Edmonton protocol. Stem cells, due to their pluripotent ability and expansion potential, provide us with a valuable tool for drug and biomedical research. Uniquely, stem cells can be used as genetic or cell-based therapies to repair, replace or replenish the cells damaged or lost due to degenerative diseases. Consequently, most recent research trends have focused on investigating newer sources to generate insulin-producing cells that are easily obtainable and usable in experimental and therapeutic settings for cell replacement therapy for diabetes. Insulin-producing cells has been isolated and extracted from various sources, including human ES cells, bone marrow or umbilical cord-derived mesenchymal cells, transdifferentiation of liver/gallbladder cells, pancreatic duct cells, islet-derived mesenchymal cells. Yet these cells produce the least amount of insulin compared to normal adult pancreas. To enhance the differentiation of adult tissue-derived progenitor cells, the major focus of present research is on growth factors, signaling molecules and processes that would promote differentiation of these cells. This chapter summarizes the advancements in stem cell biology research, problems encountered, limitations and future of stem cell therapy with the emphasis on transplantation of pancreatic islet or β cells as a therapeutic step for the cure of type 1 Diabetes.

2.1 INTRODUCTION

Diabetes is one of the major non-communicable diseases which results in inadequate insulin secretion due to destruction or loss of beta-cells which may be either from autoimmune disorder or environmental factors resulting in infiltration of inflammatory cells in pancreatic islets and destroying the beta-cells. Type 1 diabetes (T1D) is associated with heart disease and stroke as well as it results in increased morbidity and mortality [1]. Because mouse models have been insufficiently significant to human type 1 diabetes, there is a need for developing human tissue or stem cell models to investigate this disease.

T1D can be diagnosed at any age and up to 50% or more are age 18 or older at the onset of symptoms. However, this disease also occurs as one of the most chronic diseases in children in some parts of the world with the most incidences seen between the ages of 5 to 7 years [2]. The clinical onset is characterized by hyperglycemia and the resulting triad of symptoms including polydipsia, polyuria, and polyphagia. If untreated, T1D leads to ketoacidosis and death.

Diagnosis of disease can be made based on the frequently elevated plasma glucose levels during fasting or two hours after an oral glucose tolerance test (OGTT). Recently the glycated hemoglobin (HbA1c) reflecting average plasma glucose over the previous 8 to 12 weeks has been reported a new diagnostic test for diabetes [3, 4]. T1D constitutes 5–10% of the total cases worldwide [5] with the majority of patients belonging to India, China and USA [1]. T1D also increases the risk of developing other autoimmune diseases such as celiac disease and autoimmune thyroiditis [6, 7].

Although there is extensive research to investigate diabetes, there is no definitive treatment for diabetes to date. Daily administration of exogenous insulin was introduced in 1922 by Banting, Best and Collip, which is the only available therapy for T1D. In addition, newer research into transplantation of islets and stem cells capable of generating insulin-producing cells have been performed which led to the new paradigm of cellular-based therapy of diabetes. However, even with these advancements in insulin and cell-based therapies, there are several complications of diabetes which even persist such as retinopathy, nephropathy, and neuropathy which affect quality of life [8].

2.1.1   PANCREAS IN DIABETES

In type 1 diabetes, the endocrine pancreas organ undergoes autoimmune destruction or loss in which the islets of Langerhans cells progressively become dysfunctional or non-functional. A genome-wide analysis study (GWAS) shows that an adult can show clinical symptoms when there is a 40% loss of pancreatic beta-cell mass [9]. Beta-cells are destroyed by the immune cells which infiltrate islets of Langerhans. These immune cells, such as T cells are recruited to the beta cells in response to a triggering event such as viral infections [10]. Initial infection calls off the antigen presenting cells to the lymph nodes close to the pancreas where they take up the antigens processed from the beta cells. These antigens are presented to the autoreactive T cells. Autoreactive T cells are self-reactive immune cells, which escape the deletion procedure during their development in thymus. Beta cell antigen presentation to these T cells leads to migration of T-lymphocytes to islets where they induce inflammation resulting in huge beta cell death [11]. Antibodies against self-beta antigens (known as auto-antibodies) are detected in serum of patients several years before the onset of diabetes. An increasing number of autoantibodies in serum increases the risk of diabetes in patients. Hence, autoantibodies can also be used to identify patients with a high risk of developing the type 1 diabetes. The presence of serum-autoantibodies initially developed a notion that type 1 diabetes is an autoimmune disease. But now research has supported that autoantibodies develop as a result of tissue damage by infiltrating immune cells and the lost beta cells are processed and generating more intracellular epitopes or antigenic epitopes resulting in producing more autoantibodies.

Unaffected beta-cells may regain insulin secretion functions after some days of *in vitro* culturing [12]. Therefore, it is possible to restore the lost function by preserving the beta cells or restoring the functions of lost beta cells by transplanting islets to the patients which can either be done by the use of human embryonic stem cells (ESCs), adult stem or progenitor cells, pancreatic stem or progenitor cells, or mesenchymal stem cells (MSCs).

2.2   DEVELOPMENT OF PANCREAS

The pancreas is an unpaired glandular organ of about 15 cm length and located posteriorly in the abdominal cavity and having average weight 68 g [13]. Histologically and functionally pancreas is divided into endocrine and exocrine compartments. Endocrine cells of pancreas form islets or Islets of Langerhans named after Paul Langerhans, who first described them in 1869. These are minute clusters of endocrine cells scattered throughout the pancreas. Islets consist of five distinct cell types: insulin-producing beta (β) cells (50–60%), glucagon-producing alpha (α) cells (30–40%), to a lesser extent somatostatin-producing delta (δ) cells, pancreatic polypeptide-secreting PP cells, and ghrelin-producing epsilon cells (ε) cells [14, 15]. The exocrine pancreas consists of acinar, centroacinar and ductal cells, which comprise 90% of the pancreatic mass, white blood vessels, connective tissue stroma, and nerves make up the remaining 8–9%. A loose thin connective tissue layer forms a septum dividing the gland into lobules [16]. The acinar cells produce digestive enzymes and bicarbonate also known as pancreatic juice that mixes with bile and is released into duodenum via pancreatic duct. Insulin hormone has hypoglycemic action. Insulin hormone is released into the bloodstream when the blood sugar level rises which helps body cells, such as muscle cells and adipose cells absorb the blood glucose and use it for energy. Due to insufficient secretion of insulin or lack of insulin-absorption power of cells, the sugar level in the bloodstream rises. Insulin is synthesized as a prohormone precursor in beta cells of islets of Langerhans. The newly synthesized proinsulin in ribosomes found on rough endoplasmic reticulum is then transported to the Golgi apparatus where it is packaged in secretory vesicles. Later the proinsulin in the vesicles is processed by a series of proteases to form mature insulin [17–19]. In some autoimmune conditions, the body's immune system destroys the pancreas, resulting in partial or complete loss of beta cells of pancreas, which, in turn, causes building up of bloodstream sugar level indicating development of Type I diabetes mellitus (T1DM) or juvenile diabetes or insulin-dependent diabetes [20]. Whole pancreas or islets are transplanted to restore insulin production or functions of beta cells, but a long-term treatment is required to avoid or stop the immune-mediated rejection of the transplanted islet cells. Islet transplantation offers a potential alternative to whole pancreas transplantation, however, early attempts rarely showed any success. Despite the development of newer sources such as human ES cells [21, 22], induced pluripotent stem cells (iPSCs) [23], hepatocytes [24, 25], pancreatic duct cells [26, 27], and islets from human and pigs [28, 29] successful therapy requires an improved differentiation protocol to enhance differentiate these cells into fully functional, insulin-secreting and blood glucose-responsive beta cells of islets of Langerhans in the pancreas.

2.2.1 *EMBRYONIC DEVELOPMENT OF PANCREAS*

During the embryogenesis in mice, the epiblast is derived from the inner cell mass of the blastocysts during gastrulation, which gives rise to principal germ layers- ectoderm, endoderm and mesoderm. Mesendodermal cells regulate the expression of several genes such as *Brachyury (T)* and *Mixl1* which control differentiation of definitive endoderm and mesoderm progenitors. Pancreas arises from definitive endoderm (DE) as a flat sheet of cells that is specified during gastrulation [30, 31].

DE then forms a primitive gut tube (GT) along which various endoderm organ domains are specified and directed [32]. Pancreas develops from the posterior foregut, emerging as buds from dorsal and ventral sides of the gut tube. At this early stage pancreatic development depends on retinoid signaling and inhibition of hedgehog signaling [33, 34]. The developing pancreas at this stage consists of epithelial progenitors that express Pdx1 (*IPF*1) and give rise to endocrine, exocrine and ductal cells of the pancreas. Moreover, this epithelium also expresses transcription factor genes such as *Hlxb9, Hnf6, Ptf1a, and Nkx6-1*. These transcription factors together with *Pdx1* encode proteins that contribute to pancreatic development [35]. After initial bud formation, the epithelium grows, proliferates and differentiates in response to signals emanating from adjacent mesenchyme such as mesenchymal Fgf10 [36]. In addition to transcription factors, several growth factors regulate the process of gastrulation. For example, in mice Nodal, a member of transforming growth factor (TGF)-β super family, which, in turn, regulates the Wnt, fibroblast growth factor (FGF) and bone morphogenetic protein-4 (BMP-4) pathways that is important for development of anterior and posterior axes during gastrulation [31, 37].

After the pancreatic epithelium growth and proliferation, Notch signaling is inhibited in some epithelial cells, allowing the expression of pro-endocrine gene *Neurogenin-3 (Ngn3)* [35]. *Ngn3* is expressed in all endocrine progenitor cells, which initiates a cascade of transcription factor expression which, in turn, initiates differentiation of endocrine cells. This cascade includes *Nkx2-2, Neurod–1, Nkx6-1, Pax6, Pax4,* and *Isl1.*

The nascent endocrine cells migrate from epithelial branches into surrounding mesenchyme to form Islets of Langerhans. A schematic of stages of differentiation of human ESCs into pancreatic islets and transcription factors involved is shown in Figure 2.1. In mice, digestive gut is formed at embryonic day 8.5. The early specification of pancreatic buds requires signals from notochord (Activin-β and FGF-2). Between the embryonic days 9.5 and 12.5 FGF-10 helps to proliferate epithelial pancreatic progenitors. On the day 9.5, some glucagon secreting alpha (α) cells begin to appear whereas most hormones secreting cells, such as α, β, δ, ε (in mice) and pancreatic progenitors (PP) emerge from pancreatic duct epithelium during the days 13.5–14.5 [30, 38].

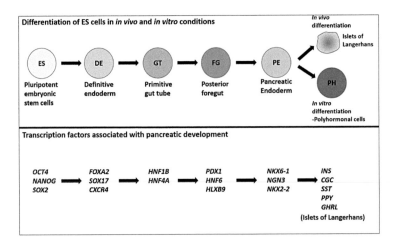

FIGURE 2.1 Schematic representation of the stages and transcription factors involved in differentiation of embryonic stem cells to insulin-producing cells. The figure depicts the differentiation of ESCs in the *in vivo* and *in vitro* conditions producing insulin-producing cells and heterogeneous cells, respectively. *In vitro* differentiation mimics development of pancreas in vertebrates. Upper panel shows the stages of endocrine pancreas development. Lower panel shows the transcription factors or markers specific to each population.

2.2.2 SIGNALING FACTORS AND TRANSCRIPTION FACTORS IN PANCREAS DEVELOPMENT

Evidences exist which show that hepatocyte growth factor (HGF) which is a mesenchymal factor which induces β-cell formation. Another mesenchymal factor epidermal growth factor (EGF) regulates the proliferation of developing pancreas. Some glucagon family members such as glucagon-like peptide 1 and 2 (GLP-1 and GLP-2), glucose insulinotrophic peptide (GIP) play a role in early differentiation of insulin-producing cells [39]. GLP-1 signaling enhances the proliferation of existing cells, induces islet neogenesis and inhibits β-cell apoptosis [40, 41]. GIP promotes growth and survival of pancreatic β-cells by increasing expression of *Bcl-2* (anti-apoptotic gene) and decreasing the pro-apoptotic gene *Bax* which results in diminished β-cell death [42]. Both GIP and GLP-1 are primary physiological incretins which are intestinal hormones that are released in response to ingested food and stimulate insulin secretion in glucose-dependent manner [43].

2.3 STEM CELLS AS REPLACEMENT THERAPY FOR DIABETES

An ideal cell therapy for replacing the lost or destroyed β-cells with the islet cells which are readily available, easily maintained, expanded, and can produce the therapeutically comparable amount of insulin when transplanted into patients.

2.3.1 HUMAN EMBRYONIC STEM CELLS

Self-renewable and pluripotent ESCs are able to produce any cell in the body. ESCs are derived from the inner cell mass (ICM) of blastocyst-stage embryos. These cells are capable of proliferating indefinitely in serum-containing media on mitotically inactivated mouse fibroblast layer (feeder cells) while retaining the potential to differentiate into any adult cell type [44]. Advancements have been made to derive and maintain hESCs on human feeder cells and feeder-free matrices in various growth factors supplemented basal media [45, 46]. These developments enable the researchers to omit nonhuman components in culture media and thus to generate clinical grade cell lines to improve the therapeutic value of stem cells. ESCs have tremendous potential to differentiate into other cell or tissues. *In vitro* differentiation of mouse and human ESCs can produce pancreatic β-like insulin-producing cells.

ES cells can be spontaneously differentiated by culturing ES cells in suspension in the absence of factors that support self-renewal. This allows them to form three-dimensional aggregates known as embryoid bodies. There are some limitations associated with the spontaneous differentiation of ESCs. Individual cell lines show much degree of differences in their propensity to differentiate into particular lineages. The ratio of specific cell populations is less and the generated cells do not produce insulin at therapeutically relevant levels.

Directed lineage specific differentiation of ES cells is the most successful method to produce a lineage of interest. In this method, ES cells are exposed to signals and growth factors that normally are present during embryonic development. Media supplemented with these factors induces ES cells into desired lineage. Directed differentiation of human ES cells in the serum-free system has been used to induce their differentiation into functional insulin-producing cells. This process initially included Activin A to induce hESCs in definitive endoderm lineage. Subsequently, all-trans retinoic acid was used to promote pancreatic differentiation followed by induction with bFGF and nicotinamide which resulted in expression of islet-specific markers [47]. Genetic engineering and transgenic mice offer a great tool to study molecular mechanisms of differentiation mechanisms of ES cells into pancreatic β-cells.

2.3.2 ADULT PANCREATIC STEM CELLS

These are clusters of progenitor cells that arise from developing foregut during embryonic development and capable of giving rise to the pancreas. They possess a tremendous potential for self-renewal and multilineage differentiation. They can generate both exocrine and endocrine cells of the pancreas. Adult pancreatic stem cells are present during fetal development of the pancreas [48] and adult pancreas (ductal, acinar and islet regions) possesses the ability to differentiate and regenerate into the endocrine phenotype of the pancreas. Highly purified epithelial cells from non-endocrine pancreas (ductal and acinar cells) differentiate into beta-cell phenotype [49]. These cells are readily accessible and therefore are good alternatives for the sources of beta cells for treatment of diabetes.

Pancreatic ductal cells offer an additional alternative for cell replacement therapy for diabetes. Islet neogenesis occurs in the pancreatic ductal epithelium during normal embryonic development and in diabetic mouse [50–53]. Both human adult ductal epithelial and fetal human pancreatic duct cells expand and proliferate *in vitro* to generate pancreatic islet-structures which respond to glucose concentration [50, 54, 55]. This growth and proliferation are promoted by the transcription factors such PDX1 and ngn-3 which is proved by the transfection of pancreatic ductal cells with adenoviral vectors which express these transcription factors [56, 57].

2.3.3   MESENCHYMAL STEM CELLS

Mesenchymal stem cells are pluripotent stromal cells that proliferate and differentiate into a variety of cell types. For example, human adipose-derived tissue from liposuction aspirates were induced to differentiate into insulin-producing cells *in vitro* with the combination of three factors: β-mercaptoethanol (BME), nicotinamide and extendin-4 [46]. Pancreatic beta-cells retain the capacity to proliferate both *in vitro* and *in vivo*. However, these cells undergo epithelial to mesenchymal transition during *in vitro* expansion. Thus, mesenchymal-like progenitor cells are also generated from beta-cells. The members of the mir-30 family of microRNAs are involved in regulation of epithelial to mesenchymal transition of human islets [58]. On the other hand, mouse beta-cells do not proliferate in culture [59]. The studies showed that certain physiochemical stimuli may induce proliferation of mouse β-cells *in vivo* [60–65]. Whether a mesenchymal intermediate is involved in this proliferation is not known [66]. Human pancreatic islet-derived progenitor cells are committed to differentiate into endocrine pancreatic lineage. In sum, islet-derived mesenchymal progenitor cells are great sources for beta-cells in diabetes replacement therapy.

2.3.4   TRANSDIFFERENTIATION

In addition to endocrine cells, pancreatic exocrine cells also transform into other mature somatic lineage via transdifferentiation. In studies performed in rodents, it has been reported that beta-cell neogenesis and proliferation occurs from acinar cells instead of ductal progenitor cells [67]. Three transcription factors PDX1, ngn-3, and Mafa could play a major role in transdifferentiation and beta-cell neogenesis from pancreatic exocrine cells [68].

Certain organs due to their structural, and embryonic origin-similarities, may also show transdifferentiation. For instance, liver and pancreas show responsiveness to glucose and expression of similar genetic transcription factors. They share characteristics, origin and their adjacent location is suggestive of the fact that liver cells are convincing extrapancreatic cells that can transdifferentiate into pancreatic β-cells and thus can be used to generate transplantable insulin-producing cells [69]. Both adult and fetal liver cells have successfully been transfected via viral agents and trans-differentiated into pancreatic β-like cells [70]. Virally mediated combined expression of Pdx1/ VP16 with neuroD/NGN3 in human hepatoma cell line, HepG2 induces insulin production and improves glucose tolerance in diabetic mice [71].

2.3.5   NOVEL SOURCES OF PANCREATIC ISLET CELLS

Other sources from which insulin-producing cells have been produced are: human umbilical cord blood stem cells [72], human neuroprogenitor cells [73], bone marrow stem cells [74], human adipose tissue-derived mesenchymal stem cells [75], and placenta-derived multipotent stem cells [76]. These adult stem cells are of great value in replacing the lost beta-cells because there is low risk of graft versus-host disease (GVHD), no risk to the donors and are of little ethical concerns. These stem cells may also be obtained from self-tissues which enables autologous stem cell therapy and therefore a platform for personalized medicine. Various types of stem cells and their advantages and disadvantages in clinical applications for therapy of diabetes type 1 are shown in Table 2.1.

Due to insufficient number of pancreatic donors, we need robust sources of pancreatic islets. In long run, we need to investigate flexible sources of pancreatic beta cells. Recently, pluripotent embryonic stem cells (ESCs) have been investigated, however, there are major ethical concerns. Additionally, teratoma formation hindered the clinical trials. Multipotent adult stem cells are more promising because they can be retrieved from the same individual and are less prone to malignant transformation compared to ES cells. Clinical trials using multipotent adult stem cells in treatment of diabetes have successfully been initiated.

Most recently, mesenchymal stem cells (MSCs) have been studied well in several diseases due to their ability to differentiate into a plethora of lineages such as adipocytes, chondrocytes, osteoblasts, myoblasts,

TABLE 2.1 Advantages and Disadvantages of Different Types of Stem Cells for Clinical Application in Type 1 Diabetes

Stem cells	Advantages	Disadvantages
Embryonic stem cells	• Pluripotent • Self-renewal and unlimited proliferation	• Ethical concerns as it requires destruction of an embryo • May be tumorigenic post transplantation
Induced pluripotent stem cells	• Pluripotent and unlimited proliferation • Autologous (patient-specific) and there is no immune-rejection after transplantation • Procedure is quick and efficient • Does not require an oocyte or destruction of an embryo	• Oncogenic and viral integration • Genome instability and clonally variable • May cause tumor after transplantation
Adult stem cells	• Found in many tissue and therefore patient-specific, no risk to donors and low risk of graft-versus host disease • Little ethical concerns as does not require destruction of an embryo	• Lineage restricted (multipotent) • Limited in proliferation, therefore, difficult to generate adequate number of specialized cells for transplantation.

cardiomyocytes, marrow stromal cells, hepatocytes, neuronal cells, renal cells, and pancreatic cells. The MSCs is nonhematopoietic multipotent self-renewing cells, which can originate from a variety of tissues such as skeletal muscle, skin and foreskin, adipose tissue, pancreas, dental pulp, salivary glands, endometrium, umbilical cord, etc. MSCs show immunomodulatory or immunosuppressive behavior as cotransplantation of MSCs with pancreatic islets in mice leads to improved beta cell function and survival along with the improved glucose homeostasis and reduced islet cell apoptosis. MSCs from various tissues have immunomodulatory effects as seen by reduced proliferation or activation of natural killer cells, dendritic cells, and T-cells in the recipient by transplanting or co-transplanting MScs. Moreover, MSCs cotransplanted with the grafts decrease secretion of inflammatory cytokines such as interferon-gamma (IFN-γ), tumor necrosis factor-alpha (TNF-α), granulocyte-macrophage-colony-stimulatingfactor (GM CSF) and monocyte chemotactic protein-1 (MCP-1). Very recently, amniotic fluid-derived stem cells have been reported to differentiate into insulin-producing cells [77].

2.4 ISLET TRANSPLANTATION FOR DIABETES

Whole pancreas transplantation restores glycemic control and hypogly-cemic awareness in patients instantly with a functional graft. However, the morbidity and mortality associated with the surgery and the adverse effects immunosuppression limit the use of this option to a small patient population. In 1960 Dr. Paul Eston Lacy pioneered the experimental techniques of beta islet cell isolation and transplantation in animals as treatment of diabetes mellitus [78–80]. Islet transplantation is a mini-mally invasive procedure with much lower morbidity. Later in 1990s Dr. James Shapiro and other collaborators at the University of Alberta in the Canadian city of Edmonton developed the protocol to isolate pancreatic islet cells and transplant the islets from a donor pancreas into a recipient. The protocol was named the Edmonton protocol and was first published in New England Journal of Medicine in the year 2000 [81]. There has been a limited success with islet transplantation in early years, but it has significantly improved after the development of the Edmonton protocol. The Edmonton protocol involved isolating islets from the cadaveric donor pancreas using a digestive enzyme mixture called liberase [82]. The protocol has been refined to achieve and maintain sustained insulin independence, enhance islet engraftment, and to reduce multiple number of islet donors. Islets are usually infused in the liver via the portal vein. Immunosuppressive medicines, sirolimus and tacrolimus, were used to control or stop recipient's immune system to maintain the transplanted islets [83–85]. Exenatide, a glucagon-like peptide–1 analog has been used to maintain survival of islet cells. Similarly, a tumor necrosis factor-alpha inhibitor, etanercept, and a T cell depleting antibody (TCDAb), antithymocyte globulin, have been used as immunosuppressive agents and improved the pancreatic islet cell transplantation [86–88]. The overall goal of islet transplantation is to infuse enough islet beta cells to control the blood glucose level and removing the need for repeated insulin injections.

2.4.1 TYPES OF ISLET TRANSPLANTATION

There are two types of pancreatic beta islet transplantation – allo-trans-plantation and auto-transplantation [89].

2.4.1.1 PANCREATIC ISLET ALLO-TRANSPLANTATION

During pancreatic allo-transplantation, the islets from the pancreas of a deceased organ donor are purified, processed and transferred into another person. Each allo-transplantation involves the use of specialized enzymes by researchers to dissociate islets from the pancreas of a single, deceased donor. Islets are processed, purified and counted in a laboratory. A patient undergoing transplantation usually receives two and sometimes more infusions each having an average of 400,000 to 500,000 islets. The beta cells of implanted islets begin to synthesize and secrete insulin.

Type 1 diabetes patients whose glucose levels are difficult to control, receive pancreatic islet allo-transplantation to bring the glucose level to normal and eliminate the symptoms of hypoglycemia without daily injections of insulin. Hypoglycemic unawareness is a dangerous situation resulting from intensified insulin injections in which Type 1 diabetes patients cannot feel the symptoms of hypoglycemia.

The procedure of allo-transplantation includes placement of a thin, flexible tube called a catheter through a small incision in the upper abdomen between the chest and the hips, into the portal vein of the liver, which is the major vein supplying blood to the liver. The islets are then slowly pushed or infused into the liver via catheter. The transplantation procedure is usually performed by a radiologist who is specialized in medical imaging who uses X rays and ultrasound to guide the placement of catheter into appropriate place. Usually, the procedure is done under local anesthesia and a sedative. This procedure is considered experimental [90, 91].

2.4.1.2 ADVANTAGES OF ALLO-TRANSPLANTATION

1. Improved blood glucose control which, in turn, slows or prevents the progression of diabetes problems such as heart disease, kidney disease and nerve or eye damage;
2. Reduction or elimination of the need for insulin injections to control disease; and
3. Prevention of hypoglycemia and hypoglycemic unawareness.

2.4.1.3   RISKS ASSOCIATED WITH PANCREATIC ISLET ALLO-TRANSPLANTATION

1. Risks associated with the transplantation procedure; especially bleeding and blood clots.
2. Transplanted islets may not function well or may stop functioning entirely.
3. Burden of lifelong immunosuppression given to protect the transplanted tissue from the host's immune system which may sometimes cause adverse effects.

2.4.1.4   PANCREATIC ISLET AUTO-TRANSPLANTATION

Patients with type 1 diabetes cannot receive islet auto-transplantation. This type of transplantation is performed in patients who have severe or untreatable pancreatitis. The patient receives general anesthesia and the surgeon removes pancreas, and then extracts and purifies islets [92]. These purified islets are then infused within hours to patient. The overall goal of the islet auto-transplantation is to give body enough healthy islets to make insulin after removal of pancreas in pancreatitis [93].

After islet transplantation, the patient receives the medication before and after the procedure until the islets are fully functional. Islets develop new blood vessels and take time until they become functional and begin to make insulin. Sometimes, an autoimmune response may destroy the transplanted islets the second time, which may result in failure of islet functions. Although, the liver is the primary site of islet infusion, research is ongoing to identify alternative sites for infusing the islets such as myogenic/muscle tissue or another organ [94]. Figure 2.2 shows the characteristics of islet-allo and -auto transplantation and risks associated.

2.4.2   ROLE OF IMMUNOSUPPRESSIVE MEDICATION IN ISLET TRANSPLANTATION

A common problem with any transplant is the rejection. Immunosuppressants are prescribed to stop host's immune system to attack transplanted or grafted cells. Immediate side effects of these medications are mouth sores and gastrointestinal problems such as upset stomach and diarrhea.

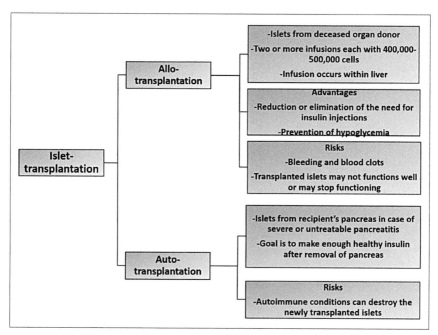

FIGURE 2.2   Representation of types of Islet transplantation.

Immunosuppressive medications have serious side effects such as high blood pressure, increased blood cholesterol or blood fat, anemia, decreased white blood cell counts, etc. These medications also increase the risk of developing certain tumors and cancers [95].

Research is being performed to develop new strategies to achieve immune tolerance in which transplanted islets are no longer recognized as foreign by the body's immune system. Immune tolerance would enable host to maintain the transplanted islets without the use of immunosuppressive medications. One of such strategies is to transplant islet cells encapsulated with a special coating, which may help to prevent rejection [96–98].

2.4.3   LIMITATIONS

There are obstacles to the use of islet allo-transplantation. The number of deceased pancreatic donors (donor pool) is usually not enough to meet the requirements of number of diabetes patients (recipient pool). In addition, many donated pancreases are not suitable for extraction of islet cells and sometimes islets cells are often damaged or destroyed during processing. Consequently, only a small number of islet transplants can be performed each year [99]. Moreover, we do not have adequate number of means to prevent rejection of transplanted islets.

Another challenge is successful islet transplantation is financial barrier. The immunosuppressive agents used in this treatment are expensive. The experimental expenses are generally not covered by health insurance companies [100]. Furthermore, immunosuppressants increase the risk for malignancies and opportunistic infections. This begs the researchers to strategize the newer ways to discover easily obtainable and expandable tissues or cells which may reduce the cost of therapy.

Alternative sources of islets such as stem cells, iPSCs, MSCs, adult pancreatic stem cells and islets from pigs are tested and screened. Islets from pigs have been transplanted into other animals, including monkeys by encapsulating the islets with a special coating to prevent rejection or by using immunosuppressive medications [97, 98]. Similarly, novel sources of islets are used such as creating islets from stem cells which will allow researchers to grow islets in laboratories indefinitely. Using autologous stem cells, it is possible to develop islets suitable for auto-transplantation.

2.4.4 POST-TRANSPLANTATION REGIMEN FOR PATIENTS

Recipients of the islet transplants should follow a meal plan worked out by a health care provider and dietitian. Immunosuppressive medications taken after a transplant can cause changes in a person's body, for example, weight gain is observed. A healthy diet and meal plan after the transplantation procedure is critically important to control blood pressure, blood cholesterol, blood glucose level, and weight gain.

2.5 CONCLUSIONS AND PROSPECTS

The pancreas is an endocrine organ in vertebrates which is essential for glucose homeostasis via secretion of insulin by beta cells. Autoimmune diseases targeting beta cells result in lack of insulin and, in turn, diabetes mellitus. Developing stem cell replacement therapy for diabetes represents a prime research interest in order to replenish or restore the functions of lost or destroyed islet beta-cells. Considerable research grew for several years, which has discovered the ways to promote differentiation of embryonic stem cells and adult stem or progenitor cells into pancreatic beta-cell lineage. The research has indicated that endocrine and exocrine pancreatic adult progenitor cells can be a potential source of islet beta-cells. The pluripotent ESCs and mesenchymal stem cells can be differentiated with the use of spontaneous or directed differentiation protocol to generate glucose-responding islet cells. The use of ESCs is associated with ethical concerns. Stem cells or mesenchymal stem cells, as well as progenitor cells recovered from readily accessible adult tissues with no or little ethical concerns such as umbilical cord blood stem cells, adipose-derived mesenchymal stem cells, fetal and adult pancreatic duct progenitor cells, are a viable source of islet cells. These tissue-derived adult stem cells can be used in autologous fashion to avoid or minimize the immunological rejection of the transplanted beta cells. The exocrine pancreatic cells, such ductal cells and acinar cells also possess the transdifferentiation capacity to produce insulin-producing cells.

Despite a multitude of advanced breakthroughs in stem cell research there has not been a promising cell tissue type to generate islet beta cells, which can produce physiological relevant levels of insulin in adult diabetes patients. Further research into testing mesenchymal stem cells, adult stem cells, and pancreatic ductal cells to analyze their differentiation power into specific islet beta-cells need to be completed. An additional pre-clinical work is to be done to identify the signaling and growth factors to direct differentiation of stem cells into insulin-producing beta-cells. A plethora of growth factors, transcriptional factors and genetic factors have been reported, however, a great deal of additional work is needed to refine Edmonton protocol and ensure differentiation parameters to obtain highly pure and multipotent stem cells to produce insulin-producing cells.

KEYWORDS

- **adult stem cells**
- **autoimmune disease**
- **stem cells**
- **type 1 diabetes**
- **islet transplantation**

REFERENCES

1. Bhartiya, D., (2016). Stemcells to replace or regenerate the diabetic pancreas: Huge potential and existing hurdles. *Indian J. Med. Res., 143*(3), 267–274.
2. Thunander, M., Petersson, C., Jonzon, K., Fornander, J., Ossiansson, B., & Torn, C., (2008). Incidence of type 1 and type 2 diabetes in adults and children in Kronoberg, Sweden. *Diabetes Res. Clin. Pract., 82*(2), 247–255.
3. World Health Organization, (2006). Definition and diagnosis of diabetes mellitus and intermediate hyperglycemia: report of a WHO/IDF consultation.
4. World Health Organization, (2011). WHO: Use of glycated hemoglobin (Hb A1c) in the diagnosis of diabetes mellitus. Abbreviated report of a WHO consultation.
5. American Diabetes Association, (2009). Diagnosis and classification of diabetes mellitus. *Diabetes Care, 32*(1), 62–67.
6. Elfström, P., Sundström, J., & Ludvigsson, J. F., (2014). Systematic review with meta-analysis: associations between coeliac disease and type 1 diabetes. *Aliment. Pharmacol. Ther., 40*(10), 1123–1132.
7. Umpierrez, G., Latif, K., Murphy, M. B., Lambeth, H., Stentz, F., & Bush, A., (2003). Thyroid dysfunction in patients with type 1 diabetes. *Diabetes Care, 26*(4), 1181–1185.
8. Nielsen, H. B., Ovesen, L. L., Mortensen, L. H., Lau, C. J., & Joensen, L. E. (2016). Type 1 diabetes, quality of life, occupational status and education level – a comparative population-based study. *Diabetes Res. Clin. Pract., 7*, 62–68
9. Klinke, D. J., (2008). Extent of beta cell destruction is important but insufficient to predict the onset of type 1 diabetes mellitus. *PLo SOne., 3*(1), 1–10.
10. Gepts, W., (1965). Pathologic anatomy of the pancreas in juvenile diabetes mellitus. *Diabetes., 14*(10), 619–633.
11. Peakman, M., (2013). Immunological pathways to beta-cell damage in Type 1 diabetes. *Diabet Med., 30*(2), 147–154.
12. Krogvold, L., Skog, O., Sundström, G., Edwin, B., Buanes, T., Hanssen, K. F., et al., (2015). Function of isolated pancreatic islets from patients at onset of type 1 diabetes: Insulin secretion can be restored after some days in a nondiabetogenic environment *in vitro* results from the Di Vid study. *Diabetes, 64*(7), 2506–2512.
13. Ogilvie, R. F. A., (1937). Quantitative estimation of the pancreatic islet tissue. *Quart. J. Med., 6*, 287–300.
14. Orci, L., & Unger, R., (1975). Functional subdivision of islets of Langerhans and possible role of D cells. *Lancet., 306* (7947), 1243–1244.
15. Bosco, D., Armanet, M., Morel, P., Niclauss, N., Sgroi, A., Muller, Y. D., et al., (2010). Unique Arrangement of alpha-and beta-cells in Human Islets of Langerhans. *Diabetes., 59*, 1202–1210.
16. Pandiri, A. R., (2014). Overview of exocrine pancreatic pathobiology. *Toxicol. Pathol., 42*, 207–216.
17. De Vos, A., Heimberg, H., Quartier, E., Huypens, P., Bouwens, L., Pipeleers, D., et al., (1995). Human and rat beta cells differ in glucose transporter but not in glucokinase gene expression. *J. Clin. Invest., 96*, 2489–2495.

18. Hellman, B., Salehi, A., Gylfe, E., Dansk, H., & Grapengiesser, E., (2009). Glucose generates coincident insulin and somatostatin pulses and antisynchronous glucagon pulses from human pancreatic islets. *Endocrinology, 150*(12), 5334–5340.

19. Steiner, D. F., Cunningham, D., & Spigelman, L. A. B., (1967). Insulin Biosynthesis: Evidence for a Precursor. *Science, 157*(3789), 697–700.

20. Klinke, D. J. (2008). Extent of beta cell destruction is important but insufficient to predict the onset of type 1 diabetes mellitus. *PLo S One, 3*(1), 1–10.

21. Stanley, E. G., & Elefanty, A. G., (2008). Building better beta cells. *Cell Stem Cel., 2*(4), 300–301.

22. Baetge, E. E. (2008). Production of beta-cells from human embryonic stem cells. *Diabetes Obes Metab., 10* Suppl, *4*, 186–194.

23. Zhang, D., Jiang, W., Liu, M., Sui, X., Yin, X., Chen, S., et al., (2009). Highly efficient differentiation of human ES cells and i PS cells into mature pancreatic insulin-producing cells. *Cell Res., 19*(4), 429–438.

24. Meivar-Levy, I., & Ferber, S., (2015). Reprogramming of liver cells into insulin-producing cells. *Best Pract Res ClinEndocrinolMetab., 29*(6), 873–882.

25. Thowfeequ, S., Li, W. C., Slack, J. M. W., & Tosh, D., Reprogramming of liver to pancreas. p. 407–418.

26. Corritore, E., Lee, Y. S., Sokal, E. M., & Lysy, P. A., (2016). β-cell replacement sources for type 1 diabetes: A focus on pancreatic ductal cells. *Ther Adv Endocrinol Metab., 7*(4), 182–199.

27. Yatoh, S., Dodge, R., Akashi, T., Omer, A., Sharma, A., Weir, G. C., et al., (2007). Differentiation of affinity-purified human pancreatic ductcells to beta-cells. *Diabetes., 56*(7), 1802–1809.

28. Montgomery, A. M., & Yebra, M., (2011). The epithelial-to-mesenchymal transition of human pancreatic β-cells: inductive mechanisms and implications for the cell-based therapy of type I diabetes. *Curr. Diabetes Rev., 7*(5), 346–355.

29. Zhu, H. T., Lu, L., Liu, X. Y., Yu, L., Lyu, Y., & Wang, B., (2015). Treatment of diabetes with encapsulated pigislets: An update on current developments. *J. Zhejiang Univ. Sci., B.16*(5), 329–343.

30. Oliver-Krasinski, J. M., & Stoffers, D. A., (2008). On the origin of the beta cell. *Genes Dev., 22*(15), 1998–2021.

31. Murry, C. E., & Keller, G., (2008). Differentiation of embryonic stem cells to clinically relevant populations: lessons from embryonic development. *Cell., 132*(4), 661–680.

32. Wells, J. M., Melton, D. A., (1999). Vertebrate endoderm development. *Annu. Rev. Cell Dev. Biol., 15*, 393–410.

33. Stafford, D., Hornbruch, A., Mueller, P. R., & Prince, V. E., (2004). A conserved role for retinoid signaling in vertebrate pancreas development. *Dev. Genes Evol., 214*, 432–441.

34. Lau, J., Kawahira, H., & Hebrok, M., (2006). Hedgehog signaling in pancreas development and disease. *Cell. Mol. Life Sci., 63*, 642–652.

35. Wilson, M. E., Scheel, D., & German, M. S., (2003). Gene expression cascades in pancreatic development. *Mech. Dev., 120*, 65–80.

36. Bhushan, A., Itoh, N., Kato, S., Thiery, J. P., Czernichow, P., Bellusci, S., et al., (2001). Fgf10 is essential for maintaining the proliferative capacity of epithelial progenitor cells during early pancreatic organogenesis. *Development, 128*, 5109–5117.

37. Zorn, A. M., & Wells, J. M., (2009). Vertebrate endoderm development and organ formation. *Annu. Rev. Cell. Dev. Biol., 25*, 221–251.

38. Cano, D. A., Hebrok, M., & Zenker, M., (2007). Pancreatic development and disease. *Gastroenterology., 132*(2), 745–762.

39. Gittes, G. K., (2009). Developmental biology of the pancreas: A comprehensive review. *Dev. Biol., 326*(1), 4–35.

40. Xu, G., Kaneto, H., Lopez-Avalos, M. D., Weir, G. C., & Bonner-Weir, S., (2006). GLP-1/exendin-4 facilitates beta-cell neogenesis in rat and human pancreatic ducts. *Diabetes Res. Clin. Pract., 73*, 107–110.

41. Tourrel, C., Bailbé, D., Meile, M. J., Kergoat, M., & Portha, B., (2001). Glucagon-like peptide-1 and exendin-4 stimulate beta-cell neogenesis in streptozotocin-treated newborn rats resulting in persistently improved glucose homeostasis at adult age. *Diabetes., 50*(7), 1562–1570.

42. Mc Intosh, C. H. S., Widenmaier, S., & Kim, S., (2009). Glucose-dependent Insulinotropic Polypeptide (Gastric Inhibitory Polypeptide, GIP), Vitamins and Hormones Elsevier. *Inc.*, p. 409–471.

43. Hansotia, T., & Drucker, D. J., (2005). GIP and GLP-1 as incretin hormones: lessons from single and double incretin receptor knockout mice. *Regul. Pept., 128*(2), 125–134.

44. Thomson, J. A., Itskovitz-Eldor, J., Shapiro, S. S., Waknitz, M. A., Swiergiel, J. J., Marshall, V. S., et al., (1998). Embryonic stem cell lines derived from human blastocysts. *Science., 282*(5391), 1145–1147.

45. Klimanskaya, I., Chung, Y., Meisner, L., Johnson, J., West, M. D., & Lanza, R., (2005). Human embryonic stem cells derived without feeder cells. *Lancet., 365*, 1636–1641.

46. Ludwig, T. E., Levenstein, M. E., Jones, J. M., Berggren, W. T., Mitchen, E. R., Frane, J. L., et al., (2006). Derivation of human embryonic stem cells in defined conditions. *Nat. Biotechnol., 24*(2), 185–187.

47. Shi, Y., (2010). Generation of functional insulin-producing cells from human embryonic stem cells *in vitro*. *Methods Mol. Biol., 636*, 79–85.

48. Lee, C. S., De León, D. D., Kaestner, K. H., & Stoffers, D. A., (2006). Regeneration of pancreatic islets after partial pancreatectomy in mice does not involve the reactivation of neurogenin-3. *Diabetes., 55*(2), 269–272.

49. Hao, E., Tyrberg, B., Itkin-Ansari, P., Lakey, J. R., Geron, I., Monosov, E. Z., et al., (2006). Beta-cell differentiation from nonendocrine epithelial cells of the adult human pancreas. *Nat. Med., 12*(3), 310–316.

50. Bonner-Weir, S., Baxter, L. A., Schuppin, G. T., & Smith, F. E., (1993). A second pathway for regeneration of adult exocrine and endocrine pancreas. A possible recapitulation of embryonic development. *Diabetes., 42*(12), 1715–1720.

51. Gu, D., & Sarvetnick, N., (1993). Epithelial cell proliferation and islet neogenesis in IFN-g transgenic mice. *Development., 118*(1), 33–46.

52. Hardikar, A. A., Karandikar, M. S., & Bhonde, R. R., (1999). Effect of partial pancreatectomy on diabetic status in BALB/c mice. *J. Endocrinol., 162*(2), 189–195.

53. Finegood, D. T., Weir, G. C., & Bonner-Weir, S., (1999). Prior streptozotocin treatment does not inhibit pancreas regeneration after 90% pancreatectomy in rats. *Am. J. Physiol., 276*(5 Pt 1), E822–827.

54. D'Alessandro, J. S., Lu, K., Fung, B. P., Colman, A., & Clarke, D. L., (2007). Rapid and efficient *in vitro* generation of pancreatic islet progenitor cells from nonendocrine epithelial cells in the adult human pancreas. *Stem Cells Dev., 16*(1), 75–89.

55. Yao, Z. X., Qin, M. L., Liu, J. J., Chen, X. S., & Zhou, D. S., (2004). *In vitro* cultivation of human fetal pancreatic ductal stem cells and their differentiation into insulin-producing cells. *World J. Gastroenterol.,* 10(10), 1452–1456.

56. Taniguchi, H., Yamato, E., Tashiro, F., Ikegami, H., Ogihara, T., & Miyazaki, J., (2003). Beta-cell neogenesis induced by adenovirus-mediated gene delivery of transcription factor pdx-1 into mouse pancreas. *Gene Ther., 10*(1), 15–23.

57. Heremans, Y., Casteele, M. V. D., in't Veld, P., Gradwohl, G., Serup, P., Madsen, O., et al., (2002). Recapitulation of embryonic neuroendocrine differentiation in adult human pancreatic duct cells expressing neurogenin 3. *J. Cell Biol.,* 159(2), 303–312.

58. Joglekar, M. V., Patil, D., Joglekar, V. M., Rao, G. V., Reddy, D. N., Mitnala, S., et al., The miR-30 family micro RNAs confer epithelial phenotype to human pancreatic cells. *Islets., 1*(2), 137–147.

59. Russ, H. A., Bar, Y., Ravassard, P., & Efrat, S., (2008). *In vitro* proliferation of cells derived from adult human beta-cells revealed by cell-lineage tracing. *Diabetes., 57*(6), 1575–1583. doi: 10.2337/db07–1283. Epub 2008 Mar 3.

60. Rieck, S., & Kaestner, K. H., (2010). Expansion of beta-cell mass in response to pregnancy. *Trends Endocrinol. Metab., 21*(3), 151–158.

61. Mercado, A. B., & Castells, S., (2006). Pancreatic beta-cell hyperactivity in morbidly obese adolescents. *Pediatr. Endocrinol. Rev., 3* Suppl 4, 560–563.

62. Dor, Y., Brown, J., Martinez, O. I., & Melton, D. A., (2004). Adult pancreatic beta-cells are formed by self-duplication rather than stem-cell differentiation. *Nature, 429*(6987), 41–46.

63. Brennand, K., Huangfu, D., & Melton, D., (2007). All beta cells contribute equally to islet growth and maintenance. *PLo S Biol., 5*(7), e163.

64. Nir, T., Melton, D. A., & Dor, Y., (2007). Recovery from diabetes in mice by beta cell regeneration. *J. Clin. Invest., 117*(9), 2553–2561.

65. Teta, M., Rankin, M. M., Long, S. Y., Stein, G. M., & Kushner, J. A., (2007). Growth and regeneration of adult beta cells does not involve specialized progenitors. *Dev. Cell., 12*(5), 817–826.

66. Cole, L., Anderson, M., Antin, P. B., & Limesand, S. W., (2009). One process for pancreatic beta-cell coalescence into islets involves an epithelial-mesenchymal transition. *J. Endocrinol., 203*(1), 19–31.

67. Lipsett, M., & Finegood, D. T., (2002). beta-cell neogenesis during prolonged hyperglycemia in rats. *Diabetes., 51*(6), 1834–1841.

68. Lima, M. J., Muir, K. R., Docherty, H. M., Mc Gowan, N. W. A., Forbes, S., Heremans, Y., et al., (2016). Generation of Functional Beta-Like Cells from Human Exocrine Pancreas. *PLo S One.,* 11(5), e0156204.

69. Yi, F., Liu, G. H., & Izpisua-Belmonte, J. C., (2012). Rejuvenatingliver and pancreas through cell transdifferentiation. *Cell Res., 22*(4), 616–619.

70. Berneman-Zeitouni, D., Molakandov, K., Elgart, M., Mor, E., Fornoni, A., Domín-guez, M. R., et al., (2014). The temporal and hierarchical control of transcription factors-induced liver to pancreas transdifferentiation. *PLo S One., 9*(2), e87812.

71. Teo, A. K., Tsuneyoshi, N., Hoon, S., Tan, E. K., Stanton, L. W., Wright, C. V., et al., (2015). PDX1 binds and represses hepatic genes to ensure robust pancreatic commitment in differentiating human embryonic stem cells. *Stem Cell Reports., 4*(4), 578–590.

72. He, B., Li, X., Yu, H., & Zhou, Z., (2015). Therapeutic potential of umbilical cord blood cells for type 1 diabetes mellitus. *J. Diabetes., 7*(6), 762–773.

73. Hori, Y., Gu, X., Xie, X., & Kim, S. K., (2005). Differentiation of insulin-producing cells from human neural progenitor cells. *PLo S Med., 2*(4), e103.

74. Xin, Y., Jiang, X., Wang, Y., Su, X., Sun, M., Zhang, L., et al., (2016). Insulin-producing cells differentiated from human bone marrow mesenchyme malstem cells *in vitro* ameliorate streptozotocin-induced diabetichy perglycemia. *PLo S One., 11*(1), e0145838.

75. Timper, K., Seboek, D., Eberhardt, M., Linscheid, P., Christ-Crain, M., et al., (2006). Human adipose tissue-derived mesenchymal stem cells differentiate into insulin, somatostatin, and glucagon expressing cells. *Biochem. Biophys. Res. Commun., 341*(4), 1135–1140.

76. Sun, N. Z., & Ji, H. S., (2009). *In vitro* differentiation of human placenta-derived adherent cells into insulin-producing cells. *J. Int. Med. Res., 37*(2), 400–406.

77. Mu, X. P., Ren, L. Q., Yan, H. W., Zhang, X. M., Xu, T. M., Wei, A. H., et al., (2017). Enhanced differentiation of human amniotic fluid-derived stem cells into insulin-producing cells *in vitro*. *J. Diabetes Investig., 8*(1), 34–43.

78. Agarwal, A., & Brayman, K. L., (2012). Update on islet cell transplantation for type 1 diabetes. *Semin. Intervent. Radiol., 29*(2), 90–98.

79. Lacy, P. E., & Kostianovsky, M., (1967). Method for the isolation of intact islets of Langerhans from the rat pancreas. *Diabetes., 16*(1), 35–39.

80. Kemp, C. B., Knight, M. J., Scharp, D. W., Ballinger, W. F., & Lacy, P. E., (1973). Effect of transplantation site on the results of pancreatic islet isografts in diabetic rats. *Diabetologia., 9*(6), 486–491.

81. Shapiro, A. M., Lakey, J. R., Ryan, E. A., Korbutt, G. S., Toth, E., Warnock, G. L., et al., (2000). Islet transplantation in seven patients with type 1 diabetes mellitus using a glucocorticoid-free immunosuppressive regimen. *N. Engl. J. Med., 343*(4), 230–238.

82. Olack, B. J., Swanson, C. J., Howard, T. K., & Mohanakumar, T., (1999). Improved method for the isolation and purification of human islets of Langerhans using Liberase enzyme blend. *Hum. Immunol., 60*(12), 1303–1309.

83. Berney, T., & Secchi, A., (2009). Rapamycin in islet transplantation: friend or foe? *Transpl. Int., 22*(2), 153–161.

84. Hyder, A., Laue, C., & Schrezenmeir, J., (2005). Effect of the immunosuppressive regime of Edmonton protocol on the long-term *in vitro* insulin secretion from islets of two different species and age categories. *Toxicol. in vitro., 19*(4), 541–546.

85. Desai, N. M., Goss, J. A., Deng, S., Wolf, B. A., Markmann, E., Palanjian, M., et al., (2003). Elevated portal vein drug levels of sirolimus and tacrolimus in islet trans-

plant recipients: local immunosuppression or islet toxicity. *Transplantation., 76*(11), 1623–1625.

86. Faradji, R. N., Tharavanij, T., Messinger, S., Froud, T., Pileggi, A., Monroy, K., et al., (2008). Long-term insulin independence and improvement in insulin secretion after supplemental islet infusion under exenatide and etanercept. *Transplantation., 86*(12), 1658–1665.

87. Rickels, M. R., Liu, C., Shlansky-Goldberg, R. D., Soleimanpour, S. A., Vivek, K., Kamoun, M., et al., (2013). Improvement in β-cell secretory capacity after human islet transplantation according to the CIT07 protocol. *Diabetes., 62*(8), 2890–2897.

88. Bellin, M. D., Barton, F. B., Heitman, A., Harmon, J. V., Kandaswamy, R., Balamurugan, A. N., et al., (2012). Potent induction immunotherapy promotes long-terminsulininde pendence after islet transplantation in type 1 diabetes. *Am. J. Transplant., 12*(6), 1576–1583.

89. Kawahara, T., Kin, T., & Shapiro, A. M., (2012). A comparison of islet auto transplantation with allotransplantation and factors elevating acute portal pressure in clinical islet transplantation. *J. Hepatobiliary Pancreat. Sci., 19*(3), 281–288.

90. Johnson, P. R., & Jones, K. E., (2012). Pancreatic islet transplantation. *Semin Pediatr Surg., 21*(3), 272–280.

91. Maruyama, M., Kenmochi, T., Akutsu, N., Otsuki, K., Ito, T., Matsumoto, I., et al., (2013). A Review of Autologous Islet Transplantation. *Cell Med., 5*(2–3), 59–62.

92. Lakey, J. R., Burridge, P. W., & Shapiro, A. M., (2003). Technical aspects of islet preparation and transplantation. *Transpl Int., 16*(9), 613–632.

93. Hermann, M., Margreiter, R., & Hengster, P., (2010). Human islet autotransplantation: the trail thus far and the highway ahead. *Adv Exp Med Biol., 654*, 711–724.

94. Cantarelli, E., & Piemonti, L., (2011). Alternative transplantation sites for pancreatic islet grafts. *Curr Diab Rep., 11*(5), 364–374.

95. Nanji, S. A., & Shapiro, A. M., (2004). Islet transplantation in patients with diabetes mellitus: choice of immunosuppression. *Bio Drugs, 18*(5), 315–328.

96. Qi, M., (2014). Transplantation of Encapsulated Pancreatic Islets as a Treatment for Patients with Type 1 Diabetes Mellitus. *Adv Med.,* 429710.

97. Dufrane, D., Goebbels, R. M., Saliez, A., Guiot, Y., & Gianello, P., (2006). Six-month survival of microencapsulated pigislets and alginate biocompatibility in primates: proof of concept. *Transplantation, 81*(9), 1345–1353.

98. Elliott, R. B., Escobar, L., Calafiore, R., Basta, G., Garkavenko, O., Vasconcellos, A., et al., (2005). Transplantation of micro- and microencapsulated piglet islets into mice and monkeys. *Transplant Proc., 37*(1), 466–469.

99. Plesner, A., & Verchere, C. B., (2011). Advances and challenges in islet transplantation: islet procurement rates and lessons learned from suboptimal islet transplantation. *J. Transplant.,* 979527.

100. Nam, S., Chesla, C., Stotts, N. A., Kroon, L., & Janson, S. L., (2011). Barriers to diabetes management: patient and provider factors. Diabetes Res Clin Pract., 93(1), 1–9.

CHAPTER 3

NOVEL ANTI-CANCER DRUGS BASED ON HSP90 INHIBITORY MECHANISMS: A RECENT REPORT

SAYAN DUTTA GUPTA

Department of Pharmaceutical Chemistry, Gokaraju Rangaraju College of Pharmacy, Hyderabad, Telangana, India,
E-mail: sayandg@rediffmail.com

3.1 INTRODUCTION

Cancer or malignant neoplasm refers to a diverse group of diseases where the atypical forms of body's own cells proliferate uncontrollably and invades neighboring tissues. Additionally, they metastasize, e.g., spreads via blood/lymph and invades healthy tissues [1]. The exact cause of cancer is not well understood as the process by which normal cells transform into abnormal ones involves multiple complex cell signaling pathways [1, 2]. Therefore, the discovery of novel anticancer chemotherapeutic agents with a high therapeutic index is a daunting task for researchers. The traditional anticancer chemotherapy involves molecules and ionic radiations that kill cancer cells along with the normal ones. The healthy cells that proliferate like cancerous ones (cells of bone marrow, gastrointestinal tract, hair follicles) are affected to a greater extent [3]. Additionally, treatment with these conventional drugs often results in failure of therapy owing to their administration in sub-optimal doses for avoiding damage to normal tissues [4]. Furthermore, resistance to the drugs also creates a bottleneck in their use. Therefore, over the last two decades anticancer research focused on identifying cancer selective macromolecules, the inhibition of which will lead to the destruction of cancer cells in a selective manner. This type of healing strategy is referred to as targeted therapy, which has revolutionized cancer treatment outcomes over the last 20 years. These cancer-specific target inhibitors have shown far better safety and efficacy profile compared to their conventional counterparts [5, 6]. However, the drawback of toxic adverse effects and resistance to chemotherapeutic agents still haunts this therapeutic approach [7, 8]. The magic antineoplastic molecule (like penicillin's for antibacterial therapy) is still elusive for cancer researchers worldwide.

The side effects of targeted therapy are due to the similar binding site morphology of target proteins and normal cellular enzymes. For example, the anticancer drug, gefitinib exhibits its action by inhibiting tyrosine kinases of cancer cells. However, this drug also has an affinity for normal cell epidermal growth factor receptor (EGFR), which is involved in the maturation of keratinocytes in the skin. Hence, folliculitis (acne-like rash) is a major side effect of gefitinib [9, 10]. Additionally, the toxic symptoms were also attributed to the presence of the target proteins in a few of the normal cells. The cancer target proteins are usually under-expressed in healthy cells. Nevertheless, in a few instances, they are able to attract the chemotherapeutic agents towards it even in lower cellular concentration, which is sufficient for normal cell destruction. An example of the aforestated adverse effect is ventricular dysfunction associated with dasatinib therapy. This drug suppresses ABL kinase of cancer cells as well as that of the cardiac myocytes. This leads to cardiac failure in approximately 2% of patients treated with dasatinib in clinical trials [11].

The other major hindrance in curing cancer via targeted therapy is the resistance to chemotherapeutic agents [12, 13]. The major causes for the development of cancer resistant drugs for treating the conditions of:

1. **Mutation:** This plays a vital role in most of the cancer resistant cases. Mutation results in the change in morphology of the target protein's active site [14]. It may also alter the amino acid sequence of the complete polypeptide. For example, the structure of Rapidly accelerated sarcoma (RAS) signaling protein is mutated in 25% of carcinoma cell lines. Hence, it is very difficult to discover effective Ras inhibitors for the treatment of cancer [15].
2. **Efflux of drug molecule:** Researchers have identified cell membrane pumps, which can expel drugs from the cytoplasm of a neoplastic cell. This prevents the cancer cells from the toxic effects of the chemotherapeutic agents. The two common pumps executing this function are multidrug resistance-associated protein (MRP) and P-glycoprotein (Pgp) [16]. These two polypeptides are also targets

of several anticancer drug discovery programs [17]. For example, resistance to imatinib in K562 CML cells was due to overexpression of Pgp protein [18].

The aforementioned lacunas are reflected in the number of marketable drugs discovered via "targeted chemotherapy." Till date, more than 500 cancer cell selective targets are available that can be explored for the discovery of novel, efficacious antineoplastic molecules with fewer adverse effects. Out of all the available targets, only 28 of them were successful in generating 33 approved anticancer compounds. These facts warrant discoveries of better cancer curative therapeutic targets than the existing ones. Hence, an ideal target for anticancer drug development should have the following attributes:

1. Over-expressed in cancer cells and completely absent in normal cells.
2. Involved in multiple oncogenic signaling networks and thus activates several biological molecules responsible for development of malignancy.
3. Stimulates diverse array of proteins responsible for the development of resistance to drug molecules via mutation. In other words, the target will not alter its morphology on prolonged treatment with the chemotherapeutic agent or during relapse of the disease.
4. The morphology of targets' catalytic site (or binding zone for its inhibitors) should be different from the binding site of normal cellular proteins.

The only cancer curative target that satisfies most of the above benchmark is heat shock protein 90 (Hsp90, molecular weight ~90 kDa) [19, 20]. Therefore, it is the most promising and encouraging protein involved in the discovery of novel chemical entities against neoplasm.

3.2 HSP90

Molecular chaperones are a group of proteins, which converts a newly synthesized immature polypeptide into a mature one [21]. Additionally, they also assist in the refolding of denatured proteins in the cell [22, 23].

They carry out their function with the help of few partner proteins, referred to as co-chaperones [24]. They are ubiquitously present and upregulated during stress conditions like inflammation, hypoxia, UV exposure, etc. [25]. The polypeptides whose structure and functions are restored or maintained by the chaperones are referred to as "client proteins." Heat shock protein 90 (Hsp90) is one such chaperone that is dependent on ATP for carrying out its repairing function, e.g., the energy required for the chaperoning function is obtained from the hydrolysis of ATP to ADP [26, 27].

Hsp90α and Hsp90β are the two isoforms of the polypeptide present in a mammalian cell [28]. They are 85% morphologically identical but differ in their mode and time of expression. The alpha form is secreted extracellularly due to external stimuli (infection, inflammation, elevated temperature, etc.) and is involved in angiogenesis, invasiveness and metastasis whereas Hsp90β is required for maturation of normal polypeptides of a cell [29]. The later is the constitutive form and is secreted under normal physiological conditions. It comprises 1–2% of total proteins of a normal healthy cell. Additionally, the beta form plays a selective role in the anti-apoptotic functions of Bcl2 and CpG-B oligodeoxynucleotide [30].

3.3   HSP90 STRUCTURE AND DYNAMICS

Heat shock protein 90 exists as homodimer polypeptide in all species (Figure 3.1) [31, 32]. Like any other normal protein, each protomer consists of the following three highly conserved and flexible domains:

1. N-terminal domain (N-domain, molecular weight ≈ 35 kDa): This segment contains the nucleotide (ATP and ADP) and inhibitor binding cleft, commonly referred to as "Bergerat fold." This catalytic site gets closed by a lid (helix-loop-helix region) during ATP/"client protein" binding state and opens up during the ADP-bound phase. This binding site is highly conserved across GHKL (gyrase, Hsp90, histidine kinase, MutL) superfamily of ATPases [33, 34].

2. Middle or the intermediate domain (M-domain, molecular weight ≈ 35 kD): It is a proteolytically resistant site meant for binding of client proteins, a few co-chaperones and γ-phosphate of ATP. The ATPase (hydrolysis of ATP) activity of Hsp90 is not quantifiable in the absence of this particular segment of the protein [35].

3. C-terminal domain (C-domain, molecular weight ≈ 12 kDa): The part consists of a pentapeptide (Met-Glu-Glu-Val-Asp or MEEVD) motif, which serves as an acceptor for tetratroicopeptide repeat (TPR) containing, co-chaperones. The dimerization of a C-terminal fragment of each monomer is one of the prerequisite conditions for ATP hydrolysis and hence the ATPase action of Hsp90 is impaired in the absence of this terminal domain. Additionally, this domain also regulates opening and closing of N-terminal's ATP binding cleft [36].

There is also a connector loop between the 'N' and 'M' domain which comprises of 30–70 charged acidic and basic amino acid residues. The morphology of this zone (amino acid sequence) is yet to be resolved at the atomic level. This connector loop does not play any significant role in the functioning of Hsp90 protein. Studies have revealed that all the three segments of the protein are highly flexible, which, in turn, affected the complete decoding of the protein's amino acid sequence. The structure of the N-terminal domain was first established in 1997 [33, 37]. The remarkable intrinsic flexibility of the protein leads to the tedious structural elucidation of the complete chaperone [38, 39]. It has taken 12 years to resolve the structure of the complete protein since its discovery as a druggable candidate [40]. However, an atomic level resolution of full-length Hsp90 is still elusive for the scientist. Additionally, the amino acid sequence of a few segments (like the lid region) is yet to be resolved.

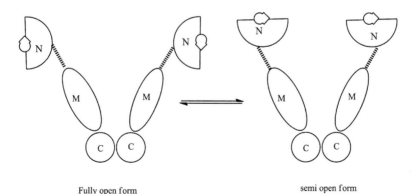

Fully open form   semi open form

FIGURE 3.1   Structural dynamics of Hsp90 chaperone.

In solution, resting stage or apostate Hsp90 is a homogenous protein and exists in two forms, e.g., semi-open and fully open conformation (Figure 3.1), Both the forms assumes a 'V' shape conformation which is often referred to as "flying sea gull" form. The clockwise rotation of the N-domain around its 2-fold axis leads to the transformation of the fully open to the semi-open form (Figure 3.1) [41, 42]. This structural dynamism is an intrinsic property of Hsp90 and is not affected by ADP/ATP binding. However, the morphological dynamics help the polypeptide in a binding wide range of client proteins and co-chaperones for carrying out its chaperoning function [43–45].

3.4   MECHANISM OF CHAPERONING FUNCTION

Hsp90 executes its chaperoning function (repair of damaged proteomes) in association with several co-chaperone proteins. The co-chaperones regulate the chaperoning function by recruiting client protein for Hsp90. They also assist in the repair and release of the matured protein from the chaperone. The energy for healing a protein is obtained from the hydrolysis of ATP to ADP and phosphate ion, which is often referred to as the ATPase activity of Hsp90 [40]. This ATPase activity can be inhibited or activated by the co-chaperones [46]. There are 18 co-chaperones identified in a human cell, which dynamically associate with Hsp90 during the chaperone cycle (Table 3.1).

These helper proteins of Hsp90 are highly conserved and performs their function in groups. They are specific for a particular client polypeptide and execute their duties via highly complex biochemical pathways, which is not fully understood by researchers [47]. The repairing of an immature protein starts with its binding to the M-domain of Hsp90 with the help of co-chaperones. This is followed by attachment of ATP to its binding site at N domain [34]. These events lead to dimerization of M- and N- segments and Hsp90 assumes a closed shape or toroidal configuration. This is followed by hydrolysis of ATP and repair of the damaged client protein (Figure 3.2). Finally, the matured protein is released along with ADP and phosphate anion into the cytosol [48]. On a few occasions, the ATP binds to the "Bergerart fold" prior to co-chaperone mediated client protein recruitment by Hsp90 (Figure 3.3). Thereafter, the usual sequence

TABLE 3.1 List of Hsp90 Co-chaperones

Serial no.	Name of cochaperone	Type of cochaperones	Function as Hsp90 co-chaperone
1	Aha1	Non-TPR domain	Stimulates Hsp90's ATPase activity and induces conformational changes in the chaperone
2	AIP	TPR domain	Complex formation with AhR, PPARα αvδ bx
3	Cdc37	Non-TPR domain	Suppression of Hsp90 ATPase activity. Recruitment of kinases for Hsp90.
4	CHIP	TPR domain	Binds ubiquitin ligase, tags protein for degradation
5	Chp1/melusin	Non-TPR domain	Forms complex with Sgt1 and Hsp90, chaperoning of NLR receptors
6	Cyp40	TPR domain	Recruitment of multiple client proteins for Hsp90
7	FKbp51	TPR domain	Recruitment of multiple client proteins for Hsp90
8	Fkbp52	TPR domain	Recruitment of multiple client proteins for Hsp90
9	Hop	TPR domain	Client protein maturation, Attenuation of Hsp90 ATPase activity, helps in Hsp90/Hsp70 interaction
10	Nudc	Non-TPR domain	Helps in mitosis and cytokinesis. Stabilizes L1S1 protein with Hsp90
11	p23	Non-TPR domain	Client protein maturation. Hsp90 ATPase inhibition
12	PP5	TPR domain	Import of kinases to Hsp90
13	Sgt1	TPR domain	Forms complex with CHORD proteins and Hsp90, chaperoning of NLR receptors
14	Tah1	TPR domain	Complex formation with Pih1 and Hsp90
15	Tom70	TPR domain	Transport of mitochondrial protein to Hsp90
16	Tpr2	TPR domain	Retrograde transfer of steroid hormone receptors from Hsp90 onto Hsp70
17	Ttc4	TPR domain	Nuclear transport proteins. Helps in transporting CDC6 DNA replication protein to Hsp90
18	Unc45	TPR domain	Transfers protein to Hsp90 for repairing

AhR = aryl hydrocarbon receptor; CDC = cell division cycle; CHORD = cysteine- and histidine-rich domains; Cyp = Cyclophilin; FKbp = FK506-binding protein; Hbx = Hepatitis B virus X protein; Hop = Hsp70-Hsp90 Organizing Protein; NLR = Nod-like receptors; NUDC = Nuclear Distribution C; Dynein Complex Regulator; PP = protein phosphatase; PPAR = Peroxisome proliferator-activated receptor; Tah = TPR (Tetratrico Peptide Repeat)-containing protein associated with Hsp90; TOM = translocase of outer membrane; Ttc = Tetratrico peptide repeat protein.

of events follows, e.g., attainment of closed conformation, breakdown of ATP, repair and release of client proteins/hydrolyzed products [49].

3.5 HSP90 AND CANCER

Hsp90 plays a major role in the development and progression of cancer. It has been hypothized with great optimism that inhibitors directed against this chaperone protein will be highly efficacious and devoid of toxic

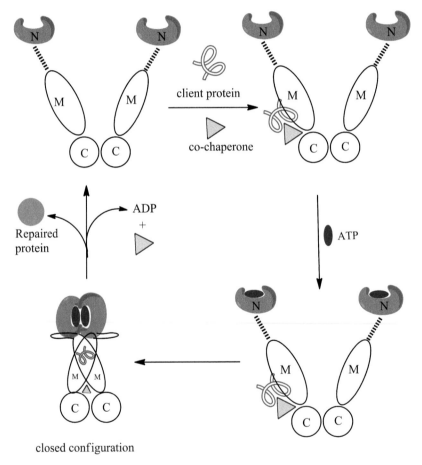

FIGURE 3.2 (See color insert.) Mechanism of repairing client protein by Hsp90 (client protein binds prior to ATP).

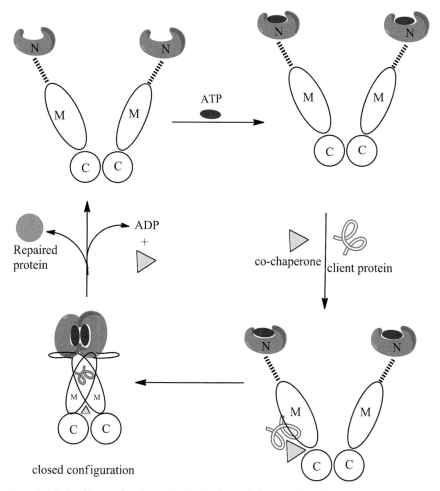

FIGURE 3.3   (See color insert.) Mechanism of chaperoning client protein by Hsp90 (ATP binds prior to client protein).

adverse effects. This assumption is based on the following experimental facts:

1. Proteins of multiple carcinogenic signaling pathways are all customers of Hsp90 and hence the attenuation of Hsp90 function will lead to the inactivation of several oncogenic proteins (Figure 3.4). This hypothesis directs us to the conclusion that compounds

against Hsp90 will be more effective than drug molecules discovered by targeting individual cancer-causing proteins [50, 51].

2. A few proteins that promote the development of resistance to chemotherapeutic agents are also dependent on Hsp90 chaperoning function for their maturation (Figure 3.4). Therefore, there is decreased chance of acquiring resistance to Hsp90 inhibitors [51].

3. Hsp90 is over translated in neoplastic cells as a co-chaperones-proteome super complex, whereas the healthy cell chaperone is under-translated (1–2% of total cellular proteins) and inhabit in an uncomplexed state [52, 53].

4. The conformation (bent shape) of the ATP binding domain of Hsp90 inhibitors is different from that of the normal cellular proteins which uses ATP for carrying out their function [54–56].

All the aforestated evidences prompted scientists worldwide to rationally design and discover novel anticancer chemical entities that will act via Hsp90 suppression.

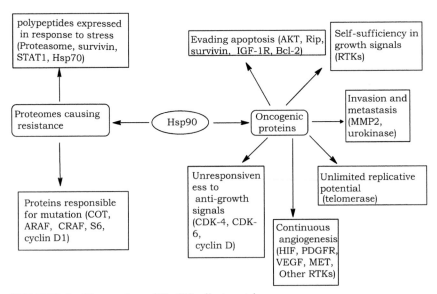

FIGURE 3.4   Diverse class of Hsp90's client proteins.

3.6   HSP90 INHIBITORS

The druggability of Hsp90 was first established in 1994 with the help of the natural product, geldanamycin (GA) [57]. Geldanamycin is a benzoquinone anamycin isolated from *Streptomyces hygroscopicus*. It was experimentally shown that GA competitively blocks the N-terminal ATP binding pocket of Hsp90, which leads to ubiqitin mediated proteosomal degradation of the client proteins [58, 59]. However, this drug molecule failed to reach clinical trials because of poor aqueous solubility, *in vivo* unstability and severe hepatotoxicity (attributed to its quinone moiety) [60]. This paved the pathway for the development of various effective geldanamycin derivatives [17-allyl amino–17-demethoxy-geldanamycin (17-AAG, Tanespimycin), 17-desmethoxy–17-N, N-dimethylaminoethyl amino geldanamycin (17-DMAG, Alvespimycin), reduced hydroquinone form of 17-AAG (Retaspimycin or IPI–504). 17 AG (17-amino–17-desmethoxygeldanamycin, IPI–493)] with improved solubility, toxicological properties and aqueous solubility (Figure 3.5) [61]. 17-AAG was the first geldanamyicn derivative to reach clinical trials alone (phase-I/II) [62, 63] and in combination with trastuzumab (phase-I) [64]. However, this chemical entity was discontinued due to poor pharmaceutical properties and patent-related concerns [65]. The water-soluble geldanamycin derivative, alvespimycin reached phase I clinical studies owing to its better safety and formulation profile [66–68]. However, the development of this Hsp90 inhibitor was halted due to huge investment required for its further clinical studies and subsequent commercialization [69]. The other oral GA analog that reached phase II/III clinical trials are retaspimycin or IPI–504 [70–72]. This chemical entity demonstrated high mortality rate among subjects and was hence not developed further [73]. IPI–493 another Hsp90 antagonist that reached phase-I clinical studies [74, 75]. This analog of GA was not taken up for advanced studies because of formulation issues and dose administration difficulties [76]. Several other potent natural GA analogs were discovered by genetic engineering techniques and mutasynthesis [77] (like macbecin, Figure 3.5). Despite showing excellent *in vitro* efficacy, these agents were found to be ineffective for testing in human subjects [76, 78].

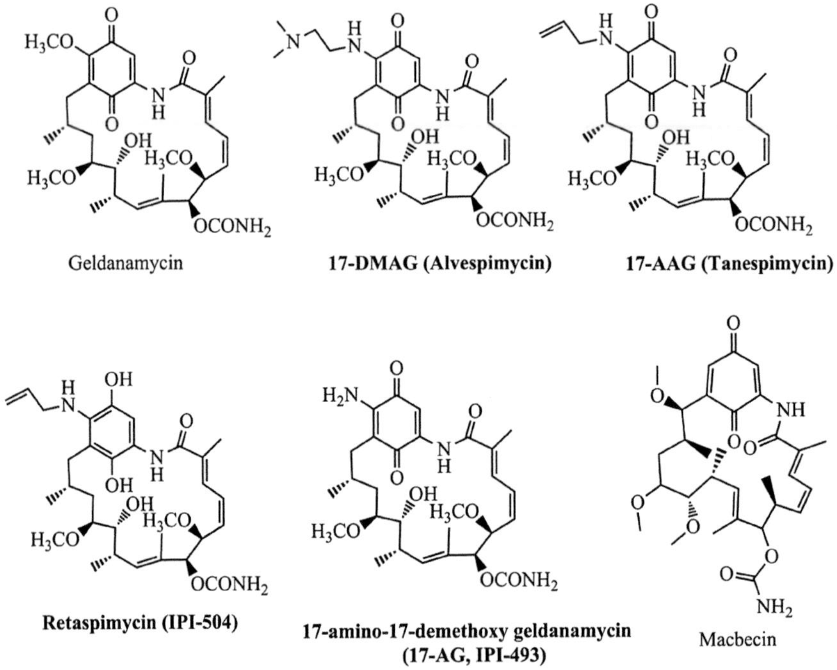

Geldanamycin   17-DMAG (Alvespimycin)   17-AAG (Tanespimycin)

Retaspimycin (IPI-504)   17-amino-17-demethoxy geldanamycin   Macbecin
                                      (17-AG, IPI-493)

FIGURE 3.5   Geldanamycin and its derivatives discovered as Hsp90 inhibitors. The compounds highlighted in bold reached clinical trials.

Radicicol (RD) is the other 1st generation, natural product N-terminal Hsp90 inhibitors, which exhibited significant *in vitro* anticancer activity. It is a macrocyclic lactone antibiotic which was first isolated from the fungus *Microsporium bonorden* in 1953 [79]. The binding orientation of radicicol at Hsp90's ATP binding cleft was found to be different from that of geldanamycin [59, 80]. It is a more potent inhibitor of the chaperone than geldanamycin (K_d of radicicol in ATPase assay = 19 nM; K_d of geldanamycin = 1.2 µM). Unfortunately, it was found to be inactive *in vivo* because of chemical instability (presence of labile allylic epoxide and α, β, γ, δ-unsaturated ketone). Therefore, it was not found to be a viable candidate for further clinical development. Hence, various derivatives of radicicol with less chemical instability were (Figure 3.6) synthesized and evaluated for their Hsp90 suppression potential [81–83]. A few of them were found to be less potent than radicicol in terms of attenuating Hsp90 ATPase function. The radicicoloximes exhibited comparable potency to that of radicicol [81]. Furthermore, they were found to be effective in the *in vitro* models wherein they reduced tumor size in a 30-day period [81, 84]. However, the stereochemical complexities associated with these oxime derivatives (mixture of E and Z isomers) restricted their further development [85]. The complexity of radicicol molecule in general also effected research on future development of this scaffold [85]. Luckily, the anticancer studies from RD and their derivatives identified a simple scaffold (resorcinol/4-chloro resorcinol) for the future discovery of Hsp90 inhibitors.

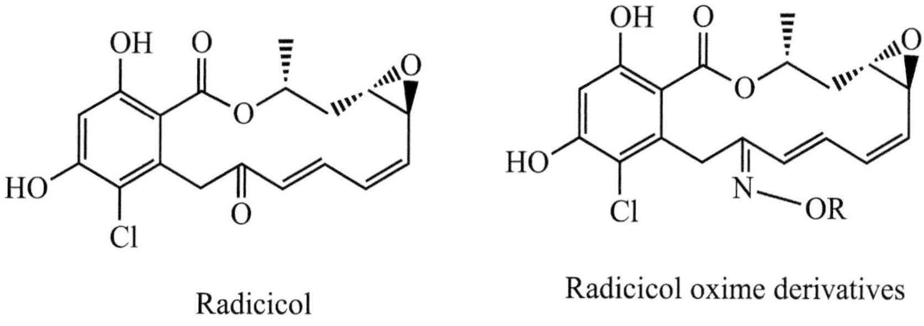

Radicicol   Radicicol oxime derivatives

FIGURE 3.6   Structure of radicicol and its derivatives.

3.6.1 2ND GENERATION HSP90 INHIBITORS

The drawbacks associated with the first generation Hsp90 antagonists led to the design of derivatives based on the structural features of 1st generation compounds [86]. The majority of the molecules for this class was discovered via structure-based drug design techniques. These classes of molecules can be mainly classified as follows:

a) chimeric inhibitors;
b) purine-4-amine scaffold containing molecules;
c) molecules with resorcinol moiety;
d) compounds with benzamide motif;
e) miscellaneous N terminal inhibitors.

3.6.1.1   CHIMERIC INHIBITORS

The toxicity associated with GD and *in vivo* failure of RD leads to the development of inhibitors by combining the structural features of GD and RD. The quinone/hydroquinone moiety of geldanamycin and radicicol's resorcinol/4-chloro resorcinol scaffold was found to play a significant role in Hsp90 binding. Hence, inhibitors were designed by retaining these two important molecular scaffolds (Figure 3.7). The first chimeric Hsp90 inhibitor discovered was radanamycin [87]. Molecular docking studies guided the design of *seco* radanamycin derivatives with a suitable linker [88]. The first approach consists of connecting the scaffolds by a two-carbon amide bond (radamide), which demonstrated significant anticancer activity in the *in vitro* and *in vivo* models [88]. In the second approach, the connector was the ester of RD along with a two-carbon chain (radester), which exhibited considerable antineoplastic activity via Hsp90 inhibition [89]. The hydroquinone moiety of the derivatives showed better activity profile than its corresponding quinone form. Subsequently, various analogs of these chimeric inhibitors were prepared and evaluated for their Hsp90 inhibition potential and antiproliferative effects [90, 91]. Unfortunately, none of them was found to be a suitable candidate for further preclinical and clinical development.

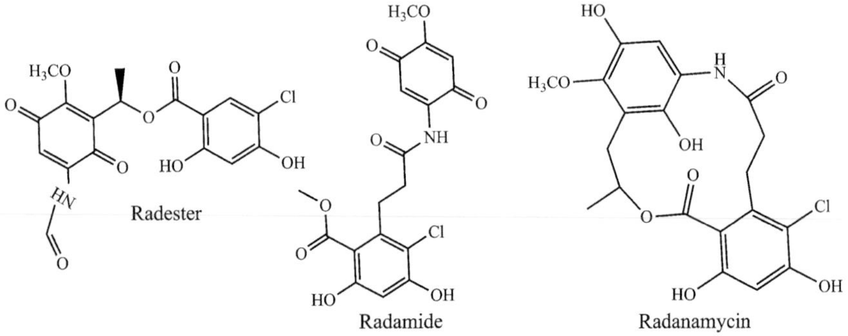

FIGURE 3.7   Chimeric inhibitors of Hsp90 based on radicicol and geldanamycin.

3.6.1.2   PURINE-4-AMINE SCAFFOLD CONTAINING MOLECULES

The ATP nucleotide has been used as a template for the design and discovery of novel small molecule Hsp90 inhibitors. The purine moiety of ATP was used for the design of new molecules by structure-dependent approach. PU-3 (Figure 3.8) was the first molecule of this class with significant Hsp90 affinity [92]. Additionally, SAR studies were carried out that resulted in analogs with improved Hsp90 inhibitory potential [92, 93]. Finally, three orally bioavailable inhibitors with Hsp90 binding affinity in nanomolar scale were discovered from this class of compounds. Ultimately, one of them (BIIB021) became the first rationally designed 2nd generation Hsp90 inhibitors to enter clinical trials [94]. PU-H71, BIIB028, CUDC–305 (Debio 0932) are the other purine analogs (Figure 3.8) that advanced into clinical trials for the treatment of advanced malignancies [95–98]. However, none of them were approved for use by regulatory authorities due to toxicities (tachycardia, respiratory failure, etc.) associated with them [92, 98].

3.6.1.3 MOLECULES WITH RESORCINOL MOIETY

The resorcinol motif plays an important role in Hsp90 inhibition. It forms important hydrogen bonding interactions with amino acids (Asp 93, Asn 51, Leu 48, Thr 184, Lys 58) and water molecules at the ATP binding site of Hsp90 chaperone [99–101]. It has been reported that compounds devoid of any one-hydroxyl group resulted in decreased Hsp90 suppression [102]. Hence, structure-based drug designing strategy was adopted by many research groups to identify novel resorcinol containing small molecules with enhanced affinity for Hsp90 protein [103]. Furthermore, the non-toxic nature of the resorcinol group prompted scientists to design novel Hsp90 inhibitors by retaining this scaffold. The resorcinol containing amides, Schiff's bases [100, 104, 105], Mannich's bases [103], pyrazoles [106], isoxazoles [107, 108], triazoles [109] were explored extensively by researchers worldwide for inhibiting Hsp90 chaperone. This study lead to the discovery of more than 2000 novel small molecule Hsp90 inhibitors with improved antiproliferative effect against cancer cells [110]. Ultimately, this group generated four molecules (Figure 3.9) that reached clinical trials [111–115].

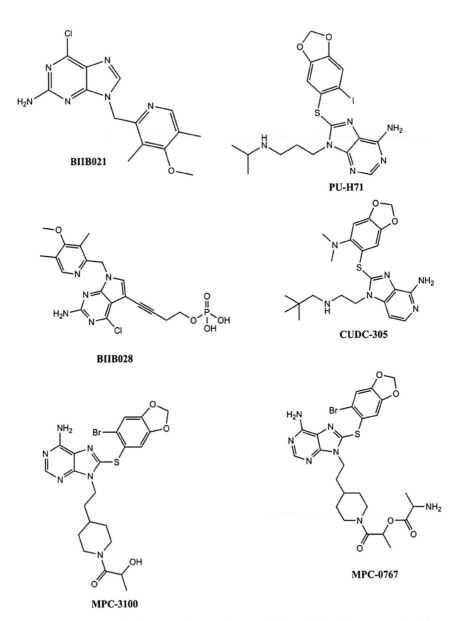

FIGURE 3.8 Chemical structure of purine based Hsp90 inhibitors that were evaluated in human beings.

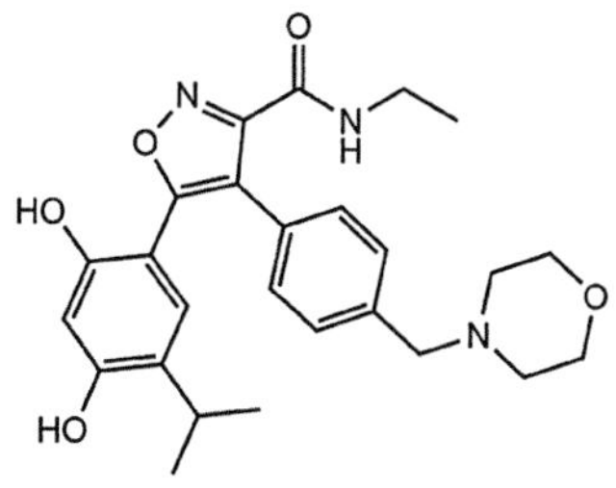

AUY922 (Luminespib, VER-52296)

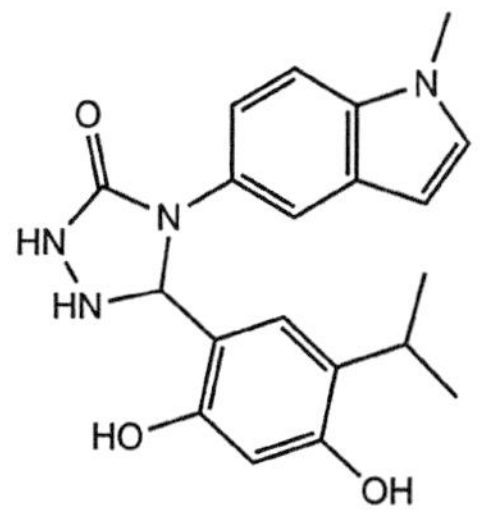

STA-9090 (Ganetespib)

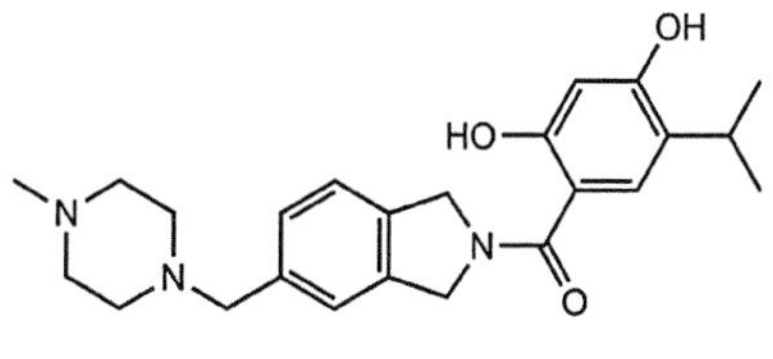

AT13387 (Onalespib)

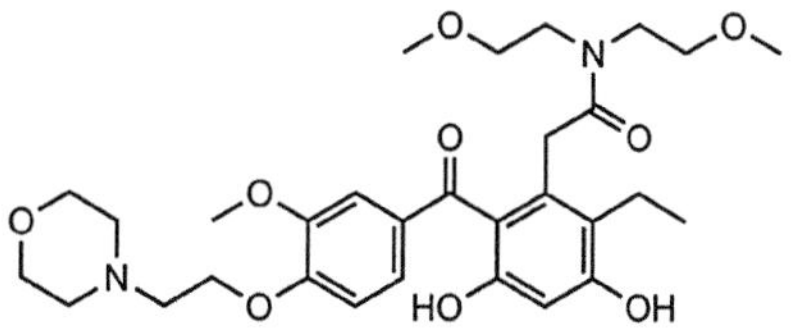

KW 2478

FIGURE 3.9 Resorcinol analogs as Hsp90 antagonist that reached clinical trials.

3.6.1.4 COMPOUNDS WITH BENZAMIDE MOTIF

The quest for the rapid discovery of novel Hsp90 inhibitor scaffold resulted in the screening of compounds by high throughput screening (HTS). In one such approach, first 2000 ATP binding proteins were adsorbed on an ATP-affinity column by loading cell lysates. Thereafter, mass spectrometry was used to identify compounds that selectively displace Hsp90 in the column. These studies identified benzamide derivatives as a new Hsp90 inhibitor chemotypes. SNX 2122 (Figure 3.10) was found to be suitable for clinical evaluation. However, this compound has variable bioavailability and produces polymorphic crystals and therefore was found unsuitable for further development [116]. Hence, a glycine prodrug of SNX 2122 (PF–04929113, SNX 5422, Figure 3.10) was developed and this molecule entered clinical trial as an oral inhibitor of Hsp90 in May 2007 [117, 118]. Additionally, the above HTS approach also generated several lead molecules, which were optimized by SAR and X-ray crystallographic studies. These research works lead to the identification of two clinically evaluable drug molecules (Figure 3.10). TAS–116 and XL–888 showed a better safety profile (like minimum ocular toxicity for TAS–116) in preclinical models [119]. Hence, these two molecules are currently in Phase I/II of clinical trials [120]. From the aforestated discussion on Hsp90 inhibitors, it is evident that 18 molecules reached various phases of clinical trials. Currently, six of them are undergoing Phase-I/II/III study either alone or in combination with other antineoplastic agents (Table 3.2) [120, 121]. The structure of one of the agents (DS–2248) is not disclosed by Daichi Sankyo Company Limited, Japan. Out of 18 molecules, only tanespimycin, retaspimycin and STA 9090 reached phase III of clinical trials [116, 121, 122].

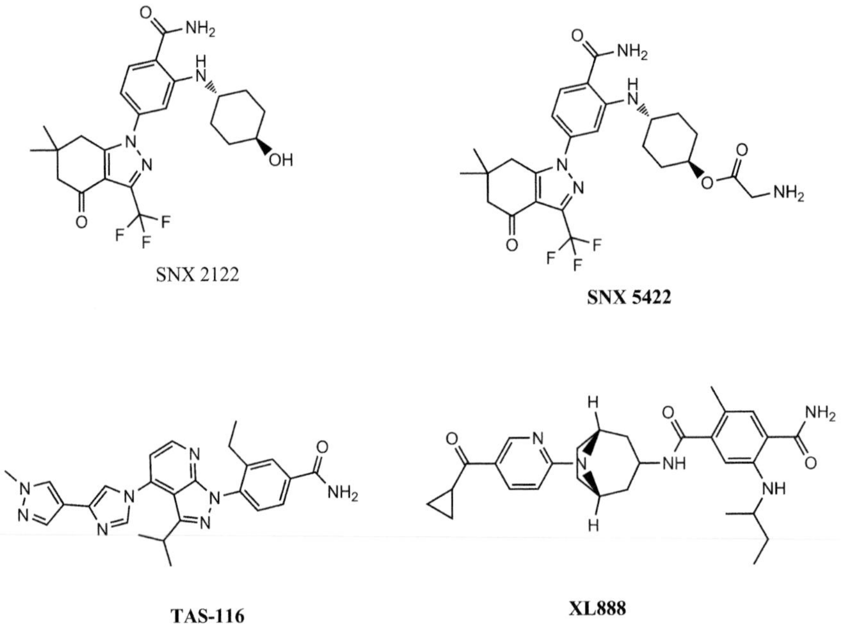

FIGURE 3.10   Benzamide derivatives as Hsp90 inhibitors that were evaluated in humans.

3.6.1.5   MISCELLANEOUS N-TERMINAL INHIBITORS

High throughput screening (HTS) and virtual screening aided in the discovery of numerous novel scaffolds with promising Hsp90 inhibitory potential [123–129]. The structure of these novel Hsp90 inhibitor moieties is highlighted in Figure 3.11. Except 7,8-dihydropyrido[4,3-*d*]pyrimidin–5(6H)-one moiety (NVP-HSP990) [98, 130], none of the scaffold was successful in generating molecules that can be evaluated in human beings. Scientists are still optimistic and are continuing their research work with these scaffolds. The scientific community is eagerly awaiting a positive result with these studies.

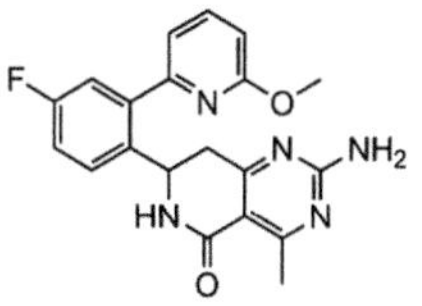

NVP HSP-990

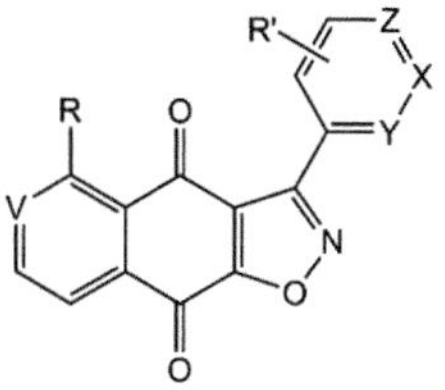

V = CH, N X,Y,Z = CH, N, N⁺CH₂Aryl

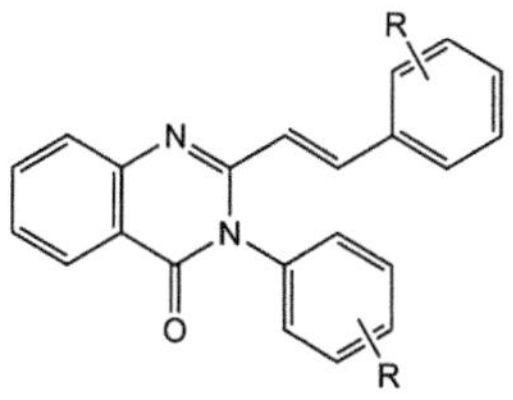

Quinazoline derivatives

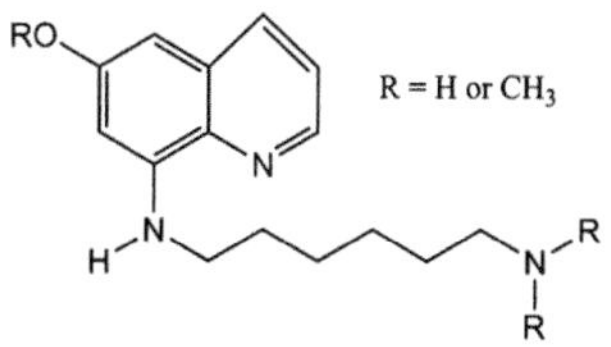

Amino quinoline derivatives

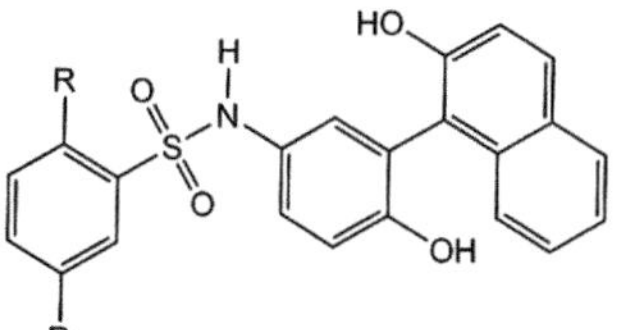

R = H or CH₃

Hydroxy napthalene aryl sulfonamide derivatives

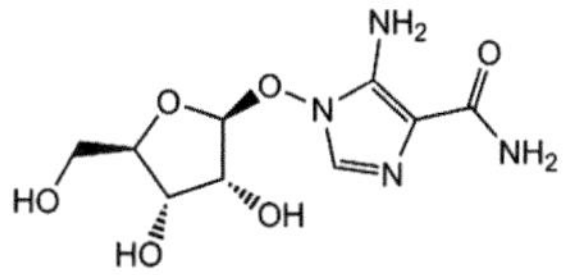

5-aminoimidazole-4-carboxamide-1-beta-D-ribofuranoside

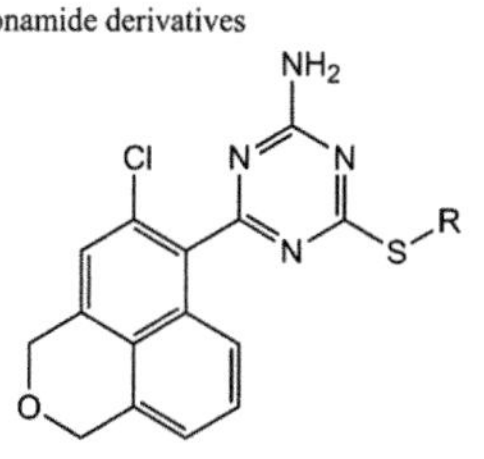

2-amino-6-(1H,3H-benzo[de]isochromen-6-yl)-1,3,5-triazines

FIGURE 3.11 Miscellaneous N-terminal inhibitors of Hsp90 chaperone. The molecule highlighted in bold reached clinical trials.

TABLE 3.2   Hsp90 Inhibitors That Reached Clinical Trials[a]

	Molecule	Chemical types	Phase	Company
1	KOS–953	Geldanamycin derivatives	II/III	BMS
2	IPI–504	Geldanamycin derivatives	II/III	Infinity
3	IPI–493	Geldanamycin derivatives	I	Infinity
4	Alvespimycin	Geldanamycin derivatives	I	BMS
5	**AUY922**	Resorcinol analogs	I/II	Novartis
6	**STA–9090**	Resorcinol analogs	I/II/III	Synta
7	KW–2478	Resorcinol analogs	I	Kyowa Hakko Kirin UK, Ltd.
8	**AT–13387**	Resorcinol analogs	I	Astex therapeutics
9	**PU-H71**	Purine scaffold	I	Samus Therapeutics, Inc.
10	BIIB021	Purine scaffold	I/II	Biogen Indec
11	BIIB028	Purine scaffold	I	Biogen Indec
12	MPC3100	Purine scaffold	I	Myriad pharmaceuticals
13	CUDC–305	Purine scaffold	I	Debiopharm International SA
14	**XL888**	Benzamide motiff	I	Exelixis
15	**SNX 5422**	Benzamide motiff	I	Serenex/Pfizer Inc
16	**TAS–116**	Benzamide motiff	I/II	Taiho pharma
17	NVPHsp990	Dihydro pyrido pyrimidine ring	I	Novartis
18	DS–2248	Structure not disclosed	I	Daiichi Sankyo Inc.

[a] Source: www.clinicaltrials.gov. Compounds in bold are presently undergoing clinical trials.

3.7   DRUG DESIGNING APPROACH: CHALLENGES AND LOOPHOLES

It has been 23 years since Hsp90 was proved as a druggable candidate. Since then, researchers have failed in getting a single Hsp90 inhibitors approved by regulatory authorities. This failure warrants urgent introspection of the strategy involved in the whole discovery process. Several scientists have tried to pass the buck on Hsp90 as a target selection. One major reason for this assumption is similar topology of the nucleotide-binding site of Hsp90 and normal ATP binding enzymes and proteins. However, it was experimentally proved that the morphology of Hsp90's ATP binding cleft is different from that of the healthy cellular enzyme's/protein's ATP binding site. In Hsp90, this cleft assumes a 'V' shaped configuration, whereas in normal protein, this site adopts a straight morphology [54–56]. Hence, blaming Hsp90 alone will be a gross injustice to this wonderful target. It can be regarded as a culprit for development and progression of cancer but cannot be a convict in anticancer drug discovery. Therefore, we tried to analyze the drug designing strategy adopted for identifying novel Hsp90 inhibitors. The two rational drug-designing approaches involved in the discovery of antagonist against Hsp90 are structure-based approach (virtual screening, molecular docking) [131–133] and ligand-dependent methodology (QSAR techniques) [128, 132–136]. The challenges with the two designing approaches are discussed in the underlying sections.

3.7.1 STRUCTURE-BASED DRUG DESIGNING (RECEPTOR-BASED STRATEGY)

Structure-dependent designing of drug molecules can be executed if the morphology of the target protein is known. The isolation of a protein in pure form and thereafter determining its amino acid sequence is a challenging task for scientists. Fortunately, the amino acid arrangements for all the three domains (N, C, and M) of Hsp90 protein have been successfully decoded. The crystal structure of Hsp90 in the free state as well as in complex with its inhibitors, cochaperones and nucleotides has been submitted to the protein data bank (PDB). A search on PDB with Hsp90 resulted in 427 entries and 276 ligands. Hence, this drug designing approach was extensively used for the discovery of novel Hsp90 antagonist. More than 95% of Hsp90 inhibitors were discovered via this receptor-based designing strategy [99, 125, 127, 137]. The software predicts the binding affinity of the compounds in terms of docking score or IC_{50} values [127]. Nevertheless, this procedure failed to bring a single molecule into the market. The limitations of this technique with respect to the design of novel Hsp90 inhibitors are listed below:

1. The binding site is considered as rigid during a docking simulation experiments. In reality, Hsp90 is a highly flexible protein in which the amino acids frequently change their orientation. Thus,

the flexibility factor is ignored in molecular docking studies which can significantly affect the results, e.g., prediction of potency in terms of binding efficiency [138]. These lacunae can be overcome by conducting molecular dynamic (MD) simulation studies [134, 139, 140] and/or induced fit docking (IFD) methodologies [141]. The disadvantage with these techniques is the huge time and cost involved in computing a result [142]. Furthermore, these procedures are also not full proof because there are instances wherein they have failed to correctly predict the potency of the compounds. For example, in one of the studies, IFD failed to correctly simulate the binding interactions of five ligands with Hsp90 protein (PDB ID: 3VHA, 2WI7, 3FT5, 3OW6, and 4YKR) [143].

2. The identification of conserved water molecules was also found to be challenging in case of Hsp90. It has been proven that water molecules play an important role during binding of ligands with the polypeptide [102, 144]. Hence, deleting all the water molecules during docking studies is not advisable in case of design of inhibitors against Hsp90. MD simulation techniques was successfully employed for the identification of these important water molecules [145, 146]. However, this technique is tedious and time-consuming. Hence, the water present within 5 angstrom from the ligand is retained during docking experimentations [133]. This saves a lot of computational time and cost. Another time and cost-saving approach is to retain all the water molecules, while running a docking engine. Finally, the waters with which the ligands forms hydrogen bonds are selected and docking machine is run again by retaining those water molecules.

3. The third limitations of structure-based designing strategy are the validation of the docking protocol. The usual method involves comparing the binding interactions (hydrogen bonding and hydrophobic) of the native protein-ligand crystal structure with that of the software-generated data [100, 147, 148]. However, very few scientists have attempted to calculate the success rate of the predicted binding affinity of the compounds by correlating them with their practical values. Hence, it is suggested that docking accuracy should be confirmed by comparing the theoretical and practical binding strengths of the compounds. Additionally, these studies should be conducted by taking into considerations a large set of compounds (minimum 100).

3.7.2   LIGAND-BASED DRUG DESIGNING

The ligand-based drug designing technique (also known as quantitative structure-activity relationship, QSAR) has been used for the discovery of novel Hsp90 inhibitors [133, 135]. However, none of the studies have generated a molecule that can be taken up for clinical trials. The major loopholes of this QSAR approach with respect to design of Hsp90 inhibitors can be summarized under the following points:

1. All the QSAR related works predicted the active structural features necessary for potent Hsp90 inhibitors. Additionally, the models were validated by external structural sets. Mere prediction with validation is enough to publish a paper in high impact journals, but not sufficient for the discovery of clinically valuable Hsp90 inhibitors. Novel chemical entities need to be designed based on the QSAR studies, followed by their synthesis and *in vitro / in vivo* pharmacological experimentations. Thereafter, the QSAR model should be validated with the newly synthesized molecules.

2. The second limitations of this approach are the prediction of only Hsp90 inhibition potential of the compounds. It has been observed that a compound exhibit poor antiproliferative effect against cancer cells in spite of being a potent Hsp90 inhibitor [103, 104, 127]. Hence a QSAR study should not only predict the Hsp90 binding efficiency of the compounds, but also their cytotoxic effects on cancer cells. In other words, a dual ligand based drug designing approach should be developed for the prediction of a potent anti-cancer agent. This strategy is challenging for a medicinal chemist. However, if implemented properly, this can lead to the successful generation of molecules with a potential to reach clinical trials.

3.8   MISCELLANEOUS MEANS OF INHIBITING HSP90

The failure of N-terminal ATP binding Hsp90 inhibitors to reach clinical trials stimulated scientists to explore other allosteric inhibitors for the chaperone. In recent years, several compounds were shown to suppress Hsp90 function by binding to sites other than the N-terminal Bergerat cleft [76, 149]. These classes of non-N domain Hsp90 inhibitors can be classified as follows:

a) C-terminal inhibitors.
b) Compounds that binds between M and N-domain.
c) Hsp90-cochaperone interaction inhibitors.
d) Hsp90-client protein interaction inhibitors
e) Hsp90 thiol oxidation inhibitors.

3.8.1 C-TERMINAL INHIBITORS

The prominent C-terminal inhibitors are coumarin derived compounds (novobiocin, chlorobiocin, coumermycin (Figure 3.12) [150]. Curcumin, epigallocatechin–3-gallate (EGCG), cisplatin, and silybin are the C-terminal Hsp90 inhibitors without the coumarin nucleus (Figure 3.13) [151]. Novobiocin is obtained from *Streptomyces niveus* and is used for the treatment of various bacterial infections. It is a DNA gyrase inhibitor, which blocks the ATPase activity of the enzyme. This coumarin antibiotic was the first carboxyl terminus Hsp90 inhibitor discovered [152]. The novobiocin binding site was revealed to be adjacent to the C-terminal dimerization domain (amino acid sequence 538–728). This site was also found to have an affinity for ATP nucleotide. Hence, it was concluded that novobiocin is a competitive inhibitor of ATP nucleotide at Hsp90's C-domain [150, 153, 154]. Further experimentation proved that novobiocin do not bind to N-terminal ATP binding cleft. However, it was practically shown that binding of novobiocin to the C-domain's ATP binding region leads to displacement of inhibitors (ATP, molecules) from the N-terminal ATP binding cleft (Bergerat's fold). This may be attributed to chaperone's flexible nature which leads to crosstalk between N and C domains of the protein. Additionally, it was also demonstrated that novobiocin binding prevents association of co-chaperones (p23, Raf1, mutant p53 and Hsc70) with Hsp90 [152]. Further, numerous derivatives of novobiocin were synthesized and evaluated for their Hsp90 inhibitory potential. These analogs were found to be 100–1000 times more potent than novobiocin in terms of Hsp90 inhibition. Additionally, structure-activity relationship studies were also carried out with these synthesized novobiocin analogs [150]. The conclusions of the SAR studies were used to design more potent and selective (will not bind to DNA gyrase) Hsp90 C-terminal inhibitors. This led to the identification of two selective and potent natural product inhibitors of Hsp90 chaperone (DHN1 and DHN2,

Figure 3.12) [151, 155]. Chlorobiocin and coumarimycin were reported to act in a similar fashion as that of novobiocin [151, 152].

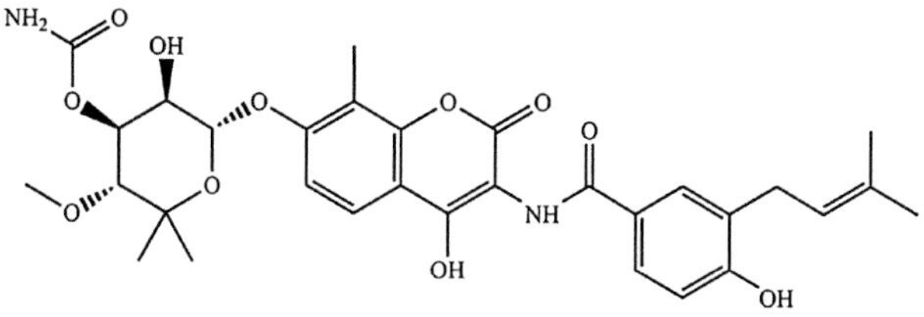

novobiocin

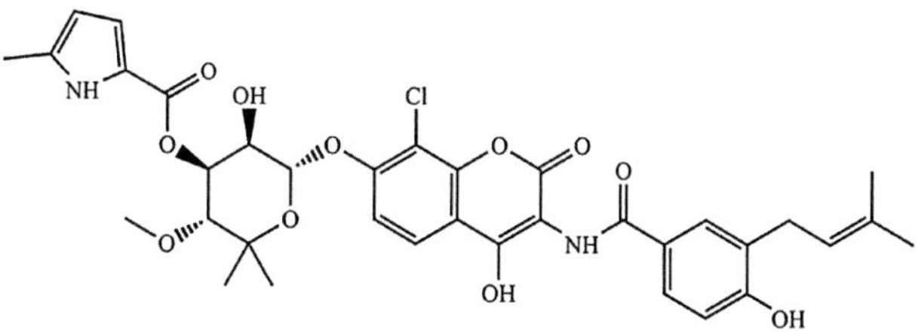

clorobiocin

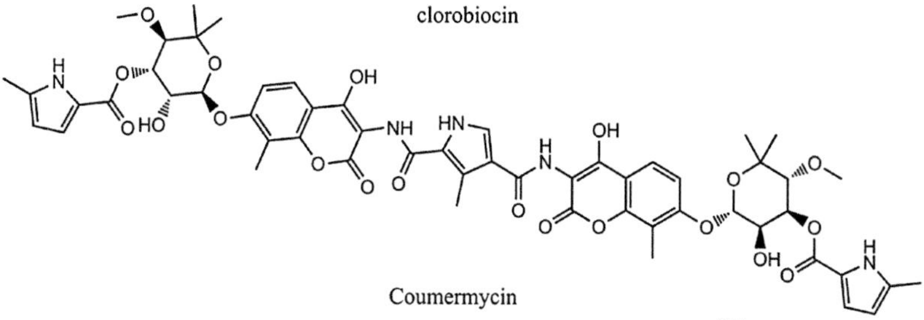

Coumermycin

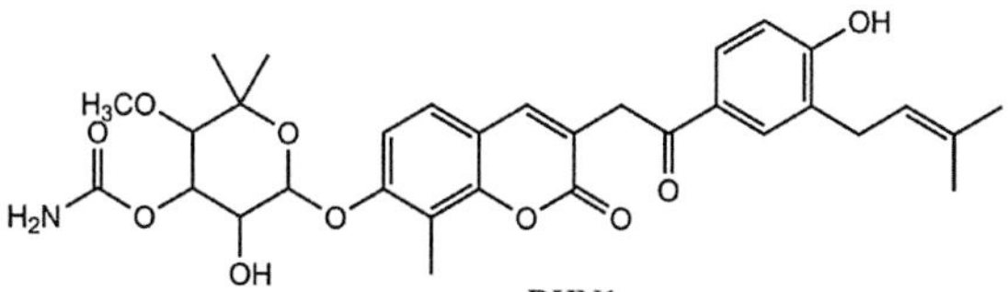

DHN1

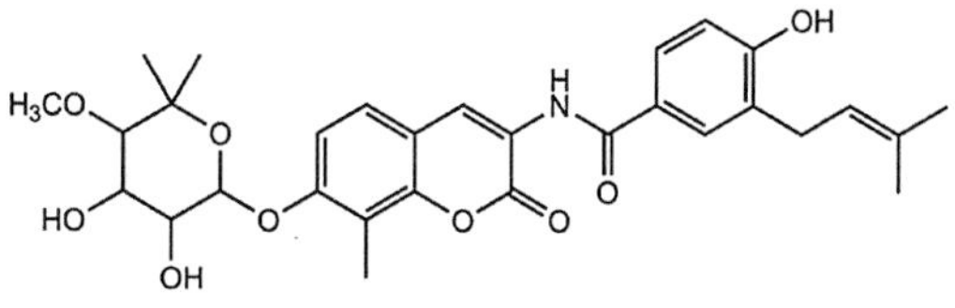

DHN2

FIGURE 3.12 Coumarin analogs as C-terminal Hsp90 inhibitors.

EGCG is a polyphenolic benzopyran ring containing compound isolated from green tea. It was experimentally shown that it binds to the amino acid residue 538–728 of Hsp90's C-terminal domain. This active site is an ATP binding cleft where novobiocin was also shown to bind [149, 156]. Currently, studies are being carried out to establish whether it's a competitive antagonist of novobiocin and cisplatin. The results for the same is still awaited [149, 157].

The natural product, curcumin is isolated from *Curcuma longa*, *Curcuma aromatica*, and *Curcuma phaeocaulis* (Zingiberaceae). Currently, it is in Phase-I/II of clinical trials for the treatment of colorectal cancer, pancreatic carcinoma, breast cancer and osteocarcinoma [158, 159]. It was experimentally found that curcumin induces degradation and down regulation of few client proteins of Hsp90 (ErbB2, p23, hTERT, bcr/abl) [160]. Additionally, curcumin and its derivatives have shown potential Hsp90 inhibitory activity by binding to the C domain [161]. However, the exact binding mode of this natural product with Hsp90 is elusive for researchers.

Cisplatin is an inorganic DNA intercalating agent that is used for the treatment of various types of cancer (Figure 3.13). It is used alone or in combination with other antineoplastic agents for the treatment of testicular, ovarian, bladder and lung cancer [162]. It has been demonstrated that cisplatin binds to a region proximal to the C domain's ATP binding pocket. The amino acid residues involved in its binding is still not elucidated [163]. Resistance and side effects associated with cisplatin treatment limit its therapeutic potential as an anticancer agent [164–166]. Hence, researchers did not get stimulation to explore further the Hsp90 binding potential of cisplatin. Nevertheless, this will provide a pathway for the design of novel cisplatin derivatives as Hsp90 inhibitors.

A flavanolignan, Silybin is another natural product C-terminal Hsp90 inhibitor isolated from *Silybum marianum* (milk thistle) seeds [167]. Western blot analysis has proven that it binds to the C domain of Hsp90 chaperone [168]. In order to improve its antiproliferative potential, various analogs were synthesized and SAR was developed [169]. All the compounds were found to bind to Hsp90's C-terminal domain. However, the exact site/mode of interaction of silybin and its derivatives with Hsp90 protein is not elucidated.

In conclusion, the C-terminal inhibitors has provided an alternative means of inhibiting the Hsp90 chaperone. Initial results with novobiocin analogs was encouraging. However, the complete binding pattern of the inhibitors with Hsp90 is still elusive. Till date, not a single crystal structure of any molecule with the chaperone's C terminal has been deposited to PDB. Structure-based drug designer are eagerly waiting for the biologists to submit such bonded crystal structures to PDB repository.

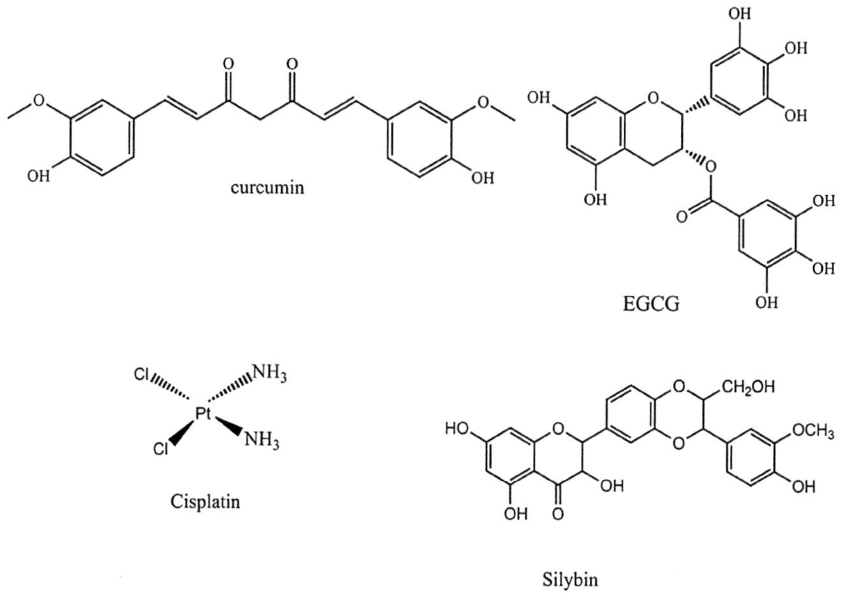

FIGURE 3.13   Non-coumarin C-terminal inhibitors of Hsp90 protein.

3.8.2   COMPOUNDS THAT BIND BETWEEN M AND N DOMAIN

A cyclic pentapeptide, sansalvamide A (isolated from marine fungus *Halodule wrightii*, Figure 3.14), exhibited potent antiproliferative activity against a wide variety of cancer cells [170]. The target of this natural product was found to be Hsp90 chaperone. Experimental studies revealed that it suppresses the chaperoning function of Hsp90 by binding at the interface between N- and M-domain. This attachment of the molecule causes a change in the orientation of the C-terminal, which, in turn, affects binding of client protein/co-chaperone to this region of the polypeptide

[171, 172]. The poor ADME properties of sansalvamide A stimulated researchers to design analogs of the peptide and evaluate their antineoplastic potential. These studies lead to the generation of sansalvamide derived lead molecules which can be further optimized for the discovery of effective Hsp90 inhibitors [173–175]. The results of the lead optimization program are still awaited.

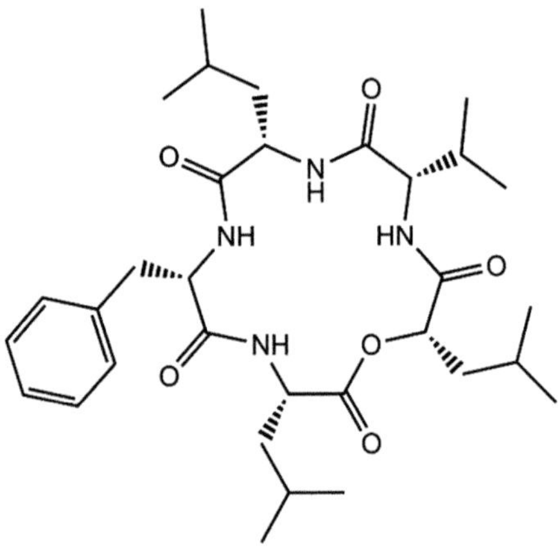

sansalvamide A

FIGURE 3.14   Structure of sansalvamide A.

3.8.3   HSP90-COCHAPERONE INTERACTION INHIBITORS

Hsp90 of cancer cells cannot heal client proteins without the help of co-chaperones [27, 46]. Therefore, disruption of Hsp90-cochaperone interactions has been a successful strategy for the degradation of client oncogenic proteins. Celastrol, gedunin, derrubone, cruentaren A, and sulforaphane were identified as Hsp90-co-chaperone interaction inhibitors (Figure 3.15). [76, 149, 176, 177] Celastrol is a triterpenoid compound isolated from the roots of *Tripterygium wilfordi* (thunder god vine plant). It exhibits its antineoplastic effects via multiple mechanisms of action [178, 179]. One of its major action was found to be Hsp90 inhibition by preventing the chaperone to interact with its co-chaperone, Cdc37 and p23. It was revealed that celastrol binds to cysteine amino acid residue at N-domain of Cdc37 co-chaperone protein. This binding interaction was found to be a Michael adduct type of covalent bond. This amino acid modification of Cdc37 by celastrol prevents its bonding with Hsp90 protein. This, in turn, blocks the recruitment of oncogenic kinases by Hsp90 as Cdc37 carries kinase enzymes to the chaperone [180, 181]. Celastrol was also reported as p23 co-chaperone-Hsp90 interaction inhibitor. Unlike Cdc37, it binds noncovalently to p23 and inactivates steroid hormone receptors as this co-chaperone is required for maintaining steroid receptor's stability and physiology [182, 183]. However, the exact mode of non-covalent binding of celastrol with p23 is not resolved.

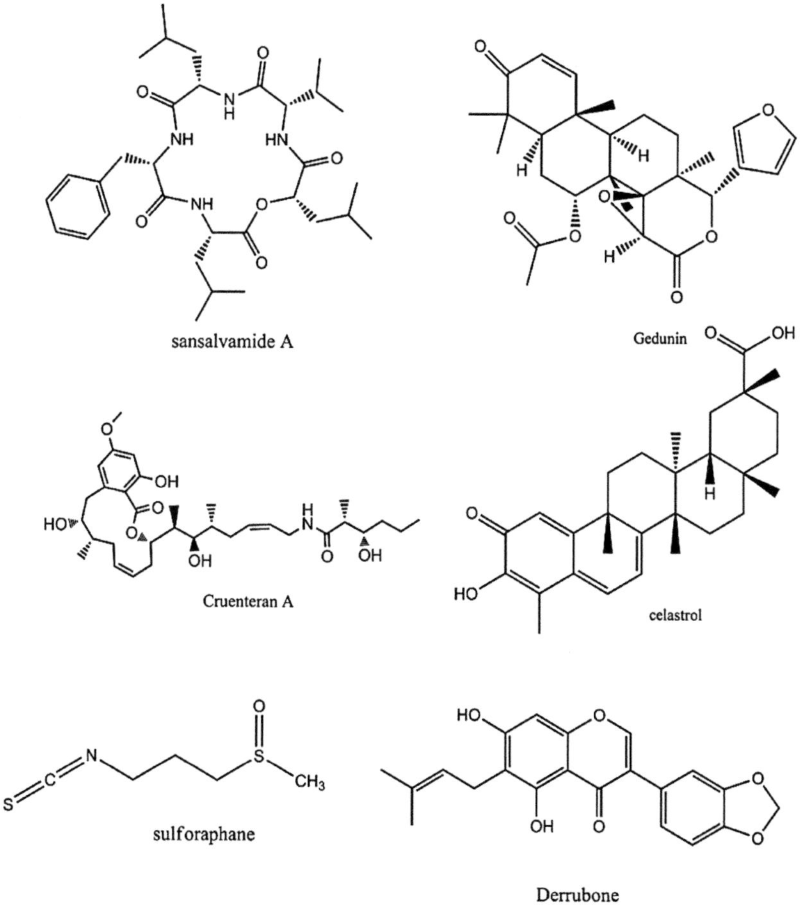

FIGURE 3.15 Chemical structure of Hsp90-cochaperone interaction inhibitors.

Gedunin is a tertratrinorterpenoid natural product isolated from the leaves of the Indian neem tree (*Azadirachta indica*). This molecule was found to exhibit potential antineoplastic activity against a wide range of cancer cells [184, 185]. Studies revealed that it induces cancer cell death primarily by binding to the Hsp90 co-chaperone, p23 [186–188]. It was proved theoretically (docking studies) and practically (mutagenesis) that gedunin forms hydrogen bonding interactions and hydrophobic contact with Thr 90/Lys 95 and Ala–94 amino acids, respectively at the C-terminal active site of p23 [186]. This interaction of gedunin with p23 prevents the co-chaperone from binding to Hsp90 protein. This, in turn, leads to inhibition of Hsp90 as the chaperone is ineffective without its co-chaperones.

Derrubone (isolfavone isolated from *Derris robusta*) is another nature-based novel Hsp90 inhibitor that executes its action by blocking Cdc37-Hsp90 interaction [189]. Cdc37 is a co-chaperone involved in the recruitment of numerous Hsp90 client peptides (mainly kinases) [190, 191]. Hence, disruption of Cdc37-Hsp90 bonding leads to the degradation of numerous client proteins of Hsp90. However, the exact mode of inhibiting Hsp90-Cdc37 bondage is still not clearly elucidated. It is proposed that it carries out its function with the help of heme-regulated eIF2a kinase (HRI) kinase enzyme, a Hsp90 client [183, 189].

Cruentaren A is a benzolactone isolated from *Myxobacterium byssovorax cruenta*. This natural product was found to exhibit potent antiproliferative effect against several cancer cell lines [192, 193]. It was demonstrated that Cruentaren A exhibits its anticancer affect by selectively inhibiting eukaryotic F_1F_0 ATP synthase enzyme (a co-chaperone of Hsp90 protein). It binds to the F_1 domain of the enzyme and do not have an affinity for any of the binding sites of other ATP synthases. This inhibition disrupts Hsp90- F_1F_0 ATP synthase interaction, which, in turn, prevents the repair of client proteins [194]. Additionally, several derivatives of cruentaren A has been synthesized and evaluated for their anticancer potential. The results of these studies revealed that three molecules were potent at nanomolar scale against L–929 mouse fibroblast cells [195]. However, an IND application is still not filed with Cruentaren A or its derivatives. This may be due to the cost involved in the development of these derivatives due to their complex stereochemical features.

A natural organosulfur compound, sulforaphane obtained from broccoli sprout was reported to disrupt Hsp90-Cdc37 binding. Additional studies revealed that it alters the topology of Hsp90's N domain, resulting in suppression of its ATPase activity. Sulforaphane promoted degradation of several client proteins (Akt, Cdk4, p53, etc.) that resulted in pancreatic cancer cell death [196–200].

3.8.4 HSP90-CLIENT PROTEIN INTERACTION INHIBITORS

The association of Hsp90 with client polypeptides can be disrupted to hinder its chaperoning function. Consequently, this alternate strategy has attracted the attention of researchers involved in the discovery of novel anticancer molecules. Studies have revealed two compounds with a potential to prevent this Hsp90-client protein interaction (Figure 3.16).

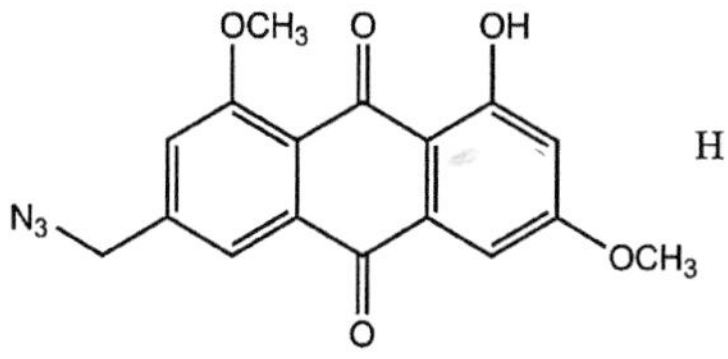

H - Lys - His - Ser - Ser - Gly - Cys - Ala - Phe - Leu - OH

shepherdin (amino acid 79-87 of survivin)

Emodin AMAD

FIGURE 3.16 Compound that prevents interaction between Hsp90 and its client proteins.

An emodin azide methyl anthraquinone derivative (AMAD) was found to prevent interaction of Her2 protein with Hsp90, which resulted in the later's degradation [201–203]. Computer-assisted docking studies revealed that this novel molecule binds to both Hsp90 and Her2 [202]. However, the aforestated hypothesis are not validated by some practical experimentation (like mutagenesis, Western blot, etc.). The results of such studies are yet to be published in the literatures.

Scientists developed a peptidomimetic molecule, Shepherdin based on survivin (client of Hsp90 responsible for cell proliferation and overexpressed in cancer cells) [204]. Shepherdin is a polypeptide whose amino acid sequence is similar to the survivin's Hsp90 binding region. It was proved experimentally that shepherdin competitively blocks the binding of survivin to Hsp90, leading to its degradation [205]. It was also shown theoretically (Molecular dynamics defined molecular docking studies) that shepherdin has affinity for the ATP binding cleft of Hsp90's N domain [129, 183]. The exact biology of the interactions of Hsp90 with its client proteins and co-chaperones is still not established. This has hindered the discovery of new molecules targeting these interactions.

3.8.5   HSP90 THIOL OXIDATION INHIBITORS

A natural product, Tubocapsenolide A (TA) opened a new window for Hsp90 inhibition (Figure 3.17). TA is a steroid obtained from the roots, stems and leaves of *Tubocapsicum anomalum* (a Chinese flora) [206]. It demonstrated significant antiproliferative activity against various cancer cell lines [206, 207]. It was proven experimentally that TA suppresses the chaperoning function of Hsp90 by inducing rapid oxidation of Hsp90's thiol group. The effect of TA was prevented by a thiol anti-oxidant, acetyl-cysteine [208]. It was also shown that TA causes proteosomal degradation of few Hsp90 client proteins like Cdk4, cyclin D1, Raf-1, Akt and mutant p53 [183, 206].

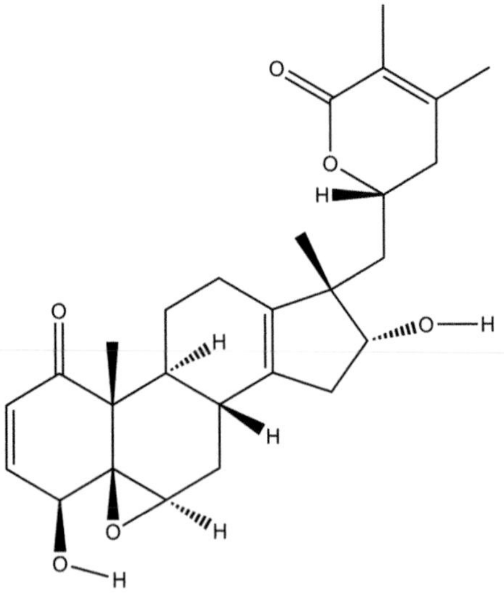

Tubocapsenolide A

FIGURE 3.17   Natural product Tubocapsenolide A.

3.9 CONCLUSION

The field of Hsp90 inhibition has shown considerable promise in anti-cancer drug discovery and development. The search for novel, efficacious antineoplastic agents based on Hsp90 inhibition started 23 years ago with natural product geldanamycin. The discovery programs were mainly spearheaded by academic institutions. Later on, several small and big pharmaceutical companies initiated Hsp90 programs by collaborations and/or acquisitions. Till date, 18 molecules have reached various stages of clinical trials either singly or in combination with other anticancer agents. Except DS–2248, the structures of all the other compounds have been disclosed by the respective organizations. However, success in the form of regulatory approval is still haunting scientists and their organizations. The N-terminal ATP binding domain was the main target area for suppressing Hsp90 and developing inhibitors against the chaperone. All the 18 molecules that were evaluated in human beings are competitive antagonist of ATP at the Hsp90's N domain. Hence, researchers tried to inhibit Hsp90 by other means (C-terminal inhibitors, Hsp90-cochaperone interaction inhibitors, etc.). All these approaches are being evaluated in preclinical level. One unexplored approach is the development of isoform(s) selective inhibitors against Hsp90. This will be a challenging task because of 80% structural similarity between Hsp90 α and β. The success achieved with selective COX-2 inhibitors should act as a stimulating factor in this research idea. The cost and time involved in the discovery process are a hindrance for fast-tracking these research projects. Hence, the need of the hour is designed, easily synthesizable small molecule Hsp90 inhibtors, which will stimulate industries and academic institutions to develop them further. Nevertheless, the new anticancer drug discovery program based on Hsp90 inhibition has not halted completely because of the involvement of Hsp90 in multiple oncogenic pathways and in every cancer type. It will always remain as one of the weapons for oncologists to fight cancer. In future, the challenge will be to design the best clinical Hsp90 inhibitors for the right indication.

KEYWORDS

- **C-terminal inhibitors**
- **geldanamycin**
- **Hsp90**

REFERENCES

1. Korohoda, W., (1988). Studies of growth, differentiation and neoplastic transformation of human and animal cells-views and trends: 1985–1988. *Postepy. Biochem., 34*(4), 269–271.
2. Hanahan, D., & Weinberg, R. A., (2000). The hallmarks of Cancer Cell, *100*(1), 57–70.
3. Kayl, A. E., & Meyers, C. A., (2006). Side-effects of chemotherapy and quality of life in ovarian and breast cancer patients. *Curr. Opin. Obstet. Gynecol., 18*(1), 24–28.
4. Carelle, N., Piotto, E., Bellanger, A., Germanaud, J., Thuillier, A., & Khayat, D., (2002). Changing patient perceptions of the side effects of cancer chemotherapy. *Cancer, 95* (1), 155–163.
5. Gerber, D. E., (2008). Targeted therapies: A new generation of cancer treatments. *Am. Fam. Physician., 77* (3), 311–319.
6. Demidenko, Z. N., & Mc Cubrey, J. A., (2011). Recent progress in targeting cancer. *Aging., 3*(12), 1154–1162.
7. Widakowich, C., de Castro, G. Jr., de Azambuja, E., Dinh, P., & Awada, A., (2007). Review: side effects of approved molecular targeted therapies in solid cancers. *The Oncologist, 12* (12), 1443–1455.
8. de Castro, G Jr., & Awada, A., (2006). Side effects of anti-cancer molecular-targeted therapies (not monoclonal antibodies). *Curr. Opin. Oncol., 18*(4), 307–315.
9. Chandra, F., Sandiono, D., Sugiri, U., Suwarsa, O., & Gunawan, H., (2017). Cutaneous Side Effects and Transepidermal Water Loss To Gefitinib: A Study of 11 Patients. *Dermatology and therapy., 7*(1), 133–141.
10. Ehmann, L. M., Ruzicka, T., & Wollenberg, A., (2011). Cutaneous side-effects of EGFR inhibitors and their management. *Skin. Therapy. Lett., 16*(1), 1–3.
11. Giles, F. J., O'Dwyer, M., & Swords, R., (2009). Class effects of tyrosine kinase inhibitors in the treatment of chronic myeloid leukemia. *Leukemia., 23*(10), 1698–16707.
12. Lovly, C. M., Iyengar, P., & Gainor, J. F., (2017). Managing Resistance to EFGR- and ALK-Targeted Therapies. *Am. Soc. Clin. Oncol. Educ. Book., 37,* 607–618.
13. Holohan, C., Van Schaeybroeck, S., Longley, D. B., & Johnston, P. G., (2013). Cancer drug resistance: An evolving paradigm. *Nat. Rev. Cancer, 13*(10), 714–726.

14. Lu, H. P., & Chao, C. C., (2012). Cancer cells acquire resistance to anticancer drugs: An update. *Biomed., J, 35*(6), 464–472.
15. Prior, I. A., Lewis, P. D., & Mattos, C. A., (2012). Comprehensive survey of Ras mutations in cancer. *Cancer Res., 72*(10), 2457–2467.
16. Teodori, E., Dei, S., Scapecchi, S., & Gualtieri, F., (2002). The medicinal chemistry of multidrug resistance (MDR) reversing drugs. *Farmaco., 57* (5), 385–415.
17. Teodori, E., Dei, S., Martelli, C., Scapecchi, S., & Gualtieri, F., (2006). The functions and structure of ABC transporters: implications for the design of new inhibitors of Pgp and MRP1 to control multidrug resistance (MDR). *Curr. Drug. Targets., 7*(7), 893–909.
18. Soverini, S., Martinelli, G., Rosti, G., Bassi, S., Amabile, M., Poerio, A., et al., (2005). ABL mutations in late chronic phase chronic myeloid leukemia patients with up-front cytogenetic resistance to imatinib are associated with a greater likelihood of progression to blast crisis and shorter survival: A study by the GIMEMA Working Party on Chronic Myeloid Leukemia. *J. Clin. Oncol., 23*(18), 4100–4109.
19. Soo, E. T., Yip, G. W., Lwin, Z. M., Kumar, S. D., & Bay, B. H., (2008). Heat shock proteins as novel therapeutic targets in cancer. *In. vivo., 22*(3), 311–315.
20. Sauvage, F., Messaoudi, S., Fattal, E., Barratt, G., & Vergnaud-Gauduchon, J., (2017). Heat shock proteins and cancer: How can nanomedicine be harnessed? *J. Control. Release: Official Journal of the Controlled Release Society, 248,* 133–143.
21. Hartl, F. U., Bracher, A., & Hayer-Hartl, M., (2011). Molecular chaperones in protein folding and proteostasis. *Nature, 475*(7356), 324–332.
22. Kim, Y. E., Hipp, M. S., Bracher, A., Hayer-Hartl, M., & Hartl, F. U., (2013). Molecular chaperone functions in protein folding and proteostasis. *Annu. Rev. Biochem., 82,* 323–855.
23. Beissinger, M., & Buchner, J., (1998). How chaperones fold proteins. *Biolo. Chem., 379*(3), 245–259.
24. Burston, S. G., & Clarke, A. R., (1995). Molecular chaperones: physical and mechanistic properties. *Essays Biochem., 29,* 125–136.
25. Zhang, X., Beuron, F., & Freemont, P. S., (2002). Machinery of protein folding and unfolding. *Curr. Opin. Struct. Biol., 12*(2), 231–238.
26. Taipale, M., Jarosz, D. F., & Lindquist, S., (2010). HSP90 at the hub of protein homeostasis: emerging mechanistic insights. *Nat. Rev. Mol. Cell. Biol., 11*(7), 515–528.
27. Li, J., Soroka, J., & Buchner, J., (2012). The Hsp90 chaperone machinery: conformational dynamics and regulation by co-chaperones. *Biochim. Biophys. Acta., 1823*(3), 624–635.
28. Sreedhar, A. S., Kalmar, E., Csermely, P., & Shen, Y. F., (2004). Hsp90 isoforms: functions, expression and clinical importance. *FEBS. Lett., 562*(1–3), 11–15.
29. Cortes-Gonzalez, C. C., Ramirez-Gonzalez, V., Ariza, A. C., & Bobadilla, N. A., (2008). Functional significance of heat shock protein 90. *Rev. Invest. Clin., 60*(4), 311–320.
30. Didelot, C., Lanneau, D., Brunet, M., Bouchot, A., Cartier, J., Jacquel, A., et al., (2008). Interaction of heat-shock protein 90 beta isoform (HSP90 beta) with cellular inhibitor of apoptosis 1 (*c*-IAP1) is required for cell differentiation. *Cell. Death. Differ., 15*(5), 859–866.

31. Pearl, L. H., & Prodromou, C., (2001). Structure, function, and mechanism of the Hsp90 molecular chaperone. *Adv. Protein. Chem., 59*, 157–186.

32. Didenko, T., Duarte, A. M., Karagoz, G. E., & Rudiger, S. G., (2012). Hsp90 structure and function studied by NMR spectroscopy. *Biochim. Biophys Acta., 1823*(3), 636–647.

33. Prodromou, C., Roe, S. M., O'Brien, R., Ladbury, J. E., Piper, P. W., & Pearl, L. H., (1997). Identification and structural characterization of the ATP/ADP-binding site in the Hsp90 molecular chaperone. *Cell., 90*(1), 65–75.

34. Richter, K., Reinstein, J., & Buchner, J., (2002). N-terminal residues regulate the catalytic efficiency of the Hsp90 ATPase cycle. *J. Biol. Chem., 277*(47), 44905–44910.

35. Meyer, P., Prodromou, C., Hu, B., Vaughan, C., Roe, S. M., Panaretou, B., et al., (2003). Structural and functional analysis of the middle segment of hsp90: implications for ATP hydrolysis and client protein and cochaperone interactions. *Mol. Cell., 11*(3), 647–658.

36. Weikl, T., Muschler, P., Richter, K., Veit, T., Reinstein, J., & Buchner, J., (2000). C-terminal regions of Hsp90 are important for trapping the nucleotide during the ATPase cycle. *J. Mol. Biol., 303*(4), 583–592.

37. Prodromou, C., Roe, S. M., Piper, P. W., & Pearl, L. H., (1997). A molecular clamp in the crystal structure of the N-terminal domain of the yeast Hsp90 chaperone. *Nat. Struct. Biol., 4*(6), 477–482.

38. Neckers, L., Mollapour, M., & Tsutsumi, S., (2009). The complex dance of the molecular chaperone Hsp90. *Trends Biochem. Sci., 34*(5), 223–236.

39. Krukenberg, K. A., Street, T. O., Lavery, L. A., & Agard, D. A., (2011). Conformational dynamics of the molecular chaperone Hsp90. *Quarterly Reviews of Biophysics, 44*(2), 229–255.

40. Pearl, L. H., & Prodromou, C., (2006). Structure and mechanism of the Hsp90 molecular chaperone machinery. *Annual Review of Biochemistry, 75*, 271–94.

41. Bron, P., Giudice, E., Rolland, J. P., Buey, R. M., Barbier, P., Diaz, J. F., et al., (2008). Apo-Hsp90 coexists in two open conformational states in solution. *Biology of the Cell, 100*(7), 413–425.

42. Mc Laughlin, S. H., Ventouras, L. A., Lobbezoo, B., & Jackson, S. E., (2004). Independent ATPase activity of Hsp90 subunits creates a flexible assembly platform. *J. Mol. Biol., 344*(3), 813–826.

43. Siligardi, G., Hu, B., Panaretou, B., Piper, P. W., Pearl, L. H., & Prodromou, C., (2004). Co-chaperone regulation of conformational switching in the Hsp90 ATPase cycle. *J. Biol. Chem., 279*(50), 51989–51998.

44. Mc Laughlin, S. H., Sobott, F., Yao, Z. P., Zhang, W., Nielsen, P. R., Grossmann, J. G., et al., (2006). The co-chaperone p23 arrests the Hsp90 ATPase cycle to trap client proteins. *J. Mol. Biol., 356*(3), 746–758.

45. Street, T. O., Lavery, L. A., & Agard, D. A., (2011). Substrate binding drives large-scale conformational changes in the Hsp90 molecular chaperone. *Mol. Cell., 42*(1), 96–105.

46. Zuehlke, A., & Johnson, J. L., (2010). Hsp90 and co-chaperones twist the functions of diverse client proteins. *Biopolymers, 93*(3), 211–217.

47. Johnson, J. L., (2012). Evolution and function of diverse Hsp90 homologs and co-chaperone proteins. *Biochim. Biophys. Acta., 1823*(3), 607–613.
48. Obermann, W. M., Sondermann, H., Russo, A. A., Pavletich, N. P., & Hartl, F. U., (1998). *In vivo* function of Hsp90 is dependent on ATP binding and ATP hydrolysis. *J. Cell. Biol., 143*(4), 901–10.
49. Richter, K., Muschler, P., Hainzl, O., & Buchner, J., (2001). Coordinated ATP hydrolysis by the Hsp90 dimer. *J. Biol. Chem., 276*(36), 33689–33696.
50. Powers, M. V., & Workman, P., (2006). Targeting of multiple signaling pathways by heat shock protein 90 molecular chaperone inhibitors. *Endocr. Relat. Cancer, 13 Suppl., 1*, 125–35.
51. Workman, P., (2004). Combinatorial attack on multistep oncogenesis by inhibiting the Hsp90 molecular chaperone. *Cancer. Lett., 206*(2), 149–157.
52. Maloney, A., & Workman, P., (2002). HSP90 as a new therapeutic target for cancer therapy: the story unfolds. *Expert. Opin. Biol. Ther., 2*(1), 3–24.
53. Chen, Y., & Ding, J., (2004). Heat shock protein 90: novel target for cancer therapy. *Chin. J. Cancer, 23*(8), 968–974.
54. Ban, C., & Yang, W., (1998). Crystal structure and ATPase activity of Mut L: Implications for DNA repair and mutagenesis. *Cell., 95*(4), 541–552.
55. Roe, S. M., Prodromou, C., O'Brien, R., Ladbury, J. E., Piper, P. W., & Pearl, L. H., (1999). Structural basis for inhibition of the Hsp90 molecular chaperone by the antitumor antibiotics radicicol and geldanamycin. *J. Med. Chem., 42*(2), 260–266,
56. Bilwes, A. M., Quezada, C. M., Croal, L. R., Crane, B. R., & Simon, M. I., (2001). Nucleotide binding by the histidine kinase Che A. *Nat. Struct. Biol., 8*(4), 353–360.
57. Stancato, L. F., Silverstein, A. M., Owens-Grillo, J. K., Chow, Y. H., Jove, R., & Pratt, W. B., (1997). The hsp90-binding antibiotic geldanamycin decreases Raf levels and epidermal growth factor signaling without disrupting formation of signaling complexes or reducing the specific enzymatic activity of Raf kinase. *J. Biol. Chem., 272*(7), 4013–4020.
58. Stebbins, C. E., Russo, A. A., Schneider, C., Rosen, N., Hartl, F. U., & Pavletich, N. P., (1997). Crystal structure of an Hsp90-geldanamycin complex: Targeting of a protein chaperone by an antitumor agent. *Cell., 89*(2), 239–250.
59. Thepchatri, P., Eliseo, T., Cicero, D. O., Myles, D., & Snyder, J. P., (2007). Relationship among ligand conformations in solution, in the solid state, and at the Hsp90 binding site: geldanamycin and radicicol. *J. Am. Chem. Soc., 129*(11), 3127–3134.
60. Neckers, L., Schulte, T. W., & Mimnaugh, E., (1999). Geldanamycin as a potential anti-cancer agent: its molecular target and biochemical activity. *Invest. New Drugs, 17*(4), 361–373.
61. Miyata, Y., (2005). Hsp90 inhibitor geldanamycin and its derivatives as novel cancer chemotherapeutic agents. *Curr. Pharm. Des., 11*(9), 1131–1138.
62. Modi, S., Stopeck, A., Linden, H., Solit, D., Chandarlapaty, S., Rosen, N., et al., (2011). HSP90 inhibition is effective in breast cancer: A phase II trial of tanespimycin (17-AAG) plus trastuzumab in patients with HER2-positive metastatic breast cancer progressing on trastuzumab. *Clin. Cancer Res., 17*(15), 5132–5139.
63. Ronnen, E. A., Kondagunta, G. V., Ishill, N., Sweeney, S. M., Deluca, J. K., Schwartz, L., et al., (2006). A phase II trial of 17-(Allylamino)-17-demethoxygeldanamycin in

patients with papillary and clear cell renal cell carcinoma. *Invest. New Drugs, 24*(6), 543–546.

64. Modi, S., Stopeck, A. T., Gordon, M. S., Mendelson, D., Solit, D. B., Bagatell, R., et al., (2007). Combination of trastuzumab and tanespimycin (17-AAG, KOS-953) is safe and active in trastuzumab-refractory HER-2 overexpressing breast cancer: A phase I dose-escalation study. *J. Clin. Oncol., 25*(34), 5410–5417.

65. Arteaga, C. L., (2011). Why is this effective HSP90 inhibitor not being developed in HER2+ breast cancer?. *Clin. Cancer Res., 17*(15), 4919–21.

66. Jhaveri, K., Miller, K., Rosen, L., Schneider, B., Chap, L., Hannah, A., et al., (2012). A phase I dose-escalation trial of trastuzumab and alvespimycin hydrochloride (KOS-1022, 17 DMAG) in the treatment of advanced solid tumors. *Clin. Cancer Res., 18*(18), 5090–5098.

67. Lancet, J. E., Gojo, I., Burton, M., Quinn, M., Tighe, S. M., Kersey, K., et al., (2010). Phase I study of the heat shock protein 90 inhibitor alvespimycin (KOS-1022, 17-DMAG) administered intravenously twice weekly to patients with acute myeloid leukemia. *Leukemia., 24*(4), 699–705.

68. Kummar, S., Gutierrez, M. E., Gardner, E. R., Chen, X., Figg, W. D., Zajac-Kaye, M., et al., (2010). Phase I trial of 17-dimethylaminoethylamino-17-demethoxygeldanamycin (17-DMAG), a heat shock protein inhibitor, administered twice weekly in patients with advanced malignancies. *Eur. J. Cancer, 46*(2), 340–347.

69. Ramanathan, R. K., Egorin, M. J., Erlichman, C., Remick, S. C., Ramalingam, S. S., Naret, C., et al., (2010). Phase I pharmacokinetic and pharmacodynamic study of 17-dimethylaminoethylamino-17-demethoxygeldanamycin, an inhibitor of heat-shock protein 90, in patients with advanced solid tumors. *J. Clin. Oncol., 28*(9), 1520–1526.

70. Sydor, J. R., Normant, E., Pien, C. S., Porter, J. R., Ge, J., Grenier, L., et al., (2006). Development of 17-allylamino-17-demethoxygeldanamycin hydroquinone hydrochloride (IPI-504), an anti-cancer agent directed against Hsp90. *Proc. Natl. Acad. Sci. U.S.A., 103*(46), 17408–17413.

71. Oh, W. K., Galsky, M. D., Stadler, W. M., Srinivas, S., Chu, F., Bubley, G., et al., (2011). Multicenter phase II trial of the heat shock protein 90 inhibitor, retaspimycin hydrochloride (IPI-504), in patients with castration-resistant prostate cancer. *Urology., 78*(3), 626–630.

72. Siegel, D., Jagannath, S., Vesole, D. H., Borello, I., Mazumder, A., Mitsiades, C., et al., (2011). A phase 1 study of IPI-504 (retaspimycin hydrochloride) in patients with relapsed or relapsed and refractory multiple myeloma. *Leuk. Lymphoma., 52*(12), 2308–2315.

73. Hanson, B. E., & Vesole, D. H., (2009). Retaspimycin hydrochloride (IPI-504): A novel heat shock protein inhibitor as an anticancer agent. *Expert. Opin. Investig. Drugs., 18*(9), 1375–1383.

74. Floris, G., Sciot, R., Wozniak, A., Van Looy, T., Wellens, J., Faa, G., et al., (2011). The Novel HSP90 inhibitor, IPI-493, is highly effective in human gastrostrointestinal stromal tumor xenografts carrying heterogeneous KIT mutations. *Clin. Cance. Res., 17*(17), 5604–5614.

75. Lee, K., Ryu, J. S., Jin, Y., Kim, W., Kaur, N., Chung, S. J., et al., (2008). Synthesis and anticancer activity of geldanamycin derivatives derived from biosynthetically generated metabolites. *Org. Biomol. Chem., 6* (2), 340–348.

76. Sidera, K., & Patsavoudi, E., (2014). HSP90 inhibitors: current development and potential in cancer therapy. *Recent. Pat. Anticancer. Drug Discov., 9*(1), 1–20.

77. Martin, C. J., Gaisser, S., Challis, I. R., Carletti, I., Wilkinson, B., Gregory, M., et al., (2008). Molecular characterization of macbecin as an Hsp90 inhibitor. *J. Med. Chem., 51*(9), 2853–2857.

78. Kim, T., Keum, G., & Pae, A. N., (2013). Discovery and development of heat shock protein 90 inhibitors as anticancer agents: A review of patented potent geldanamycin derivatives. *Expert. Opin. Ther. Pat., 2* (8), 919–943.

79. Delmotte, P., & Delmotte-Plaque, J., (1953). A new antifungal substance of fungal origin. *Nature, 171*(4347), 344.

80. Petrikaite, V., & Matulis, D., (2011). Binding of natural and synthetic inhibitors to human heat shock protein 90 and their clinical application. *Medicina (Kaunas), 47*(8), 413–420.

81. Soga, S., Neckers, L. M., Schulte, T. W., Shiotsu, Y., Akasaka, K., Narumi, H., et al., (1999). KF25706, a novel oxime derivative of radicicol, exhibits *in vivo* antitumor activity via selective depletion of Hsp90 binding signaling molecules. *Cancer Res., 59*(12), 2931–2938.

82. Proisy, N., Sharp, S. Y., Boxall, K., Connelly, S., Roe, S. M., Prodromou, C., et al., (2006). Inhibition of Hsp90 with synthetic macrolactones: synthesis and structural and biological evaluation of ring and conformational analogs of radicicol. *Chem. Biol., 13*(11), 1203–1215.

83. Shiotsu, Y., Neckers, L. M., Wortman, I., An, W. G., Schulte, T. W., Soga, S., et al., (2000). Novel oxime derivatives of radicicol induce erythroid differentiation associated with preferential G(1) phase accumulation against chronic myelogenous leukemia cells through destabilization of Bcr-Abl with Hsp90 complex. *Blood., 96*(6), 2284–2291.

84. Soga, S., Shiotsu, Y., Akinaga, S., & Sharma, S. V., (2003). Development of radicicol analogs. *Cur. Cancer Drug Targets, 3*(5), 359–369.

85. Winssinger, N., Fontaine, J. G., & Barluenga, S., (2009). Hsp90 inhibition with resorcyclic acid lactones (RALs). *Curr. Top. Med. Chem., 9*(15), 1419–1435.

86. Piper, P. W., &Millson, S. H., (2011). Mechanisms of Resistance to Hsp90 Inhibitor Drugs: A Complex Mosaic Emerges. *Pharmaceuticals, 4*(11), 1400–1422.

87. Wang, M., Shen, G., & Blagg, B. S., (2006). Radanamycin, a macrocyclic chimera of radicicol and geldanamycin. *Bioorg. Med. Chem. Lett., 16* (9), 2459–2462.

88. Clevenger, R. C., & Blagg, B. S., (2004). Design, synthesis, and evaluation of a radicicol and geldanamycin chimera, radamide. *Org. Lett., 6* (24), 4459–4462.

89. Shen, G., & Blagg, B. S., (2005). Radester, a novel inhibitor of the Hsp90 protein folding machinery. *Org. Lett., 7*(11), 2157–2160.

90. Jadhav, V. D., Duerfeldt, A. S., & Blagg, B. S., (2009). Design, synthesis, and biological activity of bicyclic radester analogs as Hsp90 inhibitors. *Bioorg. Med. Chem. Lett., 19*(24), 6845–6850.

91. Hadden, M. K., & Blagg, B. S., (2009). Synthesis and evaluation of radamide analogs, a chimera of radicicol and geldanamycin. *J. Org. Chem., 74*(13), 4697–4704.

92. Chiosis, G. (2006). Discovery and development of purine-scaffold Hsp90 inhibitors. *Curr. Top. Med. Chem., 6*(11), 1183–1191.

93. Chiosis, G., Lucas, B., Shtil, A., Huezo, H., & Rosen, N., (2002). Development of a purine-scaffold novel class of Hsp90 binders that inhibit the proliferation of cancer cells and induce the degradation of Her2 tyrosine kinase. *Bioorg. Med. Chem., 10*(11), 3555–3564.

94. Lundgren, K., Zhang, H., Brekken, J., Huser, N., Powell, R. E., Timple, N., et al., (2009). BIIB021, an orally available, fully synthetic small-molecule inhibitor of the heat shock protein Hsp90. *Mol. Cancer. Ther., 8* (4), 921–929.

95. Gallerne, C., Prola, A., & Lemaire, C., (2013). Hsp90 inhibition by PU-H71 induces apoptosis through endoplasmic reticulum stress and mitochondrial pathway in cancer cells and overcomes the resistance conferred by Bcl-2. *Biochim. Biophys. Acta., 1833*(6), 1356–1366.

96. Caldas-Lopes, E., Cerchietti, L., Ahn, J. H., Clement, C. C., Robles, A. I., Rodina, A., et al., (2009). Hsp90 inhibitor PU-H71, a multimodal inhibitor of malignancy, induces complete responses in triple-negative breast cancer models. *Proc. Natl. Acad. Sci. U.S.A., 106*(20), 8368–8373.

97. Hong, D., Said, R., Falchook, G., Naing, A., Moulder, S., Tsimberidou, A. M., et al., (2013). Phase I study of BIIB028, a selective heat shock protein 90 inhibitor, in patients with refractory metastatic or locally advanced solid tumors. *Clin. Cance. Res., 19*(17), 4824–4831.

98. Isambert, N., Delord, J. P., Soria, J. C., Hollebecque, A., Gomez-Roca, C., Purcea, D., et al., (2015). Debio0932, a second-generation oral heat shock protein (HSP) inhibitor, in patients with advanced cancer-results of a first-in-man dose-escalation study with a fixed-dose extension phase. *Ann. Oncol., 26*(5), 1005–1011.

99. Dutta Gupta, S., Snigdha, D., Mazaira, G. I., Galigniana, M. D., Subrahmanyam, C. V., Gowrishankar, N. L., et al., (2014). Molecular docking study, synthesis and biological evaluation of Schiff bases as Hsp90 inhibitors. *Biomed. Pharmacother., 68* (3), 369–376.

100. Dutta Gupta, S., Revathi, B., Mazaira, G. I., Galigniana, M. D., Subrahmanyam, C. V., Gowrishankar, N. L., et al., (2015). 2, 4-dihydroxy benzaldehyde derived Schiff bases as small molecule Hsp90 inhibitors: Rational identification of a new anticancer lead. *Bioorg. Med. Chem., 59*, 97–105,

101. Kreusch, A., Han, S., Brinker, A., Zhou, V., Choi, H. S., He, Y., et al., (2005). Crystal structures of human HSP90alpha-complexed with dihydroxyphenylpyrazoles. *Bioorg. Med. Chem. Lett., 15*(5), 1475–1478.

102. Gopalsamy, A., Shi, M., Golas, J., Vogan, E., Jacob, J., Johnson, M., et al., (2008). Discovery of benzisoxazoles as potent inhibitors of chaperone heat shock protein 90. *J. Med. Chem., 51*(3), 373–375.

103. Dutta Gupta, S., Bommaka, M. K., Mazaira, G. I., Galigniana, M. D., Subrahmanyam, C. V., Gowrishankar, N. L., et al., (2015). Molecular docking study, synthesis and biological evaluation of Mannich bases as Hsp90 inhibitors. *Int. J. Biol. Macromol., 80*, 253–259.

104. Dutta Gupta, S., Snigdha, D., Mazaira, G. I., Galigniana, M. D., Subrahmanyam, C. V., Gowrishankar, N. L., et al., (2014). Molecular docking study, synthesis and biological evaluation of Schiff bases as Hsp90 inhibitors. *Biomed. Pharmacother., 15*(14), 369–376.

105. O'Boyle, N. M., Knox, A. J., Price, T. T., Williams, D. C., Zisterer, D. M., Lloyd, D. G., et al., (2011). Lead identification of beta-lactam and related imine inhibitors of the molecular chaperone heat shock protein 90. *Bioorg. Med. Chem., 19*(20), 6055–6068.

106. Yang, Y., Liu, H., Du, J., Qin, J., & Yao, X., (2011). A combined molecular modeling study on a series of pyrazole/isoxazole based human Hsp90alpha inhibitors. *J. Mol. Model., 17*(12), 3241–3250.

107. Brough, P. A., Aherne, W., Barril, X., Borgognoni, J., Boxall, K., Cansfield, J. E., et al., (2008). 4, 5-diarylisoxazole Hsp90 chaperone inhibitors: potential therapeutic agents for the treatment of cancer. *J. Med. Chem., 51*(2), 196–218.

108. Baruchello, R., Simoni, D., Grisolia, G., Barbato, G., Marchetti, P., Rondanin, R., et al., (2011). Novel 3, 4-isoxolediamides as potent inhibitors of chaperone heat shock protein 90. *J. Med. Chem., 54*(24), 8592–8604.

109. Taddei, M., Ferrini, S., Giannotti, L., Corsi, M., Manetti, F., Giannini, G., et al., (2014). Synthesis and evaluation of new Hsp90 inhibitors based on a 1,4,5-trisubstituted 1,2,3-triazole scaffold. *J. Med. Chem., 57*(6), 2258–2274.

110. Jensen, M. R., Schoepfer, J., Radimerski, T., Massey, A., Guy, C. T., Brueggen, J., et al., (2008). NVP-AUY922: A small molecule HSP90 inhibitor with potent antitumor activity in preclinical breast cancer models. *Breast. Cancer Res., 10*(2), R33.

111. Seggewiss-Bernhardt, R., Bargou, R. C., Goh, Y. T., Stewart, A. K., Spencer, A., Alegre, A., et al., (2015). Phase 1/1B trial of the heat shock protein 90 inhibitor NVP-AUY922 as monotherapy or in combination with bortezomib in patients with relapsed or refractory multiple myeloma. *Cancer, 121*(13), 2185–2192.

112. Goyal, L., Wadlow, R. C., Blaszkowsky, L. S., Wolpin, B. M., Abrams, T. A., Mc Cleary, N. J., et al., (2015). A phase I and pharmacokinetic study of ganetespib (STA-9090) in advanced hepatocellular carcinoma. *Invest. New Drugs, 33*(1), 128–137.

113. Jhaveri, K., Chandarlapaty, S., Lake, D., Gilewski, T., Robson, M., Goldfarb, S., et al., (2014). A phase II open-label study of ganetespib, a novel heat shock protein 90 inhibitor for patients with metastatic breast cancer. *Clin. Breast. Cancer, 14*(3), 154–160.

114. Do, K., Speranza, G., Chang, L. C., Polley, E. C., Bishop, R., Zhu, W., et al., (2015). Phase I study of the heat shock protein 90 (Hsp90) inhibitor onalespib (AT13387) administered on a daily for two consecutive days per week dosing schedule in patients with advanced solid tumors. *Invest. New Drugs, 33*(4), 921–930.

115. Yong, K., Cavet, J., Johnson, P., Morgan, G., Williams, C., Nakashima, D., et al., (2016). Phase I study of KW-2478, a novel Hsp90 inhibitor, in patients with B-cell malignancies. *Br. J. Cancer, 114*(1), 7–13.

116. Biamonte, M. A., Van de Water, R., Arndt, J. W., Scannevin, R. H., Perret, D., & Lee, W. C., (2010). Heat shock protein 90: inhibitors in clinical trials. *J. Med. Chem., 53*(1), 3–17.

117. Infante, J. R., Weiss, G. J., Jones, S., Tibes, R., Bauer, T. M., Bendell, J. C., et al., (2014). 3rd, Orlemans, E. O., Ramanathan, R. K., Phase I dose-escalation studies of

SNX-5422, an orally bioavailable heat shock protein 90 inhibitor, in patients with refractory solid tumors. *Eur. J. Cancer, 50*(17), 2897–2904.

118. Reddy, N., Voorhees, P. M., Houk, B. E., Brega, N., Hinson, J. M., Jr., & Jillela, A., (2013). Phase I trial of the HSP90 inhibitor PF-04929113 (SNX5422) in adult patients with recurrent, refractory hematologic malignancies. *Clin. Lymphoma. Myeloma. Leuka., 13*(4), 385–391.

119. Ohkubo, S., Kodama, Y., Muraoka, H., Hitotsumachi, H., Yoshimura, C., Kitade, M., et al., (2015). TAS-116, a highly selective inhibitor of heat shock protein 90 alpha and beta, demonstrates potent antitumor activity and minimal ocular toxicity in preclinical models. *Mol. Cancer. Ther., 14*(1), 14–22.

120. Hendriks, L. E. L., & Dingemans, A. C., (2017). Heat shock protein antagonists in early stage clinical trials for NSCLC. *Expert. Opin. Investig. Drugs., 26*(5), 541–550.

121. Bhat, R., Tummalapalli, S. R., & Rotella, D. P., (2014). Progress in the discovery and development of heat shock protein 90 (Hsp90) inhibitors. *J. Med. Chem., 57*(21), 8718–8728.

122. Subramaniam, D. S., Warner, E. A., & Giaccone, G., (2017). Ganetespib for small cell lung cancer. *Expert. Opin. Investig. Drugs., 26*(1), 103–108.

123. Bargiotti, A., Musso, L., Dallavalle, S., Merlini, L., Gallo, G., Ciacci, A., et al., (2012). Isoxazolo(aza)naphthoquinones: A new class of cytotoxic Hsp90 inhibitors. *Eur. J. Med. Chem., 53*, 64–75.

124. Casale, E., Amboldi, N., Brasca, M. G., Caronni, D., Colombo, N., Dalvit, C., et al., (2014). Fragment-based hit discovery and structure-based optimization of aminotri-azoloquinazolines as novel Hsp90 inhibitors. *Bioorg. Med. Chem., 22*(15), 4135–4150.

125. Park, H., Kim, Y. J., & Hahn, J. S., (2007). A novel class of Hsp90 inhibitors isolated by structure-based virtual screening. *Bioorg. Med. Chem. Lett., 17*(22), 6345–6349.

126. (*d*) Hong, T. J., Park, H., Kim, Y. J., Jeong, J. H., & Hahn, J. S., (2009). Identification of new Hsp90 inhibitors by structure-based virtual screening. *Bioorg. Med. Chem. Lett., 19*(16), 4839–4842.

127. Ganesh, T., Min, J., Thepchatri, P., Du, Y., Li, L., Lewis, I., et al., (2008). Discovery of aminoquinolines as a new class of potent inhibitors of heat shock protein 90 (Hsp90): Synthesis, biology, and molecular modeling. *Bioorg. Med. Chem., 16*(14), 6903–6910.

128. Suda, A., Kawasaki, K., Komiyama, S., Isshiki, Y., Yoon, D. O., Kim, S. J., et al., (2014). Design and synthesis of 2-amino-6-(1H,3H-benzo[de]isochromen-6-yl)-1,3,5-triazines as novel Hsp90 inhibitors. *Bioorg. Med. Chem., 22*(2), 892–905.

129. Meli, M., Pennati, M., Curto, M., Daidone, M. G., Plescia, J., Toba, S., et al., (2006). Small-molecule targeting of heat shock protein 90 chaperone function: rational identification of a new anticancer lead. *J. Med. Chem., 49*(26), 7721–7730.

130. Spreafico, A., Delord, J. P., De Mattos-Arruda, L., Berge, Y., Rodon, J., Cottura, E., et al., (2015). A first-in-human phase I, dose-escalation, multicenter study of HSP990 administered orally in adult patients with advanced solid malignancies. *Br. J. Cancer, 112*(4), 650–659.

131. Lionta, E., Spyrou, G., Vassilatis, D. K., & Cournia, Z., (2014). Structure-based virtual screening for drug discovery: principles, applications and recent advances. *Curr. Top. Med. Chem., 14*(16), 1923–1938.

132. Sakkiah, S., Thangapandian, S., John, S., Kwon, Y. J., & Lee, K. W., (2010). 3D QSAR pharmacophore based virtual screening and molecular docking for identification of potential HSP90 inhibitors. *Eur. J. Med. Chem., 45*(6), 2132–2140.
133. Zhang, H., Zan, J., Yu, G., Jiang, M., & Liu, P., (2012). A combination of 3D-QSAR, molecular docking and molecular dynamics simulation studies of benzimidazole-quinolinone derivatives as in NOS inhibitors. *Int. J. Mol. Sci., 13*(9), 11210–11227.
134. Abbasi, M., Sadeghi-Aliabadi, H., & Amanlou, M., (2017). 3D-QSAR, molecular docking, and molecular dynamic simulations for prediction of new Hsp90 inhibitors based on isoxazole scaffold. *J. Biomol. Struct. Dyn.,* 1–16.
135. Winkler, D. A. (2002). The role of quantitative structure – activity relationships (QSAR) in biomolecular discovery. *Briefings. Bioinformatics, 3*(1), 73–86
136. Gaurav, A., & Singh, R., (2012). 3D QSAR pharmacophore, Co MFA and Co MSIA based design and docking studies on phenyl alkyl ketones as inhibitors of phosphodiesterase 4. *Med. Chem., 8*(5), 894–912.
137. Doddareddy, M. R., Thorat, D. A., Seo, S. H., Hong, T. J., Cho, Y. S., Hahn, J. S., et al., (2011). Structure-based design of heat shock protein 90 inhibitors acting as anti-cancer agents. *Bioorg. Med. Chem., 19*(5), 1714–1720.
138. Chen, Y. C. (2015). Beware of docking! *Trends. Pharmacol. Sci.*36 (2), 78–95.
139. Vettoretti, G., Moroni, E., Sattin, S., Tao, J., Agard, D. A., Bernardi, A., et al., (2016). Molecular Dynamics Simulations Reveal the Mechanisms of Allosteric Activation of Hsp90 by Designed Ligands. *Sci. Rep., 6*, 23830.
140. Moroni, E., Morra, G., & Colombo, G., (2012). Molecular dynamics simulations of hsp90 with an eye to inhibitor design. *Pharmaceuticals, 5*(9), 944–962.
141. Lauria, A., Ippolito, M., & Almerico, A. M., (2009). Inside the Hsp90 inhibitors binding mode through induced fit docking. *J. Mol. Graph. Model., 27*(6), 712–722.
142. Verkhivker, G. M., Dixit, A., Morra, G., & Colombo, G., (2009). Structural and computational biology of the molecular chaperone Hsp90: from understanding molecular mechanisms to computer-based inhibitor design. *Curr. Top. Med. Chem., 9*(15), 1369–1385.
143. Misini Ignjatovic, M., Caldararu, O., Dong, G., Munoz-Gutierrez, C., Adasme-Carreno, F., & Ryde, U., (2016). Binding-affinity predictions of HSP90 in the D3R Grand Challenge 2015 with docking, MM/GBSA, QM/MM, and free-energy simulations. *J. Comput. Aided. Mol. Des., 30*(9), 707–730.
144. Kung, P. P., Sinnema, P. J., Richardson, P., Hickey, M. J., Gajiwala, K. S., Wang, F., et al., (2011). Design strategies to target crystallographic waters applied to the Hsp90 molecular chaperone. *Bioorg. Med. Chem. Lett., 21*(12), 3557–3562.
145. Yan, A., Grant, G. H., & Richards, W. G., (2008). Dynamics of conserved waters in human Hsp90: implications for drug design. *J. R. Soc. Interface., 6*(5), S199–S205.
146. Hospital, A., Goni, J. R., Orozco, M., & Gelpi, J. L., (2015). Molecular dynamics simulations: advances and applications. *Adv. Appl. Bioinform. Chem., 8*, 37–47.
147. Santos-Martins, D. (2016). Interaction with specific HSP90 residues as a scoring function: validation in the D3R Grand Challenge 2015. *J. Comput. Aided. Mol. Des, 30*(9), 731–742.
148. Wang, J. C., & Lin, J. H., (2013). Scoring functions for prediction of protein-ligand interactions. *Curr. Pharm. Des., 19*(12), 2174–2182.

149. Khandelwal, A., Crowley, V. M., & Blagg, B. S., (2016). Natural Product Inspired N-Terminal Hsp90 Inhibitors: From Bench to Bedside? *Med. Res. Rev., 36*(1), 92–118.

150. Marcu, M. G., Schulte, T. W., & Neckers, L., (2000). Novobiocin and related coumarins and depletion of heat shock protein 90-dependent signaling proteins. *J. Natl. Cancer. Ins., 92*(3), 242–248.

151. Donnelly, A., & Blagg, B. S., (2008). Novobiocin and additional inhibitors of the Hsp90 C-terminal nucleotide-binding pocket. *Curr. Med. Chem., 15*(26), 2702–2717.

152. Allan, R. K., Mok, D., Ward, B. K., & Ratajczak, T., (2006). Modulation of chaperone function and cochaperone interaction by novobiocin in the C-terminal domain of Hsp90: evidence that coumarin antibiotics disrupt Hsp90 dimerization. *J. Biol. Chem., 281*(11), 7161–7171.

153. Sgobba, M., Forestiero, R., Degliesposti, G., & Rastelli, G., (2010). Exploring the binding site of C-terminal hsp90 inhibitors. *J. Chem. Inf. Model., 50*(9), 1522–1528.

154. Marcu, M. G., Chadli, A., Bouhouche, I., Catelli, M., & Neckers, L. M., (2000). The heat shock protein 90 antagonist novobiocin interacts with a previously unrecognized ATP-binding domain in the carboxyl terminus of the chaperone. *J. Biol. Chem., 275*(47), 37181–37186.

155. Burlison, J. A., Avila, C., Vielhauer, G., Lubbers, D. J., Holzbeierlein, J., & Blagg, B. S., (2008). Development of novobiocin analogs that manifest anti-proliferative activity against several cancer cell lines. *J. Org. Chem., 73*(6), 2130–2137.

156. Yin, Z., Henry, E. C., & Gasiewicz, T. A., (2009). (–)-Epigallocatechin-3-gallate is a novel Hsp90 inhibitor. *Biochem., 48*(2), 336–345.

157. Palermo, C. M., Westlake, C. A., & Gasiewicz, T. A., (2005). Epigallocatechin gallate inhibits aryl hydrocarbon receptor gene transcription through an indirect mechanism involving binding to a 90 k Da heat shock protein. *Biochem., 44*(13), 5041–5052.

158. Li, Y., Zhang, D., Xu, J., Shi, J., Jiang, L., Yao, N., et al., (2012). Discovery and development of natural heat shock protein 90 inhibitors in cancer treatment. *Acta. Pharmaceutica. Sinica B., 2*(3), 238–245.

159. Mahammedi, H., Planchat, E., Pouget, M., Durando, X., Cure, H., Guy, L., et al., (2016). The New Combination Docetaxel, Prednisone and Curcumin in Patients with Castration-Resistant Prostate Cancer: A Pilot Phase II Study. *Oncology., 90*(2), 69–78.

160. Wu, L. X., Xu, J. H., Huang, X. W., Zhang, K. Z., Wen, C. X., & Chen, Y. Z., (2006). Down-regulation of p210(bcr/abl) by curcumin involves disrupting molecular chaperone functions of Hsp90. *Acta. Pharmacol. Sin., 27*(6), 694–699.

161. Fan, Y., Liu, Y., Zhang, L., Cai, F., Zhu, L., & Xu, J., (2017). C0818, a novel curcumin derivative, interacts with Hsp90 and inhibits Hsp90 ATPase activity. *Acta Pharmaceutica Sinica. B., 7*(1), 91–96.

162. Dasari, S., & Tchounwou, P. B., (2014). Cisplatin in cancer therapy: molecular mechanisms of action. *Eur. J. Pharmacol., 740*, 364–378.

163. Itoh, H., Ogura, M., Komatsuda, A., Wakui, H., Miura, A. B., et al., (1999). A novel chaperone-activity-reducing mechanism of the 90-k Da molecular chaperone HSP90. *Biochem. J., 343*Pt *3*, 697–703.

164. Ozols, R. F., Bookman, M. A., du Bois, A., Pfisterer, J., Reuss, A., & Young, R. C., (2006). Intraperitoneal cisplatin therapy in ovarian cancer: comparison with standard intravenous carboplatin and paclitaxel. *Gynecol. Oncol., 103*(1), 1–6.

165. Galluzzi, L., Vitale, I., Michels, J., Brenner, C., Szabadkai, G., Harel-Bellan, A., et al., (2014). Systems biology of cisplatin resistance: past, present and future. *Cell Death & Disease, 5*, e1257.

166. Galluzzi, L., Senovilla, L., Vitale, I., Michels, J., Martins, I., Kepp, O., et al., (2012). Molecular mechanisms of cisplatin resistance. *Oncogene, 31*(15), 1869–83.

167. Abenavoli, L., Capasso, R., Milic, N., & Capasso, F., (2010). Milk thistle in liver diseases: past, present, future. *Phytother. Res.*24 (10), 1423–1432.

168. Verma, S., Goyal, S., Jamal, S., Singh, A., & Grover, A., (2016). Hsp90: Friends, clients and natural foes. *Biochimie., 127*, 227–240.

169. Zhao, H., Brandt, G. E., Galam, L., Matts, R. L., & Blagg, B. S., (2011). Identification and initial SAR of silybin: An Hsp90 inhibitor. *Bioorg. Med. Chem. Lett., 21*(9), 2659–2664.

170. Cueto, M., Jensen, P. R., & Fenical, W., (2000). N-Methylsansalvamide, a cytotoxic cyclic depsipeptide from a marine fungus of the genus fusarium. *Phytochem., 55*(3), 223–236.

171. Ramsey, D. M., Mc Connell, J. R., Alexander, L. D., Tanaka, K. W., Vera, C. M., & Mc Alpine, S. R., (2012). An Hsp90 modulator that exhibits a unique mechanistic profile. *Bioorg. Med. Chem. Lett., 22*(9), 3287–3290.

172. Vasko, R. C., Rodriguez, R. A., Cunningham, C. N., Ardi, V. C., Agard, D. A., & Mc Alpine, S. R., (2010). Mechanistic studies of Sansalvamide A-amide: An allosteric modulator of Hsp90. *ACS. Med. Chem. Lett., 1*(1), 4–8.

173. Sellers, R. P., Alexander, L. D., Johnson, V. A., Lin, C. C., Savage, J., Corral, R., et al., (2010). Design and synthesis of Hsp90 inhibitors: exploring the SAR of sansalvamide A derivatives. *Bioorg. Med. Chem., 18*(18), 6822–6856.

174. Alexander, L. D., Sellers, R. P., Davis, M. R., Ardi, V. C., Johnson, V. A., Vasko, R. C., et al., (2009). Evaluation of di-sansalvamide A derivatives: synthesis, structure-activity relationship, and mechanism of action. *J. Med. Chem., 52*(24), 7927–7930.

175. Pan, P. S., Vasko, R. C., Lapera, S. A., Johnson, V. A., Sellers, R. P., Lin, C. C., et al., (2009). A comprehensive study of Sansalvamide A derivatives: The structure-activity relationships of 78 derivatives in two pancreatic cancer cell lines. *Bioorg. Med. Chem., 17*(16), 5806–5825.

176. Hall, J. A., Forsberg, L. K., & Blagg, B. S., (2014). Alternative approaches to Hsp90 modulation for the treatment of cancer. *Future. Med. Chem., 6*(14), 1587–1605,

177. Seo, Y. H. (2015). Small Molecule Inhibitors to Disrupt Protein-protein Interactions of Heat Shock Protein 90 Chaperone Machinery. *J. Cancer. Prev., 20*(1), 5–11.

178. Yang, H., Chen, D., Cui, Q. C., Yuan, X., & Dou, Q. P., (2006). Celastrol, a triterpene extracted from the Chinese "Thunder of God Vine," is a potent proteasome inhibitor and suppresses human prostate cancer growth in nude mice. *Cancer Res., 66*(9), 4758–4765.

179. Pang, X., Yi, Z., Zhang, J., Lu, B., Sung, B., Qu, W., et al., (2010). Celastrol suppresses angiogenesis-mediated tumor growth through inhibition of AKT/mammalian target of rapamycin pathway. *Cancer Res., 70*(5), 1951–1959.

180. Zhang, T., Li, Y., Yu, Y., Zou, P., Jiang, Y., & Sun, D., (2009). Characterization of celastrol to inhibit hsp90 and cdc37 interaction. *J. Biol. Chem., 284*(51), 35381–35389.

181. Sreeramulu, S., Gande, S. L., Gobel, M., & Schwalbe, H., (2009). Molecular mechanism of inhibition of the human protein complex Hsp90-Cdc37, a kinome chaperone-cochaperone, by triterpene celastrol. *Angew. Chem. Int. Ed. Engl., 48*(32), 5853–58355.

182. Chadli, A., Felts, S. J., Wang, Q., Sullivan, W. P., Botuyan, M. V., & Fauq, A., (2010). Ramirez-Alvarado, M., Mer, G., Celastrol inhibits Hsp90 chaperoning of steroid receptors by inducing fibrillization of the Co-chaperone p23. *J. Biol. Chem., 285*(6), 4224–4231.

183. Li, Y., Zhang, T., Schwartz, S. J., & Sun, D., (2009). New developments in Hsp90 inhibitors as anti-cancer therapeutics: mechanisms, clinical perspective and more potential. *Drug. Resist. Updat., 12*(1–2), 17–27.

184. Uddin, S. J., Nahar, L., Shilpi, J. A., Shoeb, M., Borkowski, T., Gibbons, S., et al., (2007). Gedunin, a limonoid from Xylocarpus granatum, inhibits the growth of Ca Co-2 colon cancer cell line *In Vitro. Phytother. Res., 21*(8), 757–761.

185. Kamath, S. G., Chen, N., Xiong, Y., Wenham, R., Apte, S., Humphrey, M., et al., (2009). Gedunin, a novel natural substance, inhibits ovarian cancer cell proliferation. *Int. J. Gynecol. Cancer, 19*(9), 1564–1569.

186. Patwardhan, C. A., Fauq, A., Peterson, L. B., Miller, C., Blagg, B. S., et al., (2013). Gedunin inactivates the co-chaperone p23 protein causing cancer cell death by apoptosis. *J. Biol. Chem., 288*(10), 7313–7325.

187. Brandt, G. E. L., Schmidt, M. D., Prisinzano, T. E., & Blagg, B. S. J., (2008). Gedunin, a Novel Hsp90 Inhibitor: Semisynthesis of Derivatives and Preliminary Structure-Activity Relationships. *J. Med. Chem., 51*(20), 6495–6502.

188. Lamb, J., Crawford, E. D., Peck, D., Modell, J. W., Blat, I. C., Wrobel, M. J., et al., (2006). The Connectivity Map: using gene-expression signatures to connect small molecules, genes, and disease. *Science., 313*(5795), 1929–1935.

189. Hadden, M. K., Galam, L., Gestwicki, J. E., Matts, R. L., & Blagg, B. S., (2007). Derrubone, an inhibitor of the Hsp90 protein folding machinery. *J. Nat. Prod., 70*(12), 2014–2018.

190. Vaughan, C. K., Mollapour, M., Smith, J. R., Truman, A., Hu, B., Good, V. M., et al., (2008). Hsp90-dependent activation of protein kinases is regulated by chaperone-targeted dephosphorylation of Cdc37. *Mol. Cell., 31*(6), 886–895.

191. Vaughan, C. K., Gohlke, U., Sobott, F., Good, V. M., Ali, M. M., Prodromou, C., et al., (2006). Structure of an Hsp90-Cdc37-Cdk4 complex. *Mol. Cell., 23*(5), 697–707.

192. Kunze, B., Sasse, F., Wieczorek, H., & Huss, M., (2007). Cruentaren A, a highly cytotoxic benzolactone from Myxobacteria is a novel selective inhibitor of mitochondrial F1-ATPases. *FEBS. Lett., 581*(18), 3523–3527.

193. Kunze, B., Steinmetz, H., Hofle, G., Huss, M., Wieczorek, H., & Reichenbach, H., (2006). Cruentaren, a new antifungal salicylate-type macrolide from *Byssovorax cruenta* (myxobacteria) with inhibitory effect on mitochondrial ATPase activity. Fermentation and biological properties. *J. Antibiot., 59*(10), 664–648.

194. Hall, J. A., Kusuma, B. R., Brandt, G. E., & Blagg, B. S., (2014). Cruentaren A binds F1F0 ATP synthase to modulate the Hsp90 protein folding machinery. *ACS Chem. Biol., 9*(4), 976–985.

195. Bindl, M., Jean, L., Herrmann, J., Muller, R., & Furstner, A., (2009). Preparation, modification, and evaluation of cruentaren A and analogs. *Chem., 15*(45), 12310–12319.

196. Li, Y., Karagoz, G. E., Seo, Y. H., Zhang, T., Jiang, Y., Yu, Y., et al., (2012). Sulforaphane inhibits pancreatic cancer through disrupting Hsp90-p50(Cdc37) complex and direct interactions with amino acids residues of Hsp90. *J. Nutr. Biochem., 23*(12), 1617–1626.

197. Zhang, Y., & Tang, L., (2007). Discovery and development of sulforaphane as a cancer chemopreventive phytochemical. *Acta. Pharmacol. Sin., 28*(9), 1343–1354.

198. Conaway, C. C., Wang, C.-X., Pittman, B., Yang, Y.-M., Schwartz, J. E., Tian, D., et al., (2005). Phenethyl Isothiocyanate and Sulforaphane and their N-Acetylcysteine Conjugates Inhibit Malignant Progression of Lung Adenomas Induced by Tobacco Carcinogens in A/J Mice. *Cancer Res., 65*(18), 8548–8557.

199. Fahey, J. W., Zhang, Y., & Talalay, P., (1997). Broccoli sprouts: An exceptionally rich source of inducers of enzymes that protect against chemical carcinogens. *Proc. Natl. Acad. Sci. U.S.A., 94*(19), 10367–10372.

200. Cornblatt, B. S., Ye, L., Dinkova-Kostova, A. T., Erb, M., Fahey, J. W., Singh, N. K., et al., (2007). Preclinical and clinical evaluation of sulforaphane for chemoprevention in the breast. *Carcinogenesis, 28*(7), 1485–1490.

201. Yan, Y., Su, X., Liang, Y., Zhang, J., Shi, C., Lu, Y., et al., (2008). Emodin azide methyl anthraquinone derivative triggers mitochondrial-dependent cell apoptosis involving in caspase-8-mediated Bid cleavage. *Mol. Cancer. Ther., 7*(6), 1688–1697.

202. Yan, Y. Y., Zheng, L. S., Zhang, X., Chen, L. K., Singh, S., Wang, F., et al., (2011). Blockade of Her2/neu binding to Hsp90 by emodin azide methyl anthraquinone derivative induces proteasomal degradation of Her2/neu. *Mol. Pharm., 8*(5), 1687–1697.

203. Yan, Y. Y., Fu, L. W., Zhang, W., Ma, H. S., Ma, C. G., Liang, Y. J., et al., (2014). Emodin azide methyl anthraquinone derivative induced G0/G1 arrest in HER2/neu-overexpressing MDA-MB-453 breast cancer cells. *J. Buon., 19*(3), 650–655.

204. Gyurkocza, B., Plescia, J., Raskett, C. M., Garlick, D. S., Lowry, P. A., Carter, B. Z., et al., (2006). Antileukemic activity of shepherdin and molecular diversity of hsp90 inhibitors. *J. Natl. Cancer. Inst., 98*(15), 1068–1077.

205. Plescia, J., Salz, W., Xia, F., Pennati, M., Zaffaroni, N., Daidone, M. G., et al., (2005). Rational design of shepherdin, a novel anticancer agent. *Cancer Cell, 7*(5), 457–468.

206. Hsieh, P. W., Huang, Z. Y., Chen, J. H., Chang, F. R., Wu, C. C., Yang, Y. L., et al., (2007). Cytotoxic with anolides from *Tubocapsicum anomalum*. *J. Nat. Prod., 70*(5), 747–753.

207. Wang, H. C., Tsai, Y. L., Wu, Y. C., Chang, F. R., Liu, M. H., Chen, W. Y., et al., (2012). Withanolides-induced breast cancer cell death is correlated with their ability to inhibit heat protein 90. *Plo S One., 7*(5), e37764.

208. Chen, W. Y., Chang, F. R., Huang, Z. Y., Chen, J. H., Wu, Y. C., & Wu, C. C., (2008). Tubocapsenolide A, a novel with anolide, inhibits proliferation and induces apoptosis in MDA-MB-231 cells by thiol oxidation of heat shock proteins. *J. Biol. Chem., 283*(25), 17184–17193.

CHAPTER 4

NANOSUSPENSIONS AS NANOMEDICINE: CURRENT STATUS AND FUTURE PROSPECTS

SHOBHA UBGADE, VAISHALI KILOR, ABHAY ITTADWAR, and ALOK UBGADE

Department of Pharmaceutics, Gurunanak College of Pharmacy, Rashtrasant Tukadoji Maharaj Nagpur University, Nagpur, Maharashtra, India, Tel: +919763403953, E-mail: shobha_yadav1402@yahoo.co.in

4.1 INTRODUCTION

The "Next big thing is really small," will not be an amplification for the success of nanotechnology in multiple domains across the world. The medical field is no exception and adoption of the technology at nanoscale has led to the emergence of 'Nanomedicine.' Nanomedicine is defined as "the monitoring, repair, construction, and control of human biological systems at the molecular level, using engineered nanodevices and nano-structures" [1]. Most broadly, nanomedicine is the process of diagnosing, treating, preventing disease and traumatic injury, relieving pain, and preserving and improving human health, using molecular tools and molecular knowledge of the human body. In short, nanomedicine is the application of nanotechnology to medicine [2]. Applications of nanotechnology in medicine are potentially enormous. It is recognized that as particles get smaller, the surface area increases with a greater proportion of atoms/ molecules found at the surface compared to those inside [3]. Drug delivery of poorly soluble molecules has seen a significant change after the inception of nano-sized particles. Nanoparticle technology has become a well-established approach for formulating poorly soluble drugs. Nanonization which is a successor of the micronization process reduces the particle size of drug to sub-micron range. Development of successful drug formulations with maximum drug availability, reduced toxicity can be achieved potentially by nanoparticle technology for better clinical, and commercial benefits.

Nanosuspension is one of the nanotechnology-based drug delivery platforms which has been explored successfully for enhancing solubility and bioavailability of poorly soluble drugs. Nanosuspensions, can thus, be put in the category of nanomedicines. This chapter reviews and discusses multiple aspects of nanosuspension technology, challenges, and applications.

4.2 NANOSUSPENSION

Nanosuspensions are defined as unique liquid sub-micron colloidal dispersions of nanosized pure drug particles that are stabilized by a suitable polymer and/or surfactant and have a particle size of 1–1000 nm [4]. The formulation of active pharmaceutical ingredients (APIs) as nanosuspensions has increased in popularity in the pharmaceutical industry in the last ten years; a few nanosuspension products are already on the market, and many more novel nanosuspensions are undergoing extensive development and investigation by academia and industry [5].

A nanosuspension platform is an efficient and intelligent drug delivery system for water-insoluble drugs because these platforms increase the saturation solubility and the surface area available for dissolution [6]. Generally, the biopharmaceutical advantages of water-insoluble drugs formulated as nanosuspensions include improvements in formulation performance, such as high drug loading, reproducibility of oral absorption, improved dose-bioavailability proportionality, reduced toxicity and side effects as well as increased patient compliance via a reduction in the number of oral units that must be administered [7, 8]. This technology is also very interesting because of the reduction of excipient use, thereby reducing the risk of toxicity from these components [9, 10]. They can be administered by various routes like (parenteral, oral, ophthalmic and nasal), and exhibit substantially increased solubility and dissolution rate, hence improved bioavailability. The high ratio of drug to excipients is particularly advantageous for nanosuspensions designed to target cancer cells, particularly if a large amount of drug is delivered in each targeting event [11].

Nanosuspensions have revealed their potential to tackle the problems associated with the delivery of poorly water-soluble and poorly lipid-soluble drugs, and are unique because of their simplicity and the advantages they confer over other strategies. Nanosuspensions represent a very promising, universal formulation approach from pre-clinical to commercialization for poorly soluble drug substances. The preparation of nanosuspensions is reported to be more cost effective and technically more simpler alternative for poorly soluble drugs [12].

4.2.1   PROPERTIES OF NANOSUSPENSIONS

Nanosuspensions have emerged as a promising strategy for the efficient delivery of poorly soluble drugs because of versatile features. In the current scenario, one of the biggest challenges in drug development is to improve the solubility characteristics for the attainment of desired bioavailability of drugs. Several strategies have been employed to overcome these limitations. Particle size reduction has been a smarter approach that can be applied to the nonspecific formulation for many years. Diminution of particle size to sub-micron range is a powerful formulation approach that can increase the dissolution rate, and the saturation solubility, subsequently improves the bioavailability of poorly water-soluble drugs. Drug nanocrystals are considered as a novel approach to improve the solubility of hydrophobic drugs since the technique is simple and effective which can quickly launch product to the market. The nanocrystals were invented at the beginning of the 1990s and the first products appeared very fast on the market from the year 2000 onwards. Additionally, drug nanocrystals are a universal approach generally applied to all poorly soluble drugs for the reason that all drugs can be disintegrated into nanometer-sized particles [13]. Drug nanocrystals are nanoscopic crystals of parent compounds with the dimension of less than 1 mm. They are composed of 100% drug without carriers and typically stabilized with surfactants or polymeric steric stabilizers. A dispersion of drug nanocrystals in an outer liquid medium and stabilized by surface active agents are so-called nanosuspensions. The dispersion medium can be water, aqueous or nonaqueous media, for example, liquid polyethylene glycol (PEG) and oils. The nanosuspensions can be used to formulate compounds that are insoluble in both water and oil and to reformulate existing drugs to remove the toxic less favorable excipients [14–16].

4.2.1.1 INCREASE IN SATURATION SOLUBILITY AND DISSOLUTION VELOCITY

An outstanding feature of nanosuspensions is the increase in saturation solubility and consequently an increase in the dissolution velocity of the compound. Saturation solubility is a compound-specific constant, which depends on physicochemical properties of the compound, dissolution medium and temperature. However, this definition is only valid for drug particles with a minimum particle size in the micrometer range. Furthermore, the saturation solubility is also a function of the crystalline structure (i.e., lattice energy) and particle size. The saturation solubility increases with decreasing particle size below 1000 nm [17].

The Ostwald-Freundlich equation (Eq. 1) directly describes the relation between the saturation solubility of the drug and the particle size.

$$log\frac{C_S}{C_a} = \frac{2\sigma V}{2.303RT_{pr}} \tag{1}$$

where C_s is the saturation solubility, C_a is the solubility of the solid consisting of large particles, σ is the interfacial tension of substance, V is the molar volume of the particle material, R is the gas constant, T is the absolute temperature, ρ is the density of the solid, r is the radius.

From the above equation, it is obvious that the saturation solubility (C_s) of the drug increases with a decrease in the particle size (r). However, this effect is not substantial for larger particles, but will be pronounced for materials that have a mean particle size of less than 1–2 μm [14, 15, 18–21].

Another possible explanation for the increased saturation solubility is the creation of high-energy surfaces when disrupting the more or less ideal drug microcrystals to nanoparticles. Quantification of increased saturation solubility of the drug in nanosuspension form can be done by performing dissolution experiments [22].

Nanosuspensions possess an increased dissolution velocity that can be explained by the Noyes-Whitney equation (Eq. 2).

$$\frac{dx}{dt} = \frac{DA}{h_D} \times (C_s - C_t) \tag{2}$$

where dx/dt is the dissolution velocity, D is the diffusion coefficient, A is the surface area, h_D is the diffusional distance, C_s is the saturation solubility, C_t is the concentration around the particles.

The dissolution velocity (dx/dt) of drug particles increases due to the greater surface area (A) and the increase in saturation solubility (C_s) of the compound. The size reduction of drug particles to nano-meter size leads to an increased surface area and thus, according to the Noyes-Whitney equation the dissolution velocity is increased [23].

Another important factor is the diffusional distance h_D, as a part of the hydrodynamic boundary layer h_H, which is also strongly dependent on the particle size as shown by Prandtl equation (Eq. 3):

$$h_H = k \left(\frac{L_2^1}{V_3^1} \right)$$

(3)

where h_H is the hydrodynamic boundary layer thickness, k denotes a constant, L is the length of the particle surface in the direction of flow, V is the relative velocity of the flowing liquid surrounding the particle. In accordance with Prandtl equation, the particle size reduction leads to a decreased diffusional distance h_D and consequently an increased dissolution velocity, as described by Noyes-Whitney equation [14, 15, 18–21].

4.2.1.2 INCREASED BIO-ADHESION

Another pronounced property is the adhesiveness generally described for nanoparticles. An increased adhesiveness of nanomaterials to surface/cell membranes is usually due to an increased contact area of smaller particles. This nature of small drug nanoparticles leads to improved oral absorption of poorly soluble drugs and substantial increase in the bioavailability. The variability in absorption and patterns of erratic absorption is reduced as a result of adhesiveness imparted by drug nanoparticles [17, 24].

4.2.1.3 IMPROVED BIOLOGICAL PERFORMANCE/ BIOAVAILABILITY

Together with the permeability, the solubility behavior of a drug is a key determinant of its bioavailability [25]. In nanosuspension technology, the drug has a reduced particle size, which leads to an increased dissolution rate and therefore improved bioavailability. An increase in the dissolution velocity and saturation solubility of a drug leads to an improvement in the *in vivo* performance of the drug irrespective of the route of administration used [22].

4.2.1.4   SURFACE MODIFICATION

The *in vivo* performance of nanosuspensions can be improved further by controlling the surface modification of the drug nanoparticles using mucoadhesive polymers. To impart a certain surface property, the drug nanoparticles can be produced using special surfactants or polymers as stabilizers or else surface modifiers can be admixed to the nanosuspension. The degree of modification depends upon the affinity and adsorption on the particle surface and can be physically measured either by zeta potential measurements or hydrophobic interaction (HIC) analysis [24].

4.2.1.5   PHYSICAL STABILITY

Another special feature of nanosuspensions is the absence of Ostwald ripening, which is suggestive of their long-term physical stability [26]. Ostwald's ripening [27, 28] has been described for ultrafine dispersed systems and is responsible for crystal growth and subsequently forming of microparticles. Ostwald ripening is caused by the differences in dissolution pressure/saturation solubility between small and large particles. The lack of Ostwald's ripening in nanosuspensions is attributed to their uniform particle size, which is created by various manufacturing processes. The absence of particles with large differences in their size in nanosuspensions prevents the existence of the different saturation solubilities, which, in turn, prevents the Ostwald's ripening effect. Also, stabilizers employed for nanosuspension render them physically stable through their shelf life [22]. However, with the progress in this field, many stability issues have been faced by the formulators that are discussed later in this chapter.

4.2.1.6   VERSATILITY

Nanosuspensions can be easily incorporated in various dosage forms, such as tablets, pellets, suppositories, hydrogels as well as can be administered through various routes that makes them versatile in nature [22].

4.2.2   PREPARATION OF NANOSUSPENSIONS

Nanosuspensions are prepared by different techniques that reduce the size to sub-micron level. Two converse technologies, namely bottom-up and top-down are used for preparing nanosuspensions. The top-down processes involve breaking down of larger particles by milling or homogenization, while the bottom-up processes associated with an assembling and controlling of precipitation at nanometer scale. The combined technologies are also becoming popular nowadays wherein a pretreatment step is followed by a high energy process to get reduced particle size; such as Nanoedge Technology [17].

4.2.2.1 BOTTOM-UP PROCESSES

In bottom-up processes, the molecules are aggregated to form particles, being crystalline or amorphous. Hydrosols were developed by Sucker in 1980 using precipitation method [29]. Drug molecules are first dissolved in an appropriate organic solvent at a supersaturation concentration to allow for the nucleation of drug seeds. Drug nanoparticles are then formed by adding the organic mixture to an antisolvent in the presence of stabilizers [30]. This hydrosol technique was modified in 1990s by Sandoz, which include charged glyceryl esters, such as lecithin, as an electrostatic stabilizer. The stabilizer adsorbs onto the surface of drug nanoparticles and prevents agglomeration [29].

4.2.2.1.1 Precipitation by Liquid Solvent-Antisolvent Addition

This technique is the most commonly used technique and may be called as 'a classical precipitation method' or in latin *via humida paratum*. In this technique drug molecules are first dissolved in an appropriate organic solvent at a supersaturation concentration to allow for the nucleation of drug seeds. Drug nanoparticles are then formed by adding the organic mixture to an antisolvent in the presence of stabilizers [30]. The characteristics of nanosuspension produced is affected by various factors like drug concentration, solvent to anti-solvent ratio, flow rate of the solvent and antisolvent, the effect of the stirring speed, stabilizer used, etc. [29]. It is a very simple process and can be used on laboratory scale. The excessive use of solvents and scale-up problems hinders the application at industrial scale.

4.2.2.1.2 *Emulsions as Templates*

Emulsions can be used as templates to produce nanosuspensions. This technique is applicable to those drugs that have solubility in either volatile organic solvent or partially water-miscible solvent. These solvents are used as dispersed phase of the emulsion. Emulsification technique can be used to prepare drug nanosuspensions by two ways. In the first method, emulsion is formed by adding an organic solvent or a mixture of solvents containing drug in the aqueous phase having suitable surfactants as stabilizer. The dispersed organic phase is evaporated under reduced pressure resulting in instantaneous precipitation of drug particles. The nanosuspension so formed is stabilized by the added surfactant. The particle size can be controlled by controlling the droplet size of the emulsion. Optimizing the concentration of surfactant could allow an increased ratio of organic phase and ultimately enhancing the drug loading in emulsion [22].

Initially, organic solvents such as methylene chloride and chloroform were used [31]. However, environmental hazards and human safety concerns about residual solvents have limited their use in routine manufacturing processes. Relatively safer solvents such as ethyl acetate and ethyl formate can still be considered for use [32]. Another method uses water or partially water-miscible solvents such as butyl lactate, benzyl alcohol, and triacetin as the dispersed phase instead of hazardous solvents [33]. Emulsion is formed by conventional method and dilution of the emulsion produces drug nanosuspension. Dilution of the emulsion with water causes complete diffusion of internal phase into the external phase, leading to the formation of a nanosuspension immediately. The prepared nanosuspension is then separated from internal phase and surfactants for suitable administration. Different separation techniques like diultrafiltration, centrifugation, ultracentrifugation can be used depending upon the concentration of formulation ingredients and desired route of administration.

Use of emulsion as the template offers various advantages. Use of specialized equipment is not necessary and particle size can be controlled easily by controlling the emulsion droplet size. Ease of scale-up. However, this technique cannot be used for drugs that are poorly soluble in both aqueous and organic media. Safety concerns arise if hazardous solvents are used. The process becomes costly when separation techniques like diultrafiltration are used. Also, higher amounts of surfactants/stabilizer are required [22].

4.2.2.1.3 *Microemulsions as Templates*

Microemulsions are thermodynamically stable and isotropically clear dispersions of two immiscible liquids, such as oil and water, stabilized by an interfacial film of surfactant and co-surfactant [34]. Their advantages, such as high drug solubilization, long shelf-life and ease of manufacture, make them an ideal drug delivery vehicle. Taking advantage of the microemulsion structure, one can use microemulsions even for the production of nanosuspensions [35]. Oil-in-water microemulsions are preferred for this purpose. The internal phase of these microemulsions could be either a partially miscible liquid or a suitable organic solvent. The drug can be either loaded in the internal phase or pre-formed microemulsions can be saturated with the drug by intimate mixing. The suitable dilution of the microemulsion yields the drug nanosuspension [22]. This technique is similar to emulsion templates. The only difference is the need for less energy input for the production of nanosuspensions by virtue of microemulsions.

4.2.2.1.4 *Supercritical Fluid Process*

This process uses the unique properties of supercritical fluid (SCF). The most commonly used SCFs for a variety of applications include supercritical fluid carbon dioxide (SC-CO_2), nitrous oxide, water, methanol, ethanol, ethane, propane, n-hexane and ammonia [36]. SC-CO_2 is an attractive solvent or anti-solvent as it is safe, inexpensive, readily available, and an ideal substitute for many hazardous and toxic solvents [37].

The most widely employed technologies for preparing nanoparticles based on supercritical fluids include rapid expansion from supercritical solutions (RESS), rapid expansion from supercritical to aqueous solutions (RESAS), supercritical anti-solvent method (SAS) and supercritical anti-solvent enhanced mass transfer method (SAS-EM) [38].

In the RESS process, the solution of the drug is prepared in a supercritical fluid (SCF) and passes through a narrow nozzle. The immediate reduction in pressure changes the density of the fluid and the rapid expansion of the supercritical fluid causes supersaturation and subsequent nucleation and precipitation of the solute [29]. This process is also known as supercritical fluid nucleation (SFN) [39]. The morphology and size distribution of the precipitated drug is a function of various parameters like nature of solute, pre- and post-expansion temperature and pressure, nozzle geometry, and solution concentration. This technique is applicable to only those drugs, which exhibit fair solubility in selected supercritical fluids. Other demerits include generation of microparticles in many cases due to particle aggregation during free jet expansion [38].

Rapid expansion from supercritical to aqueous solutions (RESAS) technique has evolved in order to minimize the generation of microparticles observed due to increased coagulation rate in the free jet expansion of RESS. The process brings about expansion of supercritical solution through an orifice or tapered nozzle into aqueous solution containing a stabilizer(s) in order to minimize the particle aggregation during free jet expansion. It is more efficient in the nanosizing process than RESS process. It cannot be used for drugs exhibiting instability in aqueous environment. Nanoparticulate suspension may not exhibit long-term stability [38].

SAS process has been developed in order to achieve nanosizing of the hydrophobic drugs that cannot be processed by RESS technique owing to their poor solubility in a supercritical fluids. In this process, supercritical fluid is used as an anti-solvent for nanoengineering [39, 40, 41]. In the SAS method, the drug is dissolved in an organic solvent, which must be miscible with the supercritical antisolvent. This drug solution is then added to the supercritical antisolvent. The solvent rapidly diffuses in the antisolvent phase and the drug precipitates due to low solubility in the antisolvent [42].

Supercritical anti-solvent enhanced mass transfer (SAS-EM) method presents an innovative approach to produce very small particles in the nanometer range, having a narrow size distribution by its virtue of enhancing the mass transfer. It also provides techniques to control the particle size of the final product. Like the SAS technique, the SAS-EM process uses a supercritical fluid as an anti-solvent. Additionally, the dispersion jet of solution is deflected by a surface vibrating at an ultrasonic frequency so as to atomize the jet into nanodroplets for enhancing the mass transfer [43].

4.2.2.1.5 High Gravity Reactive Precipitation (HGRP) Technology

It represents an innovative methodology employed for the generation of nanoparticles. Its utility for the generation of inorganic nanomaterials like

$CaCO_3$ (15–40 nm) [44], $Al(OH)_3$ (1–10 nm), and $SrCO_3$ (40 nm) has been very well established. This process yields nanoparticles at significantly lower production costs than conventional production methods for nanomaterials. The method is based on the reactive precipitation method and is tailored to generate nanomaterials by employing high gravity micromixing of reactants with the help of rotating packed bed (RPB). The high gravity micromixing helps in enhancing the mass transfer and heat transfer between the reactants, thus inducing the rapid nucleation of the final product while suppressing the crystal growth. When the reactants enter the rotating packed bed, they are spread or split into very thin films or nanodroplets under the high shear created by high gravity. An intense micromixing and centrifugal force together help in enhanced mass transfer resulting in the production of nanoparticles [45].

4.2.2.1.6 Continuous Precipitation Technique

Descendant of hydrosol technique is continuous precipitation technique. In this technique, the drug is dissolved in the water-miscible solvent and dispersed into aqueous phase at a controlled rate [46]. The dispersion of the drug solution in the aqueous phase causes diffusion of the solvent leading to the generation of nanostructured drug particles. The stabilizers present in the aqueous phase adsorb on the nucleated drug particles hindering the crystal growth. The drug nanoparticles can be processed further to remove the solvent, and the dried drug nanoparticles can be obtained by suitable techniques.

4.2.2.1.7 Miscellaneous Techniques

Another bottom-up technologies includes sonocrystallization, confined impinging liquid jet precipitation and multi-inlet vortex mixing. The principle involved in sonocrystallization is creation and collapse of bubbles which results in shock waves and change in temperature and pressure for nucleation. Ultrasonic waves cause faster and more uniform nucleation through the sonicated volume, leading to smaller and more uniform sized particles. The experimental setup for sonoprecipitation is simple, comprising of an ultrasound probe in a mechanically stirred reaction tank where the anti-solvent is mixed with the drug solution to precipitate fine drug particles. Lack of suitable equipment for commercial scale production is the key barrier for sonoprecipitation process [47]. Confined impinging liquid jet precipitation method causes extreme turbulence and intense mixing. This is created by a jet of drug solution impinging a jet of anti-solvent coming through two opposing nozzles mounted in a small chamber. Precipitation of drug is caused by the shear mixing effect. It is a single-pass process, so precipitation of the compound is completed soon after mixing [47]. Multi-inlet vortex mixing allows mixing of streams of unequal volumetric flow. It provides capability to control the composition of supersaturation and solvent by varying the velocity of individual streams. This is a suitable technique for preparing nanoparticles of herbal and metallic origin that are difficult to prepare by other techniques. During scale-up, optimization of flow rates is difficult and this may alter particle size and morphology [47]. Bottom-up processes offers interesting possibilities and ways to incorporate multiple active ingredients in a single nanocarrier. However, a basic drawback of many precipitation processes is the use of organic solvent which is needed to be removed, leading to the high cost of production. Also, in case of low water and organic solvent soluble drug, the large solvent volumes are required. Hence, in pharmaceutical industry, the bottom-up processes has not been employed for the production of the marketed drug [17].

4.2.2.2   TOP-DOWN PROCESSES

The top-down process involves diminution of particles in the micrometer range to nanodimension by using milling techniques or high-pressure homogenization.

4.2.2.2.1 Media Milling

The nanosuspensions are produced using high-shear media mills or pearl mills. The media mill consists of a milling chamber, a milling shaft and a recirculation chamber. The milling chamber is charged with the milling media, water, drug and stabilizer, and the milling media or pearls then rotate at a very high shear rate. The milling process is performed under controlled temperatures. The impaction of the milling media with the drug provides the high energy and shear forces to break the microparticulate drug into nano-sized particles. The milling medium can be composed of glass, zirconium oxide or highly cross-linked polystyrene resin. The media milling process can successfully process micronized and non-micronized drug crystals. Once the formulation and the process are optimized, very little batch-to-batch variation is observed in the quality of the dispersion. This patent-protected technology was developed by Liversidge et al. [48]. Drugs that are poorly soluble in both aqueous and organic media can be easily formulated into nanosuspensions using media milling. This technique offers ease of scale-up and little batch-to-batch variation. The final formulation has narrow size distribution. The major drawback and concern is the generation of residues of milling media, which may be introduced in the final product as a result of erosion [28, 49]. Use of polystyrene resin-based milling medium has greatly reduced this problem. For this medium, residual monomers are typically 50 ppb and the residuals generated during the milling process are not more than 0.005% w/w of the final product or the resulting solid dosage form.

4.2.2.2.2 Dry Co-grinding

Nanosuspensions prepared by high-pressure homogenization and media milling using pearl-ball mill are wet–grinding processes. Recently, nanosuspensions can be obtained by dry milling techniques. Physicochemical properties and dissolution of poorly water-soluble drugs are improved by co-grinding because of an improvement in the surface polarity and transformation from a crystalline to an amorphous drug [50, 51]. Dry co-grinding can be carried out easily and economically and can be conducted without organic solvents. The co-grinding technique can reduce particles to the submicron level and a stable amorphous solid can be obtained. Successful work in preparing stable nanosuspensions using dry-grinding of poorly soluble drugs with soluble polymers and copolymers after dispersing in a liquid media has been reported [52, 53].

4.2.2.2.3 Microfluidization

Microfluidization is another milling technique which results in minimal product contamination, however, this technique has not yet been explored extensively. Besides minimal contamination, this technique can be easily

scaled up [54]. In this method, a sample dispersion containing large particles is made to pass through specially designed interaction chambers at
high pressure. The specialized geometry of the chambers along with the
high pressure causes the liquid stream to reach extremely high velocities
and these streams then impinge against each other and against the walls
of the chamber resulting in particle size reduction. The shear forces developed at high velocities due to attrition of particles against one another and
against the chamber walls, as well as the cavitation fields generated inside
the chamber are the main mechanisms of particle size reduction with this
technique [55]. Nanojet technology uses microfluidization for production
of nanosuspension.

4.2.2.2.4   *High Pressure Homogenization*

High pressure homogenization is a commonly employed method for
producing nanosuspensions of poorly soluble drugs [56–61]. This method
involves forcing a suspension, which contains drug(s) and stabilizers,
through a valve with a small orifice under pressure. High-pressure homogenization is often classified into two groups:

    i)   Dissocubes (homogenization in aqueous media),
    ii) Nanopure (homogenization in water-free media or water mixtures).

The Dissocubes technology was developed by Muller and co-workers
by employing piston gap homogenizers. Dissocubes operates at high
pressure of up to 1500 bar where a suspension passes through a small
gap. This causes an increase in the dynamic pressure with simultaneous
reduction in the static pressure, which reduces the boiling point of water
to room temperature. Consequently, at room temperature water starts
boiling creating gas bubbles. When the suspension departs the gap and
the pressure returns to atmospheric level, the gas bubbles implode. This
phenomenon is called cavitation. The combined forces of cavitation, high
shear, and collisions lead to fracture of the drug microparticles into nano-
sized particles [62]. Homogenization pressure, number of homogenization cycles, hardness of drugs, and temperature (when thermosensitive
drugs are processed) are factors that influence the physical characteristics
(such as particle size) of the resulting nanosuspensions. Metal contamination due to the erosion is less pronounced in this technique than in media

milling. High-pressure homogenization is considered as a safe technique for producing nanosuspensions. Less than 1 ppm metal contaminations were detected under processing conditions of 20 cycles and pressure of 1500 bar [62–64]. The main drawback of this method is the need for pretreatment to obtain microparticles before starting the homogenization process and the many cycles of homogenization [65, 66]. For some purposes such as dispersing drug nanocrystals in low molecular weight PEG or in oil, liquid nanosuspensions are dispersed in nonaqueous media or media with reduced water content. Because of the high boiling point and low vapor pressure of oily fatty acids and oils, the drop in pressure is not sufficient for cavitation and thus the latter is not a determining factor in this process. To compensate for the insufficient drop in pressure, the Nanopure process is conducted at low temperature which is often referred to as "deep-freeze" method. Conducting the process at 0 °C or even below the freezing point produces results comparable to those achieved using dissocubes [67].

4.2.2.3   COMBINATION TECHNOLOGIES

Milling, high-pressure homogenization, and precipitation are the main methods employed for the production of drug nanosuspensions. However, there is an intensive research for new technologies leading to many other approaches for the production of drug nanosuspensions. The technology combines, generally a pre-treatment step followed by a high-energy process, such as the NanoEdge™ technology. In the first step, crystals are precipitated; and the obtained suspension is then subjected to a high-energy process, typically high-pressure homogenization [17]. SmartCrystal technology is another technology with a number of different processes that are combined either to accelerate production by reducing the number of passes through the homogenizer or to obtain very small nanocrystals below 100 nm. Such small nanocrystals are difficult to produce via pearl milling or simple high-pressure homogenization, especially in large-scale industrial production [17]. Recently, a novel tri-combination technology as "Precipitation-lyophilization-homogenization (PLH) method" for preparation of nanocrystals had been proposed by Junyaprasert group [68]. This combined technology is composed of precipitation, lyophilization, and homogenization techniques, respectively.

4.2.3 FORMULATION DEVELOPMENT

Preparation of nanosuspension needs due consideration while selecting formulation components (excipients) and methods. Both of these factors play a major role in formulating nanosuspension with desired attributes.

4.2.3.1 CHARACTERISTICS OF ACTIVE PHARMACEUTICAL INGREDIENT

It has already been discussed that poorly soluble drugs can be easily formulated as nanosuspension. But, to specify the selection of drug following characteristics of Active Pharmaceutical Ingredient(API) should be considered [25]:

1. Water-insoluble but which are soluble in oil (high log P) or API are insoluble in both water and oils.
2. Drugs with reduced tendency of the crystal to dissolve, regardless of the solvent.
3. API with very large dose.

Furthermore, while particle size and morphology of the starting API are of less concern if nanosizing is to be employed in formulation development, the chemical form of the API needs to be considered. Typically, the neutral form is the preferred starting form. While there has been an example of a salt form-containing nanoformulation, such as Par Pharmaceutical's MEGACE® ES with its acetate salt, pharmaceutical salts are generally not preferred. Possible difficulties with the salt form are [23]:

i) Risk of disproportionation, for example, an HCl salt disproportionating to the free base form during nanomilling,
ii) Risk of aggregation due to charge based interactions in the small intestine, such as those with bile salts, and
iii) Rapid solubilization and turnover of nanoparticles of a salt form into larger particles of the neutral form due to the pH changes in the gastrointestinal (GI) tract.

APIs with ionizable groups and pKa between 2 and 7 (e.g., physiological pH range) run the risk of charge-based interactions even if not presented as a salt.

4.2.3.2 STABILIZERS

Stabilizers are indispensable in the formulation of nanosuspension. The nano-sized particles possess high surface energy and are prone to agglomeration or aggregation of drug crystals. Stabilizers prevent agglomeration or aggregation to yield a physically stable formulation by providing steric or ionic barriers. The main function of a stabilizer is to wet the drug particles thoroughly and to prevent Ostwald's ripening and agglomeration [22].

The formulation of nanosuspension requires a careful selection of stabilizers. At the nanometer domain, the attractive forces between particles, due to dispersion or van der Waals forces, come into play [69]. This attractive force increases dramatically as the particles approach each other, ultimately resulting in an irreversible aggregation. To overcome this, repulsive forces are needed. Stabilizers stabilize the nanoparticles against interparticle forces and prevent them from aggregating. There are two modes by which stabilizer imparts repulsive forces or energetic barriers—steric stabilization and electrostatic stabilization. Steric stabilization is achieved by adsorbing polymers onto the particle surface. As the particles approach each other, the osmotic stress created by the encroaching steric layers acts to keep the particles separated. Electrostatic stabilization is obtained by adsorbing charged molecules, which can be ionic surfactants or charged polymers, onto the particle surface. Charge repulsion provides an electrostatic potential barrier to particle aggregation (Figure 4.1). Typically, the use of steric stabilization alone is sufficient to stabilize the nanoparticles and prevent irreversible aggregation. However, to circumvent flocculation, steric stabilization is often combined with electrostatic stabilization for enhanced repulsive contribution.

Common pharmaceutical excipients that are suitable for use as polymeric stabilizers include the cellulosics, such as hydroxypropyl cellulose (HPC) and hydroxypropylmethyl cellulose (HPMC), povidone (PVP K30), and pluronics (F68 and F127). The surfactant stabilizers can be non-ionic, such as polysorbate (Tween 80), or anionic, such as sodium laurylsulfate (SLS) and docusate sodium (DOSS). Cationic surfactants are typically not used as stabilizers for oral formulation due to their antiseptic properties [23].

Apart from these, D-α-tocopherol polyethylene glycol 1000 succinate (TPGS), polyethylene glycols (PEGs), polyvinyl alcohols (PVAs) have also been used as stabilizers [70]. However, the nanosuspensions are not stabilized permanently by these stabilizers and aggregation may occur during

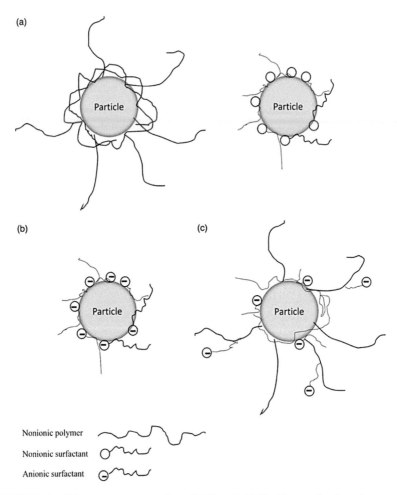

FIGURE 4.1   Schematic representation of different stabilization mechanisms in nanosuspensions: (a) steric stabilization imparted by non-ionic polymers or non-ionic surfactants, (b) electrostatic stabilization imparted by anionic surfactants, and (c) electrostatic stabilization imparted by both non-ionic polymers and anionic surfactants.

storage or when nanosuspensions are being dried. Furthermore, some of the common stabilizers raise toxicity concerns if used in large quantity for a long-term, limiting the therapeutic application of drug nanosuspensions [4, 71–74]. For example, Cremophor® EL and Tween–80 are two commercial surfactants that are widely used to solubilize poorly water-soluble drugs, but they also cause serious neuro- and nephrotoxicity as well as acute hypersensitivity reaction [75, 76]. Thus, there remains to be a demand to find new stabilizers with better stabilizing capacity and less toxicity. Food biopolymers, especially food proteins, are widely used in formulated foods because they have high nutritional values and are generally recognized as safe [77, 78]. The proteins include soybean protein isolate (SPI), whey protein isolate (WPI), β-lactoglobulin (β-lg), etc. (Table 4.1). The drug-to-stabilizer ratio in the formulation may vary from 1:20 to 20:1 and should be investigated in specific case(s). Lecithin is the stabilizer of choice if one intends to develop a parenteral nanosuspension [22].

TABLE 4.1 List of Different Stabilizers Used in Drug Nanosuspensions

Drug	Stabilizer	Manufacturing method	References
Albendazole	SLS, Carbopol, PS 80, HPMC	High Pressure Homogenization	[154]
Amphotericin – B	Tween 80, Pluronic® F68	High Pressure Homogenization	[155]
Azithromycin	Lecithin, Tween–80, Pluronic® F68	High Pressure Homogenization	[156]
Budenoside	Lecithin, Span 85, tyloxapol, cetyl alcohol	High Pressure Homogenization	[148]
Buparvaquone	Pluronic® F68 and PVA	High Pressure Homogenization	[157]
Hydroxycamptothecin	Lipoid S75, Pluronic® F68, Solutol® HS 15	High Pressure Homogenization	[158]
Nifedipine	HPMC	High Pressure Homogenization	[58]
Cinnarizine, itraconazole and phenylbutazone	TPGS 1000	Wet milling	[159]
Glipizide	SLS, PVPK–30	Wet milling	[160]
Omeprazole, albendazole and danazol	Pluronic® F108, F68	Wet milling	[161]
Carbamazepine	HPMC, PVP K17	Anti-solvent precipitation	[162]
Cyclosporin A	Tween–80	Anti-solvent precipitation	[163]
Indomethacin	Food proteins-soybean protein isolate, whey protein isolate, β-lactoglobulin	Precipitation ultrasonication	[70]

4.2.3.3   SOLVENTS AND CO-SURFACTANTS

Considering the poor water solubility of most of the chemical moieties, organic solvents are required in the formulation in most of the cases, especially when emulsion and microemulsion as a template is used. The acceptability of the organic solvents in the production of pharmaceuticals, their toxicity potential and the ease of their removal from the formulation needs to be considered when formulating nanosuspensions. The pharmaceutically acceptable and less hazardous water-miscible solvents, such as isopropanol and ethanol, and partially water-miscible solvents, such as ethyl formate, ethyl acetate, butyl lactate, triacetin, propylene carbonate and benzyl alcohol, are preferred over the conventional hazardous solvents, like dichloromethane.

Co-surfactants are generally used when using microemulsions to formulate nanosuspensions. The choice of co-surfactant is critical since co-surfactants can greatly influence phase behavior. The effect of co-surfactant on uptake of the internal phase for selected microemulsion composition and on drug loading should be investigated. Although bile salts, dipotassium glycerrhizinate are used as co-surfactants, various solubilizers, such as transcutol, glycofurol, ethanol and isopropanol can also be safely used as co-surfactants in the formulation of microemulsions [22].

4.2.3.4   OTHER ADDITIVES

Incorporation of additives in the nanosuspension formulation depends upon the desired route of administration, characteristics of the drug moiety. Common additives used include buffers, salts, polyols, osmogents, and cryoprotectant.

4.2.3.5   POST-PRODUCTION PROCESSING OF NANOSUSPENSIONS

Processing of nanosuspensions post-production becomes essential when the drug candidate is highly susceptible to hydrolytic cleavage or chemical degradation. Processing may also be required when the best possible stabilizer is not able to stabilize the nanosuspension for a longer period of time or there are acceptability restrictions with respect to the desired route.

Considering these aspects, various post-production processing techniques can be used to convert nanosuspension into more convenient and stable formulation [22].

4.2.3.5.1 Solidification Techniques

Nanosuspensions are thermodynamically unstable. Solidification techniques transform nanosuspensions into solid dosage forms such as tablets, capsules, and pellets. Solid dosage forms increase the storage stability of nanosuspensions. It is also convenient from marketing perspective and is practically important for patient convenience. Pelletization, granulation, spray drying, and lyophilization are the unit-operations of the solidification technique [79, 80]. Matrix formers (e.g., mannitol, cellulose derivatives) are usually added to the nanosuspensions before solidification to prevent destabilization of particles due to creating additional thermal stresses such as heating during spray drying or freezing during lyophilization. For example, microcrystalline cellulose was used by Bernard et al. as a matrix former during the freeze-drying process of Itraconazole nanosuspensions [81].

Nanosuspensions are typically converted to a solid dosage form for clinical formulations. Prior to drying, redispersants need to be added to the nanosuspension to ensure complete redispersion of nanoparticles into their primary, pre-drying state [82]. Sugars, such as sucrose, lactose, and mannitol, are commonly used as redispersants in oral formulations. The sugar molecules serve as "protectants" and prevent nanoparticles from aggregating as they are concentrated during drying [82].

Methods such as lyophilization or spray drying may be employed to produce a dry powder of nano-sized drug particles. Rational selection has to be made in these unit operations considering the drug properties and economic aspects. Generally, spray drying is more economical and convenient than lyophilization [22].

The nanosuspension drug delivery system can be employed as a liquid dosage form or transformed into solid dosage form such as powder, tablet, pellet, capsule, and film dosage forms. Therefore, nanosuspensions can be administrated by a variety of routes including oral, intravenous, ocular, dermal, pulmonary, and, etc. [83].

The effect of post-production processing on the particle size of the nanosuspension and moisture content of the nanosized drug should be given due consideration.

4.2.3.5.2 Surface Modification Techniques

Rapid or burst release of nanosuspensions may cause toxicity and severe side effects. Hence, surface modification is required in order to control drug release and/or prolonged residence at the site of action. The surface engineering by surface coating is important for targeted drug delivery systems. PEG is commonly used to modify the nanoparticle surface. This leads to reduced protein adsorption and opsonization of nanoparticles and leads to prolonged systematic circulation time [84]. Longer circulation time is required to allow nanoparticles sufficient time to leak out of the vasculature in infectious and inflammatory areas including cancer tissues [85, 86]. Carefully engineered nanoparticles surface can also effectively target the diseased tissue.

4.2.4 CHARACTERIZATION OF NANOSUSPENSIONS

4.2.4.1 PARTICLE SIZE AND POLYDISPERSITY INDEX

Mean particle size and polydispersity index (PI) is an essential characteristic parameter of nanosuspensions since it affects the saturation solubility, dissolution velocity, physical stability, and biological performance. It is determined by photon correlation spectroscopy (PCS). The mean particle diameter of nanosuspension can be determined rapidly and accurately using PCS. The wide particle size distribution, also known as polydispersity index can also be measured using PCS. The PI for nanosuspension should be as low as possible for long-term stability of the formulation. The PI value of 0.1–0.25 indicates a fairly narrow size distribution whereas a PI value greater than 0.5 indicates a very broad distribution. The measuring range of PCS is limited to approximately 3 nm–3 μm. Therefore, additionally laser diffractometry (LD) is used to detect any content of particles in the micrometer range or aggregates of drug nanoparticles. Laser diffractometry yields a volume size distribution and can be used to measure particles ranging from 0.05–80 μm and in certain instruments particle sizes up to 2000 μm can be measured. The typical LD characterization includes determination of diameter 50% LD [50] and diameter 99% LD [99] values, which indicate that either 50 or 99% of the particles are below the indicated size. Nanosuspensions meant for parenteral or pulmonary delivery should be analyzed by LD to minimize the risk of capillary blockade or emboli formation associated with particles having the size greater than 5–6 μm. Data obtained from these two techniques are not the same because LD data are volume based and the PCS mean diameter is the light intensity weighted size. The nanosuspensions can be suitably diluted with deionized water before carrying out PCS or LD analysis [22, 24]. Coulter counter technique is much more efficient technique than LD for particle size analysis as it helps in accurate quantification of microparticulate drug content. It is used in case of intravenous administration of nanosuspension and gives an absolute number of particles per volume unit for the different size of classes [22].

4.2.4.2   CRYSTALLINE STATE AND PARTICLE MORPHOLOGY

Nanosizing can produce polymorphic or morphological changes in the drug. Additionally, amorphous state of the drug particles can be generated when nanosuspensions are prepared. Therefore, it becomes necessary to examine the crystalline state and the morphology of the drug particle. Differential scanning calorimetry and X-ray diffraction analysis can be used to assess the crystalline structure of nanosuspensions as well as extent of changes in physical state. Scanning electron microscopy can be used to study the particle morphology [22].

4.2.4.3   PARTICLE CHARGE (ZETA POTENTIAL)

Particle charge governs the physical stability of the nanosuspension and therefore is one of the most important evaluation parameters. The zeta potential of a nanosuspension depends upon the stabilizer and the drug itself. A minimum zeta potential of ±30 mV is required for an electrostatically stabilized nanosuspension whereas a minimum zeta potential of ±20 mV is desirable in the case of a combined electrostatic and steric stabilization [22, 24].

4.2.4.4 DISSOLUTION VELOCITY AND SATURATION SOLUBILITY

These two parameters allow estimation of the systemic availability of the drug. Dissolution velocity and saturation solubility can be determined using standard pharmacopeial methods at different temperatures and in different physiological buffers [24].

4.2.4.5 BIOLOGICAL PERFORMANCE

The correlation between *in vitro* and *in vivo* performance of nanosuspension formulation and monitoring in-vivo performance is a very critical parameter. The extent of tissue distribution and interaction with plasma proteins has pronounced effect on in-vivo performance of the formulation and this, in turn, depends on the surface properties of the drug like surface hydrophobicity [22]. Techniques such as hydrophobic interaction chromatography can be used to determine surface hydrophobicity [87], whereas 2-D PAGE [88] can be employed for the quantitative and qualitative measurement of protein adsorption after intravenous injection of drug nanosuspensions.

4.2.4.6 TOXICOLOGY STUDIES

In recent years, a concern of nanotoxicity of nanoparticles is increased due to the fact that nanoparticles have the ability to enter the cell and cause damage to single cells. Therefore, nanotoxicity of drug nanocrystals cannot be neglected and hence, the interaction of nanoparticles with cells and its uptaking should be considered when the nanoparticles are developed. Normally, the nanoparticles in range of 100 nm up to 1000 nm can only be taken up by quite limit number of cells with phagocytic activity and are not easy to access limiting the toxicity risk. The nanoparticles with size below 100 nm can be taken up by all cells by endocytosis, which leads to higher risk of toxicity. All the nanosized formulations must undergo cytotoxicity studies before their use to minimize the associated risks [17].

4.3   STABILITY OF NANOSUSPENSIONS

4.3.1   STABILITY ISSUES

Stability issues are inevitably encountered during the development of nanosuspension technology and in pharmaceutical applications, thus representing the limiting step in the development of such platforms [89, 90]. Despite progress in this field, nanosuspensions possess instability due to nucleation and particle growth [91]. The high surface area of drug nanosuspensions that offers unique biopharmaceutical characteristics also makes them thermodynamically unstable and promotes agglomeration and crystal growth [92]. The high surface energy of the produced nano-sized crystals results in particle size growth, a phenomenon known as Ostwald's ripening, in the absence of appropriate stabilizers. Crystal growth or flocculation during the production process or shelf life of the nanosuspensions may directly affect dissolution and *in vivo* performance due to the formation of larger particles with a reduced specific surface area [93]. Some of the common nanosuspension stability issues are discussed in the following subsections.

4.3.1.1   AGGREGATION

Nanosuspensions that are not appropriately stabilized may aggregate during storage or the solidification process. Unsuitable stabilizers result in the agglomeration of smaller particles in the nanosuspensions due to the Ostwald's ripening phenomenon [94]. During Ostwald's ripening, coarse particles grow at the expense of the redissolution of fine particles [95]; because smaller particle-sized nanocrystals are more soluble than large ones, mass transfer occurs from the fine to coarse particles [96]. A nanosuspension is a thermodynamically unstable system. Aggregation in nanosuspension occurs as a result of Ostwald's ripening and the nanosized system tends to reduce the Gibbs free energy [97]. Therefore, limiting or inhibiting this aggregation is an important aspect of nanosuspension development. The prepared nanosuspensions should have a relatively homogeneous particle size to avoid large differences in the saturation solubility of different-sized crystals [56]. The use of an appropriate stabilizer is essential to produce a stable nanosuspension and the efficiency of a particular stabilizer depends on its interaction with the drug compound [22].

Aggregation may occur either during preparation process or storage. In case of top-down approach for nanosuspension production, as the surface area increases, nano-sized drug particles aggregate due to thermodynamic effects, ultimately reducing the process efficiency [98]. Thus, the use of a stabilizer to cover the surface of nanoparticles during milling or homogenizing is important. During storage, aggregation becomes even more important. The selection of appropriate stabilizers and concentrations are critical parameters of stability [99]. Functional co-polymers bearing alkynyl or azido groups have been reported to prevent aggregation by "click-chemistry"-mediated crosslinking [100]. The successful crosslinking of the polymers around the drug nanosuspensions may prevent aggregation and control the dissolution rate, thus stabilizing the drug nanosuspensions.

4.3.1.2   SEDIMENTATION

Nanosuspensions are colloidal dispersions and are intermediate in size between true solutions and coarse dispersions. Sedimentation is the extreme presentation of nanosuspension instability [101]. Generally, aggregation and Ostwald's ripening are the primary steps to instability. However, when the gravity of the drug particle is greater than the buoyant force provided by the dispersion system, sedimentation is inevitable. This process is irreversible, and the nanosuspension cannot be redispersed. Therefore, predicting and preventing sedimentation are the chief considerations in the nanosuspension formulation [102].

Flocculation, a type of sedimentation, results from attractive interactions between particles. Flocculation of nanosuspensions may occur by polymer bridging, charge neutralization, polymer–particle surface complex formation or depletion flocculation, or by a combination of these mechanisms [103]. Depending on the nature of the polymer and the character of its interaction with the nanocrystal surface, there are two main possible mechanisms of flocculation: [1] bridging as a result of adsorption of a macromolecule onto the surface of nanocrystals and [2] surface charge neutralization [104].

4.3.1.3   CRYSTALLINE TRANSFORMATION

The presence of insoluble drugs in amorphous state fully or partially in nanocrystals is theoretically an ideal method to improve the dissolution rate [4]. However, amorphous state is thermodynamically unstable relative to the crystalline state. Therefore, although the amorphous form of a formulated nanosuspension is more soluble and has a higher dissolution rate than the crystalline state, amorphous solids are thermodynamically unstable and tend to transform to a crystalline state, limiting their extensive commercialization [11]. Therefore, a crystalline nanosuspension drug is more stable and therefore often more favorable [105]. There are no appropriate control methods for this transformation. Therefore, once the production process generates an amorphous material, there is a risk of crystalline transformation during storage or release, and crystalline transformation is an inevitable problem during nanosuspension storage.

4.3.1.4   CHEMICAL INSTABILITY

In addition to physical stability, chemical stability is an important aspect of drug delivery platforms in the pharmaceutical sciences [106]. Generally, nanosuspensions have been prepared in a dispersion medium of water or water-mixture environments, although a few processes have used non-aqueous media [107]. Therefore, chemical reactions, such as hydrolysis and oxidation, are considerable problems in the formulation of nanosuspensions and limit drug chemical stability in nanosuspensions [61]. The preparation of chemically stable nanosuspensions to prevent the hydrolysis of labile compounds remains challenging.

4.3.1.5   IN VIVO STABILITY

When a nanosuspension is infused into the plasma, the environment, including the pH and ionic strength, of the nanosuspension changes significantly which could alter the surface charge and zeta potential of the nanosuspensions and may induce the aggregation of the nanosuspensions. However, no study has specifically focused on the factors that influence *in vivo* stability [108].

4.3.2   FACTORS INFLUENCING STABILITY

The basic and essential drug attributes are safety, efficiency and stability. Moreover, safety and efficiency are at least partially dependent on stability; good stability is a warranty for reliable safety and efficiency. The stability of nanosuspensions in drug delivery is a very important and critical aspect of this technology [108]. Therefore, there is an urgent need to address these drawbacks of nanosuspensions. For enhancing the efficiency and stability of nanosuspension development it is crucial to define the most influential factors affecting the critical quality attributes (CQAs) of such formulations, importance of physicochemical characteristics of drugs and stabilizers, as well as formulation and manufacturing process parameters that affect nanosuspensions CQAs.

4.3.2.1   STABILIZER SELECTION AND USE

Stabilizers are vital for preventing the aggregation of high-energy nano-suspensions [109]. However, different drugs require different polymeric stabilizers, and no single stabilizer system is appropriate for all drugs and preparation processes [110]. Unfortunately, there are no systematic empirical or theoretical guidelines for stabilizer selection and optimization. Empirical knowledge of the relationship between stabilizer efficacy and nanosuspension stability is also limited. Furthermore, the type and amount of stabilizer also have a significant effect on the physical stability and *in vivo* behaviors of the nanosuspensions [111, 112]. In addition to the problems mentioned above, potential toxicity concerns about the long-term use of large quantities of stabilizers have also limited the therapeutic application of drug nanosuspensions [113].

Commonly used stabilizers and their mode of stabilization has already been discussed earlier in this chapter. However, some of the contributing factors related to stabilizers will be discussed here. The stabilizers used in the preparation of nanosuspensions should adsorb onto the surfaces of drug nanosuspensions and provide a steric or electrostatic stabilization effect due to the hydrophobic moieties present in the stabilizers. Thus, the surface of the nanosuspensions particles could be covered by the adsorbed stabilizers. To establish successful stabilization effects within a reasonable processing time, strong and fast adsorption at full coverage and a long desorption time scale are required, in addition to steric repulsion [114]. The adsorption rate decreases as the molecular weight of the steric stabilizer increases [96]. The length and molecular weight of the polymer are the thermodynamic driving forces for physical adsorption on the surface of the particle. In general, polymer-type steric stabilization is suitable for ensuring drug stability during processing because it does not usually destroy the crystal structure of the drug particles, in contrast to conventional low-molecular-weight surfactants such as sodium lauryl sulfate (SLS) [114]. The stability and robustness of nanosuspensions are mainly manipulated by various formulation and process variables [115]. Therefore, the stabilization effect of the stabilizers is a critical parameter, and consequently, the selection of proper steric and electrostatic stabilizers and the optimization of their concentration play a major role in nanosuspension formulation. Developing and discovering new stabilizers are other important aspects of nanosuspension technology [116].

A stabilizer, which covers the surface of the nanosuspension to produce ionic or steric stabilization, is an essential component of the formulation. The hydrophobic chains of stabilizer adsorbed on the surface of the nanosuspension undergo a continuous thermal motion that results in a dynamic rough surface preventing coalescence via repulsive entropic forces. However, diffusion, convection, adsorption, and desorption of stabilizers occur simultaneously with mechanical fracturing and the generation of the new surface area of the nanosuspension system takes place [117]. This is a dynamic state and therefore, the rate of surface area formation represents a competition between the net adsorption of the stabilizers and the generation of the new surface. A higher rate of surface adsorption results in more efficient preparation methods and improved nanosuspension stability. Polymer length and molecular weight of a polymer acts as the driving force for the physical adsorption of the stabilizer on the surface of the particle. The higher the molecular weight of a polymeric stabilizer, the slower will be the adsorption [96]. Through the use of Atomic Force Microscopy (AFM), direct visualization of the adsorbed stabilizer directly on the drug surface would facilitate the elucidation of the extent of the interaction between the drug and the stabilizer and provides vital information in predicting the wettability, aggregation, crystal growth, and Ostwald ripening of the prepared nanosuspensions. Therefore, AFM methods could be employed to screen and predict the most suitable stabilizers for the preparation of nanosuspensions [108].

The stabilizer concentration is of utmost importance for the stabilization of nanosuspensions, and thus the amount of stabilizer added to the formulation should be optimized. Surfactant-type stabilizers should be used at concentrations below the critical micelle concentration (CMC). However, an inadequate amount of stabilizer will not provide complete coverage of the drug molecule surface, which is necessary to provide a steric repulsion between the nanoparticles in suspension [93]. A concentration higher than the CMC could actually result in less surfactant adsorption, which would further destabilize the nanosuspensions and thereby contributed to the increased particle size [118, 119]. Furthermore, at high stabilizer concentrations high above the plateau of the adsorption isotherm, electrostatic stabilizers could cause a decrease in the diffuse layer, leading to a decreased zeta potential and, consequently, decreased physical stability [120]. Also, nanosuspension dispersions are usually stabilized by increasing the polymer concentration to approximately 1%.

Moreover, high concentration of long chain polymers may lower the rate of dissolution especially for poorly water-soluble drugs. Further, stabilizers like sodium lauryl sulfate, Pluronic at high concentration, sometimes offer challenges in producing patient friendly dosage form especially for pediatric group due to local gastric irritation [96].

In addition to stabilizer concentration, temperature also plays an important factor in determining nanosuspension stability. The formation of hydrophobic interactions between the nanosuspension and stabilizer is a negative entropic process. Thus, the higher the temperature of the nanosuspension, the more thermodynamically unfavorable the system stability becomes. Thus, in hydrophobic nanosuspension systems, the tendency of aggregation is enhanced at higher temperature [121].

The solubility of the formulated drug in the stabilizer solution plays a significant role in the increase in particle size during storage, indicating Ostwald's ripening. For use in nanosuspensions, a stabilizer should have little effect on drug solubility. In addition, a drug that is soluble in the stabilizer in the solid state could also form an amorphous solid dispersion at the interface of the formulated nanosuspensions [5].

Because the zeta potential is one of the factors determining the physical stability of a dispersion system, its value can be used to predict the physical stability of the prepared nanosuspensions [122]. For some candidate compounds, the selection of a suitable zeta potential stabilizer and a sufficiently high zeta potential is challenging. The investigation of the influence of the other stabilizers on the zeta potential changes is also important for formulating nanosuspensions.

4.3.2.2   PHYSICAL PROPERTIES OF DRUG CANDIDATE

APIs with low enthalpy values are poor candidates for formulation as nanosuspensions, irrespective of the stabilizer used. APIs with low log P values also forms the least stable nanosuspensions [123].

4.3.2.3   FORMULATION APPROACHES

The type of preparation method chosen has great influence on the stability of nanosuspensions. Significant advances have been made in nanosuspension preparation technologies over the last several decades. Milling, high-pressure homogenization (HPH), impinging jet, electro-spraying, liquid-based methods, and supercritical fluid processes are the technologies that are currently available and are being actively developed for the preparation of nanosuspensions [110]. Furthermore, the bottom-up approach, such as nano-precipitation, is also an effective technique for preparing nanosuspensions [124]. As the particle size is reduced, there is a dramatic increase in surface energy (extra Gibbs free energy) that needs to be compensated for surface stabilization. So, all nanosuspension preparation methods must reduce thermodynamic instability and kinetic instability. However, different methods involve different theories on decreasing particle size and increasing surface area. Therefore, the potential influence of different nanosuspension preparation techniques on stability is a critical factor in the selection of a preparation technique.

In general, nanosuspension morphology and stability are influenced by the manufacturing approach and operating parameters. For example, the HPH process may yield short, needle-shaped nanosuspensions that increase in length during storage due to Ostwald's ripening. Furthermore, the HPH method does not induce crystal form transformation. By contrast, media milling can significantly damage the crystallinity of particles, resulting in spherical nanocrystals. Similar to the media milling approach, the bottom-up technique can also induce crystal form transformation, although needle-shaped nanosuspensions are obtained. Furthermore, the bottom-up method is more liable to Ostwald's ripening and increases in particle size.

In the HPH method, a suspension is forced through a small gap, making miniaturization of this technology less straightforward [125]. HPH-prepared nanosuspensions can protect chemically labile drugs from degradation [56]. This technique is a suitable alternative approach for preparing drug nanosuspension formulations to protect chemically labile drugs from degradation. During the HPH process, the homogenization pressure is the main parameter that determines the final particle size and crystal form of the prepared nanosuspensions [126]. In addition to the influence of homogenization pressure in the crystalline form, the combination of freeze-drying with HPH also influences the effectiveness with which homogenization reduces particle size [127].

Among the different techniques used for the preparation of nanocrystals, wet media milling is considered an attractive approach that permits relatively easy scale-up with respect to industrial pharmaceutical nanosuspension production [120]. During the milling process, the drug crystals break into smaller particles, and thus fresh surfaces are continuously generated [93]. Therefore, the time gap between the rate of new surface generation and the time of stabilizer absorption onto the new surface determine the stability and the efficiency of the nanosuspensions in the wet milling. In the media milling process, the degree of compensation depends on the interactions between the drugs and stabilizers [114]. In addition, retaining the crystallinity of drug particles during processing is beneficial for drug stability [80]. However, media milling can occasionally and significantly damage the crystallinity of the particles. Therefore, the greatest concern in media milling is shear-induced amorphous drug formation. Compared to its crystalline form, the amorphous drug form is characterized by enhanced solubility and Ostwald's ripening [123]. Caution is needed in the generalization of active pharmaceutical ingredients because each drug crystalline form has different physical stability [110]. Crystal defects and the re-crystallization of amorphous drug are potential sources of instability in nanosuspensions during preparation and storage. In the media-milling processes, considerable heat is generated, which may cause degradation of heat-sensitive APIs. The media-milling approach may also cause mechanical activation of drug particle surfaces.

Similar to the potential instability induced by the top-down approach, the bottom-up technique may adversely influence nanosuspension formulations by generating various unstable polymorphs, hydrates and solvates during processing [128]. Furthermore, the bottom-up approach involves the use of solvents, which are usually difficult to remove completely [129, 130]. Therefore, proper control of residual solvent levels is required to limit the potential toxicity of organic solvents. The use of organic solvents is also associated with a number of challenges, such as the proper and safe handling of flammable and/or explosive solvents, which requires special facilities and equipment [127]. Moreover, the bottom-up approach usually results in needle-shaped particles due to the rapid growth in one direction, which influences the physical stability of the nanosuspensions [131].

4.3.3 STABILITY IMPROVEMENT STRATEGIES

The stability of nanosuspensions can be improved by considering all the influential factors discussed above. The most important factors include selection of appropriate stabilizer and method for nanosuspension preparation. Also, compared with aqueous nanosuspensions, the solidified state is preferred because aggregation and other instability factors are significantly decreased [101]. Therefore, prepared nanosuspensions are commonly solidified.

4.3.3.1 USE OF APPROPRIATE STABILIZER AND METHOD

The type of stabilizer used, its amount and mechanism of stabilization is a crucial parameter when nanosuspension formulation is to be developed. Although there are various types of stabilizers available, but depending upon the desired application and drug properties, suitable stabilizer should be incorporated into nanosuspension. Also, a suitable method for production of nanosuspension should be used.

4.3.3.2 USE OF SOLIDIFIED FORMS OF NANOSUSPENSION

Prepared nanosuspensions are commonly converted to the solid state to avoid aggregation and other instability factors [101]. The solidified powder is then formulated as other dosage forms, such as sterile powder for injection, tablets and capsules for oral delivery and nebulized for pulmonary delivery [91, 132–134]. The end dosage forms of the nanosuspensions prevent aggregation, hydrolysis, and other sources of instability [117]. The solidification process is a critical step in the formulation of the final product. The selection of a suitable method for nanosuspension solidification is a critical aspect of nanosuspension preparation [135]. Moreover, nanosuspensions are always diluted, and have to be administered in large volumes to achieve therapeutic levels in the circulation [136]. To overcome these problems, it is desirable to formulate nanosuspensions into solid dose forms [4, 137]. To reduce time and energy consumption, spray drying is generally preferred by the pharmaceutical industry to transform liquid nanosuspensions into a dry product [135, 138]. Similar to spray drying, the electro-spray drying technique also could be used as a novel strategy for the preparation of redispersible drug nanosuspension formulations without aggregation [139].

Nanosuspension solidification can also be achieved using other established methods, such as freeze-drying [80]. Freeze-drying is the most suitable approach for the solidification of thermolabile drugs, such as protein drugs and vaccines [140]. Freeze-drying is also commonly used to retain the redispersibility of drug nanoparticles after solidification. Freezing rate and the type of cryoprotectants are two critical factors in the freeze-drying process. A cryoprotectant is usually added to the freshly prepared nanosuspensions to avoid freeze damage due to ice formation and to avoid particle aggregation [141]. A series of polymer cryoprotectants, such as polyatomic alcohols, are usually selected to stabilize nanosuspensions during the freeze-drying process. The most frequently used polyatomic alcohols include mannitol, lactose, sucrose, trehalose, and glucose.

Some solidification methods have also been developed to convert the liquid nanosuspensions into other solid products such as fluid-bed coating, casting method, hydrogel method and wet granulation. The fluid-bed coating and extrusion-spheronization are the two main methods to convert liquid nanosuspension into solid pellets [83]. Fluid-bed coating of pellets is a "one-step" technique that can add a drug film coating onto a nonpareil pellet core using a fluid-bed [142]. Fluid bed coating is more scalable. In the casting method, the liquid nanosuspension is mixed with some excipients, casted and dried to form a film [143]. In hydrogel method, liquid nanosuspension is incorporated into hydrogel. It is a more practical way to convert liquid nanosuspensions into dry products and to improve dosage stability [61]. Wet granulation is defined as a process to form drug granulates by sticking powder (binder or drug in powder form) with liquid (drug or binder in liquid) using a mesh screen, high shear mixer, or fluid bed [144]. After a solidification process of liquid nanosuspenison, the resultant drug nanoparticle powder can be mixed with excipients and then be converted into a tablet dosage form using a press machine [145]. Also, drug nanoparticle-loaded capsule dosage can be prepared by filling dry drug nanoparticle powder into a capsule [146]. Drug nanosuspension can also be transformed into polymeric films offering more economical and scalable option [147].

4.4   NANOSUSPENSIONS IN DRUG DELIVERY APPLICATIONS

Drug delivery using nanosuspension as a novel drug delivery system has opened new ways for administration of poorly soluble and poorly bioavailable drugs through various routes. Nanosized drug in the form of nanosuspension can be used for achieving increased oral absorption, increased oral bioavailability. Apart from improving oral absorption, nanosuspensions also improve dose proportionality, reduced fed/fasted state variability and reduced inter-subject variability. Nanosuspensions are also advantageous in achieving quick onset of action for drugs that are completely but slowly absorbed. Drug nanosuspensions can also be incorporated into dosage forms such as tablets, capsules, oral thin films, pellets, etc. by means of standard manufacturing techniques [22].

Nanosuspension is an ideal drug delivery system for the parenteral route. It becomes easy to process nanosized drug particles of almost all drugs for parenteral administration. Moreover, the absence of any harsh solvents/co-solvents and/or any potentially toxic ingredient in nanosuspensions enables them to bypass the limitations of parenteral administration. Nanosuspensions can be administered via different parenteral routes, ranging from intra-articular to intraperitoneal to intravenous injection [22].

Nanosuspensions may prove to be an ideal approach for delivering drugs that exhibit poor pulmonary absorption. Current formulations like suspension aerosols, inhalers shows a limited diffusion and dissolution of the drug at the site of action because of its poor solubility and microparticulate nature, which may affect the bioavailability of the drug, rapid clearance of the drug from the lungs because of ciliary movements [148], less residence time for the drugs, leading to absence of prolonged effect, unwanted deposition of the drug particles in pharynx and mouth. Nano-suspensions can solve the problems associated with conventional systems because of their versatile nature. The nanoparticulate nature of the drug allows the rapid diffusion and dissolution of the drug at the site of action. At the same time, the increased adhesiveness of the drug to mucosal surfaces [149] offers a prolonged residence time for the drug at the absorption site. They prevent unwanted deposition of particles in the mouth and pharynx, leading to decreased local and systemic side effects of the drug. Nanosus-pensions could be used in all available types of nebulizer. However, the extent of influence exerted by the nebulizer type as well as the nebuliza-tion process on the particle size of nanosuspensions should be ascertained.

Nanosuspensions can prove to be a boon for drugs that exhibit poor solubility in lachrymal fluids. Nanosuspensions, by their inherent ability to improve the saturation solubility of the drug, represent an ideal approach for ocular delivery of hydrophobic drugs. Moreover, the nanoparticulate nature of the drug allows its prolonged residence in the cul-de-sac, giving sustained release of the drug. Likewise, nanosuspensions can be used for targeted delivery as their surface properties and *in vivo* behavior can easily be altered by changing either the stabilizer or the composition of stabi-lizers or polymers. Their versatility and ease of scale-up and commercial production enables the development of commercially viable nanosuspen-sions for targeted delivery [22].

Nanosuspensions afford a means of administering increased concen-trations of poorly water-soluble drugs to the brain with decreased systemic effects and have been successfully employed for brain targeting. Several concepts of targeting of nanosuspension dosage forms for the treatment of bioweapon-mediated diseases have been developed at the Baxter Health-care Corporation [153]. Summary of benefits of some drug nanosuspen-sions have been shown in Table 4.2.

In future, it would be possible to engineer nanosuspensions by using the agents that enhance permeation [150] and/or minimize gut-related metabolic issues [151, 152]. This amalgamated approach would facilitate delivery of the compounds belonging to BCS Class IV that exhibit poor water solubility and poor membrane permeability. Currently, studies are in progress to identify strategies for manipulating the surface properties, size, and shape of drug nanosuspensions in order to eliminate sequestra-tion by the phagocytic cells of MPS-enriched organs whenever desired.

TABLE 4.2 Summary of Various Applications of Drug Nanosuspensions

Drug	Indication	Delivery route	Benefits of nanosuspension	References
Atovaquone	Antibiotic	Oral	Increased oral absorption and bioavailability	[164]
Danazol	Gonadotropin inhibitor	Oral	Drastic improvement in bioavailability and reduced intersubject variability and fed/fasted ratio	[165]
Naproxen	NSAID	Oral	Quick onset of action, increased Area Under Curve	[166]
Ketoprofen	NSAID	Oral	Sustained release	[167]
Paclitaxel	Anti-cancer	Parenteral	Improved tolerable dose of drug, reduced therapy cost and improved therapeutic performance	[168]
Clofazimine	Anti-leprotic	Parenteral	Improved stability and efficacy as compared to lipo-somal form	[169]
Flurbiprofen and ibuprofen	NSAID	Ocular	Sustained release and improved *in vivo* performance as compared to existing marketed formulations	[170, 171]
Budesonide	Corticosteroid	Pulmonary	Increased diffusion and dissolution of drug and mucoadhesion	[148]
Dalargin	Peptide drug	Targeted drug delivery	Brain targeting using surface modification	[172]
Buparvaquone	Antibiotic	Targeted drug delivery	Surface modified mucoadhesive nanosuspension with prolonged and targeted action	[61, 173]

4.5   SCALE-UP AND COMMERCIALIZATION

The possibility and ability for large-scale production of a delivery system or a dosage form is the essential prerequisite for its introduction to the pharmaceutical market. In addition, the production technology needs to be low-cost as extremely expensive technology will not find its way to the market. Another important point is to have a production line which can produce on large-scale and has acceptance by the regulatory authorities for qualification and validation [24].

Nanosuspension are simple systems and the chances are higher to launch products on the market. There are already some products available in the market. The list is shown in Table 4.3. Successful scale-up depends upon the simplicity, ease and consistency of production process as well as products. Not only have these, but also the convenient form of delivery enhanced the consumer acceptability. Nanosuspensions can be conveniently transformed into suitable dosage forms and hence can be easily accepted by the patients. Nanosuspensions appear to be a unique and yet commercially viable approach to combat problems associated with the delivery of poorly bioavailable hydrophobic drugs. Production techniques such as media milling and high-pressure homogenization have been successfully employed for large-scale production of nanosuspensions [22].

4.6   CONCLUSION

The nanosuspension technology for delivering nanosized drug particle has shown tremendous potential in improving solubility and bioavailability characteristics of poorly soluble drugs. Ease of formulation and transformation into suitable dosage forms has made them versatile carriers for various administration routes. Still, some stability and toxicity issues need more attention of researchers to prove the suitability of this technology for commercial production of solid dosage forms of nanosized drugs. Many multinational pharmaceutical companies have already filed the New Drug Application (NDA) based on nanosuspension techniques for drug categories like anticancer, immunosuppressant, anti-inflammatory and antimicrobials where varied routes of administration like oral, intravenous, pulmonary, topical have been mentioned. This technique is gaining

TABLE 4.3   List of Commercially Available Drug Nanosuspensions

Product	Drug compound	Indication	Company	Nanoparticle technology
RAPAMUNE®	Sirolimus	Immunosuppressant	Wyeth	Élan Drug Delivery Nanocrystals®
EMEND®	Aprepitant	Anti-emetic	Merck	Élan Drug Delivery Nanocrystals®
TriCor®	Fenofibrate	Hypo-cholesteremic	Abbott	Élan Drug Delivery Nanocrystals®
MEGACE ES®	Megestrol Acetate	Appetite stimulant	PAR Pharmaceutical	Élan Drug Delivery Nanocrystals®
TRIGLIDE™	Fenofibrate	Hypo-cholesteremic	First Horizon Pharmaceutical	SkyePharma IDD®-P technology

popularity in pharmaceutical industries to address the issues like poor solubility, dose reduction, low bioavailability due to its simplicity, scalability, and stability. Major challenges like achieving impurity profiling can be overcome by adopting the nanosuspension formulation approach, since the requirement of excipients is very less. This technology provides a ray of light for formulators to overcome the poor solubility of drug molecules. Thus, nanosuspensions can be successfully used as a nanomedicine in the near future.

KEYWORDS

- bioavailability characteristics
- nanosuspension technology
- new drug application
- pharmaceutical industries

REFERENCES

1. Morrow, K. J., Bawa, R., & Wei, C., (2005). Recent advances in basic and clinical nanomedicine. *Medical Clinics of North America, 2007,* 91(5), 805–843.

2. Freitas, R. A., (2005). What is nanomedicine? *Nanomedicine, 1*(1), 2–9.

3. Chan, V. S. W., (2006). Nanomedicine: an unresolved regulatory issue. *Regulatory Toxicology and Pharmacology, 46*(3), 218–224.

4. Rabinow, B., (2004). Nanosuspensions in drug delivery. Nat. Rev. *Drug Discov., 3,* 785–796.

5. Kayaert, P., (2012). Van den Mooter G. Is the amorphous fraction of a dried nanosuspension caused by milling or by drying? A case study with naproxen and cinnarizine. *Eur. J. Pharm. Biopharm., 81,* 650–656.

6. Kassem, M., Abdel Rahman, A., Ghorab, M., Ahmed, M., & Khalil, R., (2007). Nanosuspension as an ophthalmic delivery system for certain glucocorticoid drugs. *Int. J. Pharm., 340,* 126–133.

7. Shegokar, R., & Muller, R., (2010). Nanocrystals: industrially feasible multifunctional formulation technology for poorly soluble actives. *Int. J. Pharm., 399,* 129–139.

8. Das, S., & Suresh, P., (2011). Nanosuspension: a new vehicle for the improvement of the delivery of drugs to the ocular surface. Application to amphotericin. *B. Nanomedicine: Nanotechnology, Biology, and Medicine, 7,* 242–247.

9. Barle, E., Cerne, M., Peternel, L., & Homar, M., (2013). Reduced intravenous toxicity of amiodarone nanosuspension in mice and rats. *Drug Chem. Toxicol., 36,* 263–269.

10. Mc Kee, J., Rabinow, B., Cook, C., & Gass, J., (2010). Nanosuspension formulation of itraconazole eliminates the negative inotropic effect of SPORANOX in dogs. *J. Med. Toxicol., 6,* 331–336.

11. Liu, F., Park, J., Zhang, Y., Conwell, C., Liu, Y., Bathula, S., & Huang, L., (2010). Targeted cancer therapy with novel high drug-loading nanocrystals. *J. Pharm. Sci., 99,* 3542–3551.

12. Kumar, G., & Krishna, K., (2011). Nanosuspensions: The Solution to Deliver Hydrophobic Drugs. *International Journal of Drug Delivery, 3,* 546–557.

13. Muller, R. H., Gohla, S., & Keck, C. M., (2011). State of the art of nanocrystals-special features, production, nanotoxicology aspects and intracellular delivery. *Eur J Pharm Biopharm., 78,* 1–9.

14. Moschwitzer, J., & Muller, R. H., (2007). Drug nanocrystals-the universal formulation approach for poorly soluble drugs. In *Nanoparticulate Drug Delivery Systems,* Thassu, D., Deleers, M., Pathak, Y., Ed., Informa Healthcare: New York, 71–88.

15. Junghanns, JUAH., & Muller, R. H., (2008). Nanocrystal technology, drug delivery and clinical applications. *Int J Nanomedicine, 3*(3), 295–309.

16. Kocbek, P., Baumgartner, S., & Kristl, J., (2006). Preparation and evaluation of nanosuspensions for enhancing the dissolution of poorly soluble drugs. *Int J Pharm., 312,* 179–186.

17. Junyaprasert, V., & Morakul, B., (2015). Nanocrystals for enhancement of oral bioavailability of poorly water soluble drugs. *AJPS., 10,* 13–23.

18. Buckton, G., & Beezer, A. E., (1992). The relationship between particle size and solubility. *Int J Pharm., 82,* 7–10.

19. Moschwitzer, J., & Muller, R. H., (2006). New method for the effective production of ultrafine drug nanocrystals. *J Nanosci Nanotech.*, *6*, 3145–3153.
20. Gulsun, T., Gursoy, R. N., Oner, L., (2009). Nanocrystal technology for oral delivery of poorly water-soluble drugs. *FABAD J Pharm Sci.*, *34*, 55–65.
21. Keck, C. M., & Muller, R. H., (2010). *Smart Crystals-review of the second generation of drug nanocrystal.* In *Handbook of materials for nanomedicine*, Torchilin, V. P., Amiji, M. M., Ed., Pan Stanford: Singapore, 555–580.
22. Date, A., Kulkarni, R., & Patravale, V., (2004). Nanosuspensions: A promising drug delivery. *Journal of Pharmacy & Pharmacology, 56*, 827–840.
23. Kesisoglou, F., Panmai, S., & Wu, Y., (2007). Nanosizing oral formulation development and biopharmaceutical evaluation. *Adv. Drug Deliv. Rev.*, *59*(7), 631–644.
24. Muller, R. H., Jacobs, C., & Kayser, O., (2001). Nanosuspensions as particulate drug formulations in therapy: Rationale for development and what we can expect for the future. *Adv. Drug. Deliv. Rev.*, *47*, 3–19.
25. Paun, J. S., & Tank, H. M., (2012). Nanosuspension: An emerging trend for bioavailability enhancement of poorly soluble drugs. *Asian J. Pharm. Tech., 2*(4), 157–168.
26. Peters, K., & Muller, R. H., (1996). Nanosuspensions for the oral application of poorly soluble drugs. In: Proceedings European Symposium on Formulation of Poorly-available Drugs for Oral Administration. *APGI,* Paris, 330–333.
27. Rawlins, E. A. (1982). *Bentley's Textbook of Pharmaceutics. 8th ed.*, Bailliere Tindall, London.
28. Muller, R. H., & Bohm, B. H. L., (1998). *Nanosuspensions.* In *Emulsions and nanosuspensions for the formulation of poorly soluble drugs*, Muller, R. H., Benita, S., Bohm, B. H. L. Ed., Medpharm Scientific Publishers .Stuttgart, 149–174.
29. Koradia, K., Koradia, H., Sheth, N., & Dabhi, M., (2015). The impact of critical variables on properties of Nanosuspension: A Review. *Int. J. Drug Dev. & Res., 7*(1), 150–161.
30. Chen, H., Khemtong, C., Yang, X., Chang, X., & Gao, J., (2011). Nanonization strategies for poorly water-soluble drugs. *Drug discovery today., 16*, 354–60.
31. Bodmeier, R., & Mc Ginity, J. M., (1998). Solvent selection in the preparation of poly (DL-lactide) microspheres prepared by solvent evaporation method. *Int. J. Pharm., 43*, 179–186
32. Sah, H. (1997). Microencapsulation technique using ethyl acetate as a dispersed solvent: effects on its extraction rate on the characteristics of PLGA microspheres. *J. Control Release, 47*, 233–245
33. Trotta, M., Gallarate, M., Pattarino, F., & Morel, S., (2001). Emulsions containing partially water miscible solvents for the preparation of drug nanosuspensions. *J. Control Release, 76*, 119–128
34. Eccleston, G. M. (1992). *Microemulsions. In Encyclopedia of Pharmaceutical Technology,* Swarbrick, S., Boylan, J. C., Ed., Marcel Dekker: New York, 375–421.
35. Trotta, M., Gallarate, M., Carlotti, M. E., & Morel, S., (2003). Preparation of griseofulvin nanoparticles from water-dilutable microemulsions. *Int. J. Pharm., 254*, 235–242
36. Wang, S., & Kienzle, F., (2000). The syntheses of pharmaceutical intermediates in supercritical fluids. *Ind. Eng. Chem. Res., 39*, 4487–4490.

37. Shekhon, B. S., (2010). Supercritical fluid technology: An overview of Pharmaceutical applications. *Int. J. Pharm Tech Res., 2*(1), 810–826.
38. Date, A. A., Patravale, V. B., (2004). Current strategies for engineering drug nanoparticles. *Current Opinion in Colloid & Interface Science, 9*, 222–235.
39. Vasukumar, K., & Bansal, A., (2003). Supercritical fluid technology in pharmaceutical research, *CRIPS., 4*, 8–12.
40. Reverchon, E., (1999). Supercritical antisolvent precipitation of micro- and nanoparticles. *J. Supercrit. Fluids, 15*, 1–21.
41. Reverchon, E., & Porta, G. D., (1999). Production of antibiotic micro- and nanoparticles by supercritical antisolvent precipitation. *Powder Technol., 106*, 23–29.
42. Sinha, B., Müller, R. H., & Möschwitzer, J. P., (2013). Bottom-up approaches for preparing drug nanocrystals: Formulations and factors affecting particle size. *International Journal of Pharmaceutics, 453*, 126–41.
43. Chattopadhyay, P., & Gupta, R. B., (2001). Production of antibiotic nanoparticles using supercritical CO_2 as antisolvent with enhanced mass transfer. *Ind. Eng. Chem. Res., 40*, 3530–3539.
44. Chen, J. F., Wang, Y. H., Guo, F., Wang, X. M., & Zheng, C., (2000). Synthesis of nanoparticle with novel technology: high gravity reactive precipitation. *Ind. Eng. Chem. Res., 39*, 948–954.
45. Chen, J. F., Zhou, M. Y., Shao, L., Wang, Y. Y., Yun, J., Chew, N., et al., (2004). Feasibility of preparing nanodrugs by high gravity reactive precipitation. *Int. J. Pharm., 269*, 267–274.
46. Hitt, J., Curtis, C., Fransen, K., Kupperblatt, G., Rogers, T., Scherzer, B., et al., (2002). Nanoparticles of poorly water soluble drugs made via a continuous precipitation process. Proceedings, AAPS Annual Meetings, Toronto, Canada.
47. Singh, S. K., Vaidya, Y., Gulati, M., Bhattacharya, S., Garg, V., & Pandey, N. K., (2016). Nanosuspension: Principles, perspectives and practices. *Curr Drug Deliv., 13*(8), 1222–1246.
48. Liversidge, G. G., Cundy, K. C., Bishop, J. F., & Czekai, D. A., (1992). Surface modified drug nanoparticles. *US Patent. 5*, 145, 684.
49. Buchmann, S., Fischli, W., Thiel, F., & Alex, R., (1996). Aqueous suspension, an alternative intravenous formulation for animal studies. *Eur. J. Pharm. Biopharm., 42*, S10.
50. Yonemochi, E. et al., (1999). Physicochemical properties of amorphous clarithromycin obtained by grinding and spray drying. *Eur J Pharm Sci., 7*, 331–8.
51. Watanabe, T., et al. (2002). Stabilization of amorphous indomethacin by co-grinding in a ternary mixture. *Int J Pharm., 241*, 103–11.
52. Wongmekiat, A., et al., (2002). Formation of fine drug particles by cogrinding with cyclodextrin. I. The use of β-cyclodextrin anhydrate and hydrate. *Pharm Res., 19*, 1867–72.
53. Mura, P., et al., (2002). Investigation of the effects of grinding and cogrinding on physicochemical properties of glisentide. *J Pharm Biomed Anal., 30*, 227–37.
54. Illig, K. J., Mueller, R. L., Ostrander, K. D., & Swanson, J. R., (1996). Use of microfluidizer processing for preparation of pharmaceutical suspensions. *Pharm. Technol., 20*, 78–88.

55. Gruverman, I. J., (2003). Breakthrough ultraturbulent reaction technology opens frontier for developing life-saving nanometer-scale suspensions and dispersions. *Drug Deliv. Technol., 3*, 52.

56. Moschwitzer, J., Achleitner, G., Pomper, H., & Muller, R. H., (2004). Development of an intravenously injectable chemically stable aqueous omeprazole formulation using nanosuspension technology. *European Journal of Pharmaceutics and Biopharmaceutics, 58*(3), 615–619.

57. Chen, Y. J., Yang, X. L., Zhao, X. L., & Xu, H. B., (2006). Preparation of oleanolic acid nanosuspension. *Zhongguo Yaoxue Zazhi., 41, 12*, 924–927.

58. Hecq, J., Deleers, M., Fanara, D., Vranckx, H., & Amighi, K., (2005). Preparation and characterization of nanocrystals for solubility and dissolution rate enhancement of nifedipine. *International Journal of Pharmaceutics, 299*, 1–2, 167–177.

59. Kumar, M. P., Rao, Y. M., & Apte, S., (2007). Improved bioavailability of albendazole following oral administration of nanosuspension in rats. *Current Nanoscience, 3*(2), 191–194.

60. Langguth, P., Hanafy, A., & Frenzel, D., (2005). Nanosuspension formulations for low-soluble drugs: pharmacokinetic evaluation using spironolactone as model compound. *Drug Development and Industrial Pharmacy, 31*(3), 319–329.

61. Muller, R. H., & Jacobs, C., (2002). Buparvaquone mucoadhesive nanosuspension: preparation, optimisation and long-term stability. *International Journal of Pharmaceutics, 237*(1–2), 151–161.

62. Gao, L., Zhang, D., & Chen, M., (2008). Drug nanocrystals for the formulation of poorly soluble drugs and its application as a potential drug delivery system. *Journal of Nanoparticle Research., 10*(5), 845–862.

63. Muller, R. H., Jacobs, C., & Kayser, O., (2000). Nanosuspensions for the formulation of poorly soluble drugs. *Pharmaceutical Emulsions and Suspensions, 105*, 383–407.

64. Krause, K. P., Kayser, O., Mader, K., Gust, R., & Muller, R. H., (2000). Heavy metal contamination of nanosuspensions produced by high-pressure homogenization. *International Journal of Pharmaceutics., 196*(2), 169–172.

65. Kumar, A., Sahoo, S. K., & Globale, P., (2011). Review on solubility enhancement techniques for hydrophobic drugs. *Pharmacie Globale, 3*(3), 1–7.

66. Reddy, G. A. (2012). Nanosuspension technology: A review. *IJPI's Journal of Pharmaceutics and Cosmetology, 2*(8), 47–52.

67. Keck, C. M., & Muller, R. H., (2006). Drug nanocrystals of poorly soluble drugs produced by high pressure homogenization. *European Journal of Pharmaceutics and Biopharmaceutics, 62*(1), 3–16.

68. Morakul, B., Suksiriworapong, J., & Leanpolchareanchai, J., (2013). Precipitation-lyophilization-homogenization (PLH) for preparation of clarithromycin nanocrystals: influencing factors on physicochemical properties and stability. *Int J Pharm., 457*, 187–196.

69. Hunter, R. J., (2001). *Foundations of Colloid Science*, 2nd edn., Oxford University Press, New York.

70. Wei, H., Yi, Lu., Jianping, Q., Lingyun, Chen., Fuqiang, Hu., & Wei, W., (2013). Food proteins as novel nanosuspension stabilizers for poorly water soluble drugs. *International journal of Pharmaceutics, 441*, 269–278.

71. He, W., Tan, Y., Tian, Z., Chen, L., Hu, F., & Wu, W., (2011). Food protein-stabilized nanoemulsions as potential delivery systems for poorly water-soluble drugs: preparation, *in vitro* characterization, and pharmacokinetics in rats. *Int. J. Nanomed., 6*, 521–533.

72. Izquierdo, P., Esquena, J., Tadros, T. F., Dederen, C., Garcia, M. J., Azemar, N., & Solans, C., (2001). Formation and stability of nano-emulsions prepared using the phase inversion temperature method. *Langmuir, 18*, 26–30.

73. Jiao, J., (2008). Polyoxyethylated nonionic surfactants and their applications in topical ocular drug delivery. Adv. *Drug Deliv. Rev., 60*, 1663–1673.

74. Liu, D., & Zhang, N., (2010). Cancer chemotherapy with lipid-based nanocarriers. *Crit. Rev. Ther. Drug Carrier Syst., 27*, 371–417.

75. Gelderblom, H., Verweij, J., Nooter, K., & Sparreboom, A., (2001). Cremophor EL: the drawbacks and advantages of vehicle selection for drug formulation. *Eur. J. Cancer., 37*, 1590–1598.

76. Hawkins, M. J., Soon-Shiong, P., & Desai, N., (2008). Protein nanoparticles as drug carriers in clinical medicine. *Adv. Drug Deliv. Rev., 60*, 876–885.

77. Chen, L., Remondetto, G. E., & Subirade, M., (2006). Food protein-based materials as nutraceutical delivery systems. *Trend Food Sci. Technol., 17*, 272–283.

78. Ma Ham, A., Tang, Z. W., Wu, H., Wang, J., & Lin, Y. H., (2009). Protein-based nanomedicine platforms for drug delivery, *5*, 1706–1721.

79. Muller, R. H., Moschwitzer, J., & Bushrab, J., (2006). Manufacturing of nanoparticles by milling and homogenization techniques in Nanoparticle Technology for Drug Delivery, *Drugs and the Pharmaceutical Sciences., 159*, 21–51, CRC Press, New York, NY, USA.

80. Eerdenbrugh, B., Mooter, G., & Augustijns, P., (2008). Top-down production of drug nanocrystals: nanosuspension stabilization, miniaturization and transformation into solid products. *International Journal of Pharmaceutics., 364*(1), 64–75.

81. Bernard, V. E., Sofie, V., & Johan, M., (2008). Microcrystalline cellulose, a useful alternative for sucrose as a matrix former during freeze-drying of drug nanosuspensions—a case study with itraconazole, *European Journal of Pharmaceutics and Biopharmaceutics, 70*(2), 590–596.

82. Abdelwahed, W., Degobert, G., & Stainmesse, S., (2006). Fessi, H.Freeze-drying of nanoparticles: formulation, process and storage considerations, *Adv. Drug Deliv, Rev., 58*(15), 1688–1713.

83. Chen, A., Shi, Y., Yan, Z., Hao, H., Zhang, Y., Zhang, J., et al., (2015). Dosage form developments of Nanosuspension drug delivery system for oral administration route. *Curr Pharm Des., 21*(29), 4355–4365.

84. Alexis, F., Pridgen, E., Molnar, L. K., & Farokhzad, O. C., (2008). Factors affecting the clearance and biodistribution of polymeric nanoparticles. *Molecular Pharmaceutics, 5*(4), 505–515.

85. Lode, J., Fichtner, I., Kreuter, J., Berndt, A., Diederichs, J. E., & Reszka, R., (2001). Influence of surface-modifying surfactants on the pharmacokinetic behavior of 14C-poly (methylmethacrylate) nanoparticles in experimental tumor models. *Pharmaceutical Research, 18*(11), 1613–1619,

86. Wu, N. Z., Da, D., Rudoll, T. L., Needham, D., Whorton, A. R., & Dewhirst, M. W., (1993). Increased microvascular permeability contributes to preferential accumulation of stealth liposomes in tumor tissue. *Cancer Research, 53*(16), 3765–3770.

87. Wallis, K. H., & Muller, R. H., (1993). Determination of the surface hydrophobicity of colloidal dispersions by mini-hydrophobic interaction chromatography. *Pharm. Ind., 55,* 1124–1128.

88. Blunk, T., Hochstrasser, D. F., Sanchez, J. C., & Muller, B. W., (1993). Colloidal carriers for intravenous drug targeting: Plasma protein adsorption patterns on surface-modified latex particles evaluated by two-dimensional polyacrylamide gel electrophoresis. *Electrophoresis, 14,* 1382–1387.

89. Dodiya, S., Chavhan, S., Korde, A., & Sawant, K. K., (2013). Solid lipid nanoparticles and nanosuspension of adefovir dipivoxil for bioavailability improvement: formulation, characterization, pharmacokinetic and biodistribution studies. *Drug Dev. Ind. Pharm., 39,* 733–743.

90. Li, X., Gu, L., Xu, Y., & Wang, Y., (2009). Preparation of fenofibrate nanosuspension and study of its pharmacokinetic behavior in rats. *Drug Dev. Ind. Pharm., 35,* 827–833.

91. Dolenc, A., Kristl, J., Baumgartner, S., & Planinsek, O., (2009). Advantages of celecoxib nanosuspension formulation and transformation into tablets, *Int. J. Pharm., 376,* 204–212.

92. Gao, L., Zhang, D., Chen, M., Zheng, T., & Wang, S., (2007). Preparation and characterization of an oridonin nanosuspension for solubility and dissolution velocity enhancement. *Drug Dev. Ind. Pharm., 33,* 1332–1339.

93. Ghosh, I., Schenck, D., Bose, S., & Ruegger, C., (2012). Optimization of formulation and process parameters for the production of nanosuspension by wet media milling technique: effect of vitamin E TPGS and nanocrystal particle size on oral absorption. *Eur. J. Pharm. Sci., 47,* 718–728.

94. Ali, H. S., York, P., & Blagden, N., (2009). Preparation of hydrocortisone nanosuspension through a bottom-up nanoprecipitation technique using microfluidic reactors. *Int. J. Pharm., 375,* 107–113.

95. Hu, J., Ng, W. K., Dong, Y. C., Shen, S. C., & Tan, R. B. H., (2011). Continuous and scalable process for water-redispersible nanoformulation of poorly aqueous soluble APIs by antisolvent precipitation and spray-drying, *Int. J. Pharm., 404,* 198–204.

96. Ghosh, I., Bose, S., Vippagunta, R., & Harmon, F. (2011). Nanosuspension for improving the bioavailability of a poorly soluble drug and screening of stabilizing agents to inhibit crystal growth. *Int. J. Pharm., 409,* 260–268.

97. Xia, D., Quan, P., Piao, H., Sun, S., Yin, Y., & Cui, F., (2010). Preparation of stable nitrendipine nanosuspensions using the precipitation-ultrasonication method for enhancement of dissolution and oral bioavailability, *Eur. J. Pharm.Sci., 40,* 325–334.

98. Xia, D., Ouyang, M., Wu, J. X., Jiang, Y., Piao, H., Sun, S., et al., (2012). Polymer-mediated anti-solvent crystallization of nitrendipine: monodispersed spherical crystals and growth mechanism, *Pharm. Res., 29,* 158–169.

99. Detroja, C., Chavhan, S., & Sawant, K., (2011). Enhanced antihypertensive activity of candesartan cilexetil nanosuspension: formulation, characterization and pharmacodynamics study. *Sci. Pharm., 79,* 635–651.

100. Fuhrmann, K., Gauthier, M. A., & Leroux, J. C., (2010). Cross-linkable polymers for nanocrystal stabilization. *J. Control Release, 148,* e12–e13.

101. Gao, Y. A., Qian, S. A., & Zhang, J. J., (2010). Physicochemical and pharmacokinetic characterization of a spray-dried cefpodoxime proxetil nanosuspension. *Chem. Pharm. Bull., 58,* 912–917.

102. Gao, Y., Li, Z., Sun, M., Guo, C., Yu, A., Xi, Y., Cui, J., Lou, H., & Zhai, G., (2011). Preparation and characterization of intravenously injectable curcumin nanosuspension. *Drug Deliv., 18,* 131–142.

103. Nasser, M. S., Twaiq, F. A., & Onaizi, S. A., (2013). Effect of polyelectrolytes on the degree of flocculation of papermaking suspensions. *Sep. Purif. Technol., 103,* 43–52.

104. Barany, S., (2004). Szepesszentgyorgyi, A. Flocculation of cellular suspensions by polyelectrolytes. *Adv. Colloid Interface Sci., 111,* 117–129.

105. Bose, S., Schenck, D., Ghosh, I., Hollywood, A., Maulit, E., & Ruegger, C., (2012). Application of spray granulation for conversion of a nanosuspension into a dry powder form. *Eur. J. Pharm. Sci., 47,* 35–43.

106. Wu, L. B., Zhang, J., & Watanabe, W., (2011). Physical and chemical stability of drug nanoparticles. *Adv. Drug Deliv. Rev., 63,* 456–469.

107. Dodiya, S. S., Chavhan, S. S., Sawant, K. K., & Korde, A. G., (2011). Solid lipid nanoparticles and nanosuspension formulation of saquinavir: preparation, characterization, pharmacokinetics and biodistribution studies. *J. Microencapsul., 28,* 515–527.

108. Wang, Y., Zheng, Y., Zhang, L., Wang, Q., & Zhang, D., (2013). Stability of nanosuspensions in drug delivery. *Journal of Controlled Release, 172,* 1126–1141.

109. Sun, W., Mao, S. R., Shi, Y., Li, L. C., & Fang, L., (2011). Nanonization of Itraconazole by High Pressure Homogenization: stabilizer Optimization and Effect of Particle Size on Oral Absorption. *J. Pharm. Sci., 100,* 3365–3373.

110. Lee, J., Lee, S. J., Choi, J. Y., Yoo, J. Y., & Ahn, C. H., (2005) Amphiphilic amino acid copolymers as stabilizers for the preparation of nanocrystal dispersion. *Eur. J. Pharm. Sci., 24,* 441–449.

111. Lou, H. Y., Gao, L., Wei, X. B., Zhang, Z., Zheng, D. D., Zhang, D. R., et al., (2011). Oridonin nanosuspension enhances anti-tumor efficacy in SMMC-7721 cells and H22 tumor bearing mice. *Colloids Surf., B, 87,* 319–325.

112. Teeranachaideekul, V., Junyaprasert, V. B., Souto, E. B., & Muller, R. H., (2008). Development of ascorbyl palmitate nanocrystals applying the nanosuspension technology. *Int. J. Pharm., 354,* 227–234.

113. Gao, L., Liu, G. Y., Ma, J. L., Wang, X. Q., Zhou, L., & Li, X., (2012). Drug nanocrystals: *in vivo* performances. *J. Control Release, 160,* 418–430.

114. Choi, J. Y., Yoo, J. Y., Kwak, H. S., Nam, B. U., & Lee, J., (2005). Role of polymeric stabilizers for drug nanocrystal dispersions. *Cur. Appl. Phys., 5,* 472–474.

115. Chiang, P. C., Ran, Y., Chou, K. J., Cui, Y., & Wong, H., (2011). Investigation of utilization of nanosuspension formulation to enhance exposure of 1,3-dicyclohexylurea in rats: preparation for PK/PD study via subcutaneous route of nanosuspension drug delivery. *Nanoscale Res. Lett., 6,* 413.

116. Ezhilarasi, P. N., Karthik, P., Chhanwal, N., & Anandharamakrishnan, C., (2013). Nanoencapsulation techniques for food bioactive components: a review. *Food Bioprocess Technol., 6*, 628–647.

117. Kim, J. H., Jang, S. W., Han, S. D., Hwang, H. D., & Choi, H. G., (2011). Development of a novel ophthalmic ciclosporin A-loaded nanosuspension using top-down media milling methods. *Pharmazie., 66*, 491–495.

118. Lo, C. L., Lin, S. J., Tsai, H. C., Chan, W. H., Tsai, C. H., Cheng, C. H., et al., (2009). Mixed micelle systems formed from critical micelle concentration and temperature-sensitive diblock copolymers for doxorubicin delivery. *Biomaterials., 30*, 3961–3970.

119. Wang, H., Pan, Q., & Rempel, G. L., (2011). Micellar nucleation differential microemulsion polymerization. *Eur. Polym., J. 47*, 973–980.

120. Singh, S. K., Srinivasan, K. K., Gowthamarajan, K., Singare, D. S., Prakash, D., & Gaikwad, N. B., (2011). Investigation of preparation parameters of nanosuspension by top-down media milling to improve the dissolution of poorly water-soluble glyburide. *Eur. J. Pharm. Biopharm., 78*, 441–446.

121. Deng, J. X., Huang, L., & Liu, F., (2010). Understanding the structure and stability of paclitaxel nanocrystals. *Int. J. Pharm., 390*, 242–249.

122. Rachmawati, H., Al Shaal, L., Muller, R. H., & Keck, C. M., (2013). Development of curcumin nanocrystal: physical aspects. *J. Pharm. Sci., 102*, 204–214.

123. George, M., & Ghosh, I., (2013). Identifying the correlation between drug/stabilizer properties and critical quality attributes (CQAs) of nanosuspension formulation prepared by wet media milling technology. *Eur. J. Pharm. Sci., 48*, 142–152.

124. Liu, G. P., Zhang, D. R., Jiao, Y., Zheng, D. D., Liu, Y., Duan, C. X., et al., (2012). Comparison of different methods for preparation of a stable riccardin D formulation via nano-technology. *Int. J. Pharm., 422*, 516–522.

125. Xiong, R. L., Lu, W. G., Li, J., Wang, P. Q., Xu, R., & Chen, T. T., (2008). Preparation and characterization of intravenously injectable nimodipine nanosuspension. *Int. J. Pharm., 350*, 338–343.

126. Sun, M., Gao, Y., Pei, Y., Guo, C., Li, H., Cao, F., et al., (2010). Development of nanosuspension formulation for oral delivery of quercetin. *J. Biomed. Nanotechnol., 6*, 325–332.

127. Salazar, J., Heinzerling, O., Muller, R. H., & Moschwitzer, J. P., (2011). Process optimization of a novel production method for nanosuspensions using design of experiments (Do E). *Int. J. Pharm., 420*, 395–403.

128. Verma, S., Gokhale, R., & Burgess, D. J., (2009). A comparative study of top-down and bottom-up approaches for the preparation of micro/nanosuspensions. *Int. J. Pharm., 380*, 216–222.

129. Kakran, M., Sahoo, N. G., Li, L., Judeh, Z., Wang, Y., Chong, K., & Loh, L., (2010). Fabrication of drug nanoparticles by evaporative precipitation of nanosuspension. *Int. J. Pharm., 383*, 285–292.

130. Kakran, M., Sahoo, N. G., Li, L., Judeh, Z., & Panda, P., (2011). Artemisinin-polyvinylpyrrolidone composites prepared by evaporative precipitation of nanosuspension for dissolution enhancement. *J. Biomater. Sci. Polym. Ed., 22*, 363–378.

131. Chen, X., Matteucci, M. E., Lo, C. Y., Johnston, K. P., & Williams, R. O., (2009). Flocculation of polymer stabilized nanocrystal suspensions to produce redispersible powders. *Drug Dev. Ind. Pharm.*, *35*, 283–296.

132. Ahuja, M., Dhake, A. S., Sharma, S. K., & Majumdar, D. K., (2011). Diclofenac-loaded Eudragit S100 nanosuspension for ophthalmic delivery. *J. Microencapsul.*, *28*, 37–45.

133. Chiang, P. C., Alsup, J. W., Lai, Y., Hu, Y., Heyde, B. R., & Tung, D., (2009). Evaluation of aerosol delivery of nanosuspension for pre-clinical pulmonary drug delivery. *Nanoscale Res. Lett.*, *4*, 254–261.

134. Stark, B., Pabst, G., & Prassl, R., (2010). Long-term stability of sterically stabilized liposomes by freezing and freeze-drying: effects of cryoprotectants on structure. *Eur. J. Pharm. Sci.*, *41*, 546–555.

135. Niwa, T., Miura, S., & Danjo, K., (2011). Design of dry nanosuspension with highly spontaneous dispersible characteristics to develop solubilized formulation for poorly water soluble drugs. *Pharm. Res.*, *28*, 2339–2349.

136. D'Addio, S. M., & Prud'homme, R. K., (2011). Controlling drug nanoparticle formation by rapid precipitation. *Adv Drug Deliv Rev. 63*(6), 417–426.

137. Zhang, H. F., Wang, D., & Butler, R., (2008). Formation and enhanced biocidal activity of water-dispersable organic nanoparticles. *Nat Nanotechnol.*, *3*(8), 506–511.

138. Mezhericher, M., Naumann, M., Peglow, M., Levy, A., Tsotsas, E., & Borde, I., (2012). Continuous species transport and population balance models for first drying stage of nanosuspension droplets.*Chem. Eng. J.*, *210*, 120–135.

139. Ho, H., & Lee, J., (2012). Redispersible drug nanoparticles prepared without dispersant by electro-spray drying.*Drug Dev. Ind. Pharm.*, *38*, 744–751.

140. Van Eerdenbrugh, B., Froyen, L., Van Humbeeck, J., Martens, J. A., Augustijns, P., & Van den Mooter, G., (2008). Drying of crystalline drug nanosuspensions—the importance of surface hydrophobicity on dissolution behavior upon redispersion. *Eur. J. Pharm. Sci.*, *35*, 127–135.

141. Mauludin, R., Muller, R. H., & Keck, C. M., (2009). Kinetic solubility and dissolution velocity of rutin nanocrystals. *Eur. J. Pharm. Sci.*, 36, 502–510.

142. Dixit, R., & Puthli, S., (2009). Fluidization technologies: Aerodynamic principles and process engineering. *J Pharm Sci.*, *98*, 3933–3960.

143. Shen, B., Shen, C., Yuan, X., Bai, J., Lv, Q., Xu, H., et al., (2013). Development and characterization of an orodispersible film containing drug nanoparticles. *Eur J Phar Biopharm.*, *85*, 1348–1356.

144. Liu, Y., Scharf, D., Graule, T., & Clemens, F. J., (2014). Granulation processing parameters on the mechanical properties of diatomite-based porous granulates. *Powder Tech.*, *263*, 159–167.

145. Sarnes. A., Kovalainen, M., Häkkinen, M. R., Laaksonen, T., Laru, J., Kiesvaara, J., et al., (2014). Nanocrystal-based per-oral itraconazole delivery: Superior *in vitro* dissolution enhancement versus Sporanox® is not realized in *in vivo* drug absorption. *J Control Rel.*, *180*, 109–116.

146. Bose, S., Schenck, D., Ghosh, I., Hollywood, A., Maulit, E., & Ruegger, C., (2012). Application of spray granulation for conversion of a nanosuspension into a dry powder form. *Eur J Pharm Sci.*, *47*, 35–43.

147. Sievens-Figueroa, L., Bhakay, A., Jerez-Rozo, J. I., Pandya, N., Romanach, R. J., Michniak-Kohu, B., et al., (2012). Preparation and characterization of hydroxypropyl methyl cellulose films containing stable BCS class-II drug nanoparticles for pharmaceutical applications. *International Journal of Pharmaceutics, 423*, 496–508.

148. Muller, R. H., & Jacobs, C., (2002). Production and characterization of a budesonide nanosuspension for pulmonary administration. *Pharm. Res., 19*, 189–194.

149. Ponchel, M., Montisci, J., Dembri, A., Durrer, C., & Duchene, D., (1997). Mucoadhesion of colloidal particulate systems in the gastrointestinal tract. *Eur. J. Pharm. Biopharm., 4*, 25–31.

150. Aungst, B. J. (1993). Novel formulation strategies for improving oral bioavailability of drugs with poor membrane permeation or presystemic metabolism. *J. Pharm. Sci., 82*, 979–986.

151. Kusuhara, H., Suzuki, H., & Sugiyama, Y., (1998). The role of p-glycoprotein and canalicular multispecific organic anion transporter in the hepato-biliary excretion of drugs. *J. Pharm. Sci., 87*, 1025–1040.

152. Benet, L. Z., Izumi, T., Zhang, Y., Silverman, J. A., & Wacher, V. J., (1999). Intestinal MDR transport proteins and P-450 enzymes as barriers to oral drug delivery. *J. Control Release, 62*, 25–31.

153. Jain, K. K. (2008). *The Handbook of Nanomedicine, 1st edn.*; Humana Press: New York.

154. Kumar, M. P., Rao, Y. M., & Apte, S., (2008). Formulation of nanosuspensions of albendazole for oral administration. *Curr. Nanosci., 4*, 53–58.

155. Kayser, O., Olbrich, C., Yardley, V., Kinderlen, A. F., & Croft, S. L., (2003). Formulation of amphotericin B as nanosuspension for oral administration. *Int. J. Pharm., 254*, 73–75.

156. Zhang, D., Tan, T., Gao, L., Zhao, W., & Wang, P., (2007). Preparation of azithromycin nanosuspensions by high pressure homogenization and its physicochemical characteristics studies. *Drug Dev. Ind. Pharm., 33*, 569–575.

157. Hernandez-Trejo, N., Kayser, O., Steckel, H., & Muller, R. H., (2005). Characterization of nebulized buparvaquone nanosuspensions-effect of nebulization technology. *J. Drug Target., 13*, 499–507.

158. Zhao, Y. X., Hu, H. Y., Chang, M., Li, W. J., Zhao, Y., & Liu, H. M., (2010). Preparation and cytotoxic activity of hydroxyl camptothec in nanosuspensions. *Int. J. Pharm., 392*, 64–71.

159. Eerdenbrugh, B. V., Froyen, L., Humbeeck, J. V., Martens, J. A., Augustijns, P., & Mooter, G. V. D., (2008). Alternative matrix formers for nanosuspension solidification dissolution performance and X-ray microanalysis as an evaluation tool for powder dispersion. *Eur. J. Pharm. Sci., 35*, 344–353.

160. Mahesh, K. V., Singh, S. K., & Gulati, M., (2014). A comparative study of top-down and bottom-up approaches for the preparation of nanosuspensions of glipizide. *Powder Technol., 256*, 436–449.

161. Tanaka, Y., Inkyo, M., Yumoto, R., Nagai, J., Takano, M., & Nagata, S., (2009). Nanoparticulation of poorly water-soluble drugs using a wet-mill process and physicochemical properties of the nanopowders. *Chem. Pharm. Bull., 57*, 1050–1057.

162. Douroumis, D., & Fahr, A., (2007). Stable carbamazepine colloidal systems using the cosolvent technique. *Eur. J. Pharm. Sci.*, *30*, 367–374.

163. Tam, J. M., Mcconville, J. T., Williams III, R. O., & Johnston, K. P., (2008). Amorphous cyclosporin nanodispersions for enhanced pulmonary deposition and dissolution. *J. Pharm. Sci.*, *97*, 4915–4933.

164. Looareesuwan, S., Chulay, J. D., Canfield, C. J., & Hutchinson, D. B., (1999). Atovaquone and proguanil hydrochloride followed by primaquine for treatment of Plasmodium vivax malaria in Thailand. *Trans. R. Soc. Trop. Med. Hyg.*, *93*, 637–640.

165. Liversidge, G. G., & Cundy, K. C., (1995). Particle size reduction for improvement of oral bioavailability of hydrophobic drugs. I Absolute oral bioavailability of nanocrystalline danazole in beagle dogs. *Int. J. Pharm.*, *127*, 91–97.

166. Liversidge, G. G., & Conzentino, P., (1995). Drug particle size reduction for decreasing gastric irritancy and enhancing absorption of naproxen in rats. *Int. J. Pharm.*, *125*, 309–313.

167. Remon, J. P., Vergote, G. J., Vervaet, C., Driessche I., Hoste, S., Smedt, S., et al., (2001). An oral controlled release matrix pellet formulation containing nanocrystalline ketoprofen. *Int. J. Pharm.*, *219*, 81–87.

168. Merisko-Liversidge, E., Liversidge, G. G., & Cooper, E. R., (2003). Nanosizing: a formulation approach for poorly-water-soluble compounds. *Eur. J. Pharm. Sci.*, *18*, 113–120.

169. Peters, K., Muller, R. H., Borner, K., Hahn, H., Leitz, K. S., Diederichs, J. E., et al., (2000). Preparation of a clofazimine nanosuspension for intravenous use and evaluation of its therapeutic efficacy in murine Mycobacterium avium infection. *J. Antimicrob. Chemother.*, *45*, 77–83.

170. Pignatello, R., Bucolo, C., Spedalieri, G., Maltese, A., & Puglisi, G., (2002). Flurbiprofen-loaded acrylate polymer nanosuspensions for ophthalmic application. *Biomaterials.*, *23*, 3247–3255.

171. Pignatello, R., Bucolo, C., Ferrara, P., Maltese, A., Puleo, A., & Puglisi, G., (2002). Eudragit RS100 nanosuspensions for theophthalmic controlled delivery of ibuprofen. *Eur. J. Pharm. Sci.*, *16*, 53–61.

172. Kreuter, J., Petrov, V. E., Kharkevich, D. A., & Alyautdin, R. N., (1997). Influence of the type of surfactant on the analgesic effects induced by the peptide dalargin after 1st delivery across the blood–brain barrier using surfactant coated nanoparticles. *J. Control Release, 49*, 81–87.

173. Kayser, O., (2001). A new approach for targeting to Cryptosporidiumparvum using mucoadhesive nanosuspensions: research and applications. *Int. J. Pharm.*, *214*, 83–85.

174. Patel, M., Shah, A., & Patel, K. R., (2011). Nanosuspension: A novel approach for drug delivery system. *JPSBR, 1*(1), 1–10.

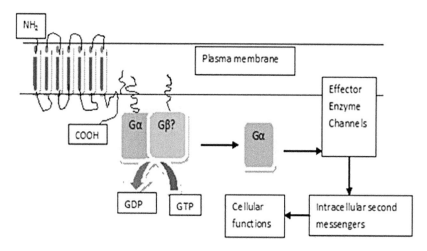

FIGURE 1.1 G-protein-coupled receptor (GPCR).

	A/B (AF1)	C (DBD)	D (Hinge)	E (LBD)	F (AF2)	
Amino end						Carboxyl end

FIGURE 1.3 Nuclear receptor and its structure.

AF-1: Activation factor 1, it is a ligand-independent factor and responsible for gene activation

DBD: DNA binding domain is composed of two highly conserved zinc fingers and responsible for targeting the receptor to highly specific DNA sequences comprising a hormone response element (RE), it also includes the hinge region.

LBD: Ligand binding domain, capable of binding to small lipophilic molecules such as steroids, retinoids, and vitamins which regulate the activity of these receptors.

AF-2: Activation factor 2, it is a legend dependent and responsible for the gene activation function.

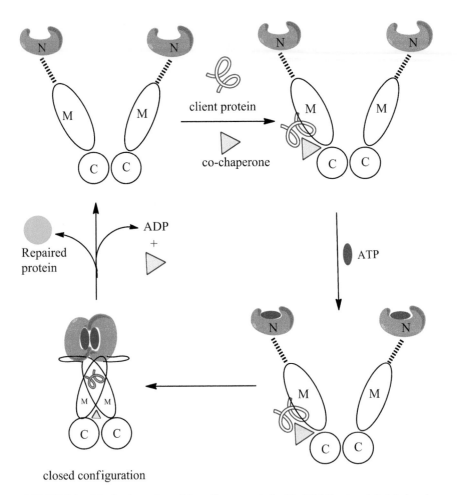

FIGURE 3.2   Mechanism of repairing client protein by Hsp90 (client protein binds prior to ATP).

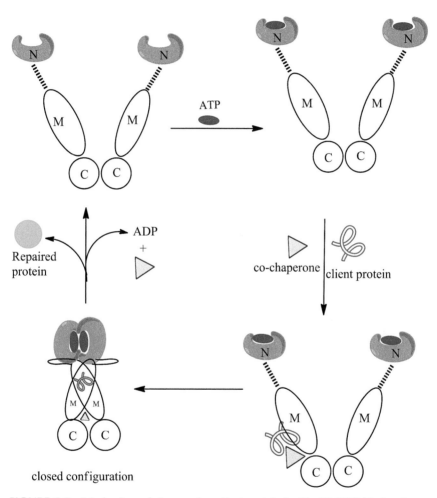

FIGURE 3.3 Mechanism of chaperoning client protein by Hsp90 (ATP binds prior to client protein).

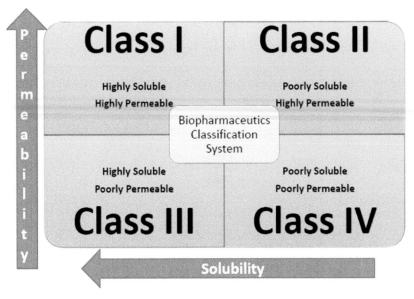

FIGURE 5.1   Biopharmaceutics classification system.

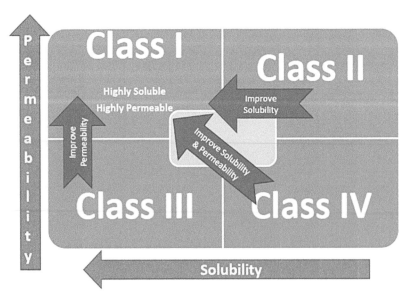

FIGURE 5.2   Conversion to BCS class I.

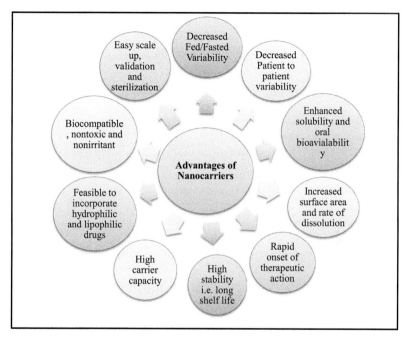

FIGURE 5.3   Advantages of nanocarrier systems.

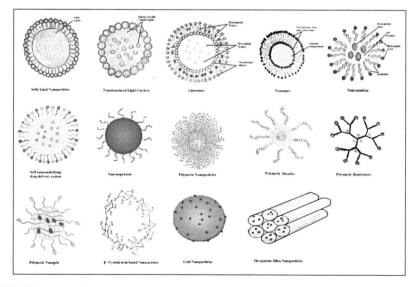

FIGURE 5.5   Different nanocarriers for oral solubility and bioavailability enhancement.

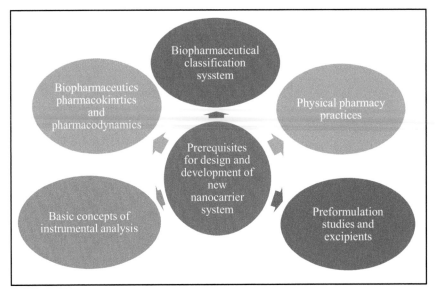

FIGURE 5.6   Prerequisites for design and development of new nanocarrier system.

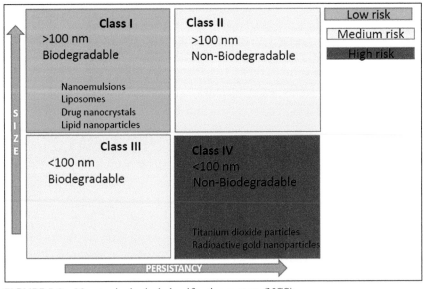

FIGURE 5.9   Nanotoxicological classification system (NCS).

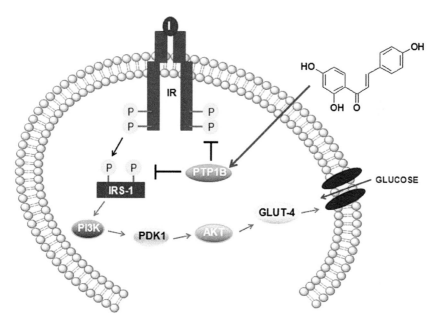

FIGURE 6.1   The physiological role of PTP-1B in glucose metabolism.

CHAPTER 5

NANOCARRIER TECHNOLOGIES FOR ENHANCING THE SOLUBILITY AND DISSOLUTION RATE OF API

ASHWINI DESHPANDE and TULSHIDAS S. PATIL

SVKM's NMIMS, School of Pharmacy and Technology Management, Shirpur, Dist. Dhule, Maharashtra, India, E-mail: Ashwini.deshpande@ nmims.edu; tulshidaspatil01@gmail.com

Enteral route of drug administration, though conventional, is still a preferred route as it is non-invasive, convenient and has better patient compliance. However, many promising drug candidates become unsuitable to be administered by this conventional, a popular route because of low solubility. The reason behind the inconsistent absorption of the BCS Class II drugs, is poor aqueous solubility. Generally, the rate-limiting step for absorption of BCS Class II drugs is the dissolution rate. Moreover, in the era of reverse engineering, a large number of new drug molecules with varied therapeutic potentials are being discovered through combinatorial screening programs, but more than 40% of them are poorly soluble in water and that is why most of them are still in the developmental process [1, 2]. By improving the solubility of such promising candidates, both, Pharma industry and consumers will be benefited. Achieving the minimum required solubility to apply for the regulatory waiver will reduce the time to reach the drug to the market. Consumers will be benefited with the therapeutic potential of the drug. Moreover, chances of dose reduction due to improved solubility cannot be overlooked.

The prerequisite for a drug to absorb in the systemic circulation is, it must be in solution form. The solubility of a drug can be defined as its amount that has passed into solution when equilibrium is attained between the solution and excess undissolved drug at a given temperature and pressure. The extent of solubility of a drug in a given solvent is measured as

TABLE 5.1   Descriptive Terms for Solubility

Descriptive Term	Parts of Solvent Required for 1 Part of Solute
Very soluble	Less than 1
Freely soluble	From 1 to 10
Soluble	From 10 to 30
Sparingly soluble	From 30 to 100
Slightly soluble	From 100 to 1000
Very slightly soluble	From 1000 to 10,000
Practically insoluble, or insoluble	Greater than or equal to 10,000

the saturation concentration where further addition of solid drug does not increase its concentration in the solution. Descriptive solubility terms as given in USP are mentioned in Table 5.1.

Solubility should not be confused with dissolution. These two are separate phenomena. The former is static and the later one dynamic. It has been well explained that solubility, dissolution and gastrointestinal permeability are fundamental parameters that control bioavailability [3]. First-pass metabolism and susceptibility to efflux mechanisms, are also important parameters influencing oral bioavailability of drugs. Modification in drug's permeability is not possible as it requires the addition or deletion of some functional groups in drugs' structure ultimately leading to changes in activity. This suggests that to be bioavailable, a drug must exhibit optimum solubility as well as permeability. Based on solubility and permeability, the drugs, e.g., therapeutic agents can be classified into four groups as per the Biopharmaceutics Classification System (BCS) as shown in Figure 5.1.

So BCS Class II drugs are targeted candidates to enhance solubility. BCS Class IV drugs also exhibit poor aqueous solubility, but they are poorly permeable too.

The drug is said to be highly soluble when the highest strength of the drug is soluble in 250 mL or less of the aqueous medium over the pH range of 1–7.5. The drug is said to be highly permeable when the extent of intestinal absorption is demonstrated to be 90% or higher based on the mass balance or in comparison to an intravenous reference dose. Various strategies have been employed to enhance the solubility of the drug and can be classified as chemical and physical modifications. Prodrugs and

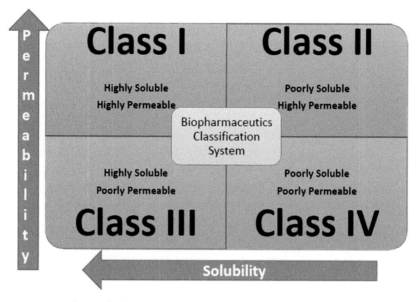

FIGURE 5.1   (See color insert.) Biopharmaceutics classification system.

salt formation are examples of chemical modification. Physical modification includes particle size techniques such as micronization, nanonization, polymorphism/pseudo polymorphism, inclusion complexes (cyclodextrins, conjugation to dendrimers), solubilization (using surfactants or an addition of co-solvents) and solid solutions or solid dispersions using hydrophilic carriers [4–8].

Each method enlisted above has its own proven pros and cons. An abundant literature is available on these methods. There are three major mechanisms involved in these methods:

- increasing the effective surface area of the particles;
- changing highly organized crystal structure; or
- surrounding the hydrophobic molecule with hydrophilic moieties.

Increasing the effective surface area of the hydrophobic drugs may sometimes lead to development of static charges on the particles or in some cases may increase the reactivity of the drug leading to its deactivation. Restricted use of organic solvents in recrystallization process and stability conditions required for metastable polymorphs limits crystal structure changes. One can use hydrophilic carriers and surfactant micelles provided they are in GRAS and IIG guidelines.

Enhancing the solubility may solve the problem of BCS II drugs, but still BCS III & IV drugs having poor permeability could not be used for the patients in spite of their best effects.

Looking at the above scenario, it is always better to encapsulate the drug in biodegradable and biocompatible carriers of nano sizes. It will also protect the drug against degradation by gastrointestinal fluids moreover absorption through the gastrointestinal epithelium or lymphatic transport can be enhanced. In short, there is a need to convert BCS Class II, III & IV to BCS Class I (see Figure 5.2) with judicial selection of excipients and modification of particle size of the formulation.

Before selecting the effective carrier system for drug, one should consider the important factors regarding API and oral route of administration:

1. A molecular mass should not be greater than 500 Daltons;
2. An octanol-to-water partition coefficient (log P) value should not be greater than 5;

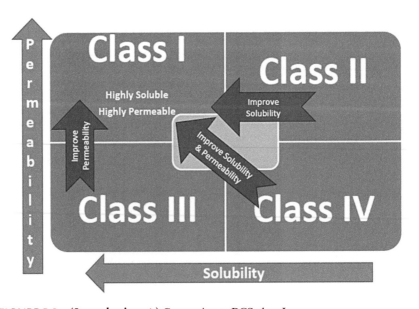

FIGURE 5.2   (See color insert.) Conversion to BCS class I.

3. pH of the GI tract: there is a significant change in pH according to location, pH of 1.2 in the empty stomach, 5–7 in the small intestine and 6 to 7.5 in the colon.

4. The physical barriers of intestinal epithelium which consists of absorptive cells (enterocytes), secretory cells (goblet cells and paneth cells) and M cells of Peyer's patches. The physical integrity of these intestinal epithelial cells is exhibited by tight junctions between the contiguous epithelial cells.

5.1 NANOTECHNOLOGY

The term Nano is derived from a Latin word meaning dwarf. It is 1/1,000,000,000th part of a meter. One can better correlate this with angstrom unit. One nanometer equals 10 angstroms.

The National Nanotechnology Initiative (NNI) defines nanotechnology as follows:

Nanotechnology is science, engineering, and technology conducted at the nanoscale, which is about 1 to 100 nanometers. Nanoscience and nanotechnology are the study and application of extremely small things and can be used across all the other scientific fields, such as chemistry, biology, physics, materials science, and engineering.

While reading this chapter, some related technical terms will come across.

Let us know more about these terms.

1. Nanotechnology can be defined as the controlled modification of at least one-novel/superior characteristics of the material/device/system by reducing the size of the material to nanometer scale – featuring between 10^{-9} to 10^{-7} meters.

2. Nanomedicines: Application of nanotechnology in the interest of health and wellbeing of life. This branch aims at a comprehensive diagnosis, monitoring, repair and improvement of all human/animal biological systems, working from the molecular level using engineered devices and nanostructures to achieve medical benefit.

European Science Foundation defines Nanomedicine as 'the science and technology of diagnosing, treating and preventing disease and traumatic injury, of relieving pain, and of preserving and improving human health, using molecular tools and molecular knowledge of the human body.'

This definition was revised by the US NIH as: 'nanomedicine refers to highly specific medical intervention at the molecular scale for curing diseases or repairing damaged tissues, such as bone, muscle, or nerve.'

3. Pharmaceutical nanotechnology: Application of nanotechnology in design development of drug delivery devices, diagnostics, imaging and biosensors. Pharmaceutical nanotechnology is an emerging area now-a-days, providing cost and time effective diagnostic tools and focused treatment of disease at a cellular level. The field also offers various avenues for disease treatment, detection of the antigen associated with different diseases such as cancer, diabetes mellitus, neurodegenerative diseases, and detection of the microorganisms and viruses associated with infections.

Based on the material used for the preparation, shapes/forms the material is transforming and their use, some potential advantages of nanocarriers systems are discussed in Figure 5.3.

Various classes of nanocarriers for oral drug delivery to enhance solubility and bioavailability are described in Figure 5.4.

Although mainstream nanotechnology expresses particles between 1 to 100 nm in diameter, the size of the individual particles tested for drug delivery of therapeutic and imaging agents may range from 2 to 1000 nm (Figure 5.5).

The current research is primarily focusing on the nanoformulations with particle size less than 200 nm because of following limitations:

1. Tumor capillaries occasionally exceed the diameter of 300 nm.
2. The nanoparticles having the size greater than 200 nm are able to activate the human complement system and may be cleared off from blood by kupffer cells.
3. Particles larger than 150 nm can get filtered in the liver via fenestrae in the sinus endothelium.

5.2 PREREQUISITES FOR DESIGN AND DEVELOPMENT OF NEW NANOCARRIER SYSTEM

Different prerequisites for the design and development of the new nanocarrier system are depicted in Figure 5.6.

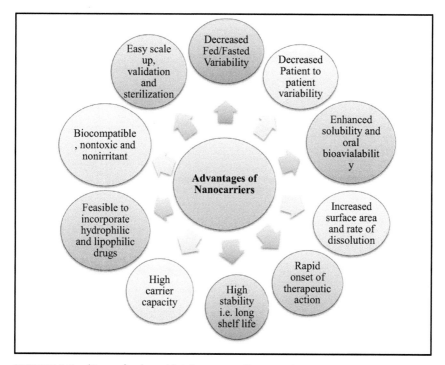

FIGURE 5.3 (**See color insert.**) Advantages of nanocarrier systems.

5.3 LIPID AND SURFACTANT BASED SYSTEMS

5.3.1 INTRODUCTION

The slow dissolution process of the poorly water-soluble drugs is the major factor of its poor bioavailability. Lipid-based formulations enhance the poor bioavailability of these drugs by following mechanisms:

i) preferably by keeping these drugs in dissolved state throughout transit state in the GI tract;
ii) avoid chemical and enzymatic degradation in aqueous environment of the GI tract; and
iii) promote lymphatic transport and avoid systemic first-pass metabolism.

Classification of nanocarriers for oral drug delivery to enhance solubility and bioavailability

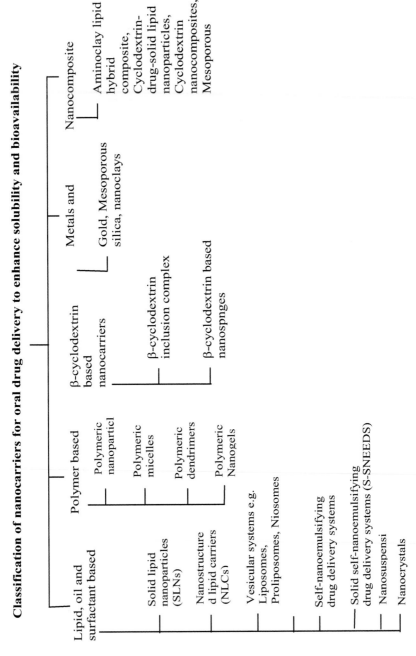

FIGURE 5.4 Classification of nanocarriers for oral drug delivery to enhance solubility and bioavailability.

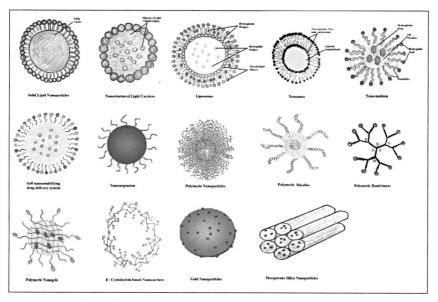

FIGURE 5.5 **(See color insert.)** Different nanocarriers for oral solubility and bioavailability enhancement.

5.3.2 FORMULATION ASPECTS

From formulation point of view, to develop lipid-based formulations following factors should be taken into account before selection of the excipients:

- miscibility of the excipients
- the ability of self-dispersion of the excipients
- digestibility of the lipids
- regulatory related issues like purity, toxicity, chemical stability, irritancy, and cost.

Lipid-based formulations primarily involves the excipients like:

- pure triglyceride oils, mono- and diglycerides;
- substantial proportion of hydrophilic or lipophilic surfactants; and
- cosolvents (Reviewed in reference [9–13]).

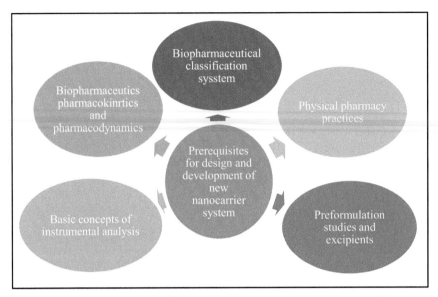

FIGURE 5.6   (See color insert.) Prerequisites for design and development of new nanocarrier system.

5.3.3   OIL PHASE: MONO, DI, AND TRI-GLYCERIDES

Triglycerides include long, medium and short chain triglycerides. Long chain triglycerides (fixed oil) are able to enhance the lymphatic transport of the drugs and bypass the hepatic first-pass metabolism as compared to medium chain tri, di and monoglycerides. Medium chain mono and diglycerides possess greater solubilization and absorption potential as compared to long chain triglycerides. Also, the medium chain glycerides are less prone to oxidation and have a high solvent capacity and promote emulsification as compared to long chain triglycerides. Thus, the mixture of long chain and medium chain triglycerides (mixed glycerides) can be used to balance the drug loading and emulsification abilities. Mixed glycerides can be obtained from the partial hydrolysis of the vegetable oils. Polar oils (e.g., span 85) also enhances the solvent capacity and promotes emulsification.

Examples of oily phases include:

- Long chain triglycerides; e.g., hydrogenated soybean oil, hydrogenated vegetable oil, corn oil, peanut oil, soybean oil, olive oil, sesame oil.

- Medium chain triglycerides include caprylic/capric triglycerides; e.g., Miglyol 810, 812, Labrafac CC, Crodamol GTCC, Captex 300, 355, triacetin (Captex 500).
- Medium chain mono and di-glycerides include caprylic/capric acids; e.g., Capmul MCM, Imwitor 742, Akoline MCM.
- Long chain monoglycerides ex. glyceryl monooleates, propylene fatty acid esters, and fatty acid esters. D-α-Tocopherol (vitamin E) was used to solubilize the drugs like paclitaxel, saquinavir, itraconazole, etc. [12].

5.3.4 WATER-INSOLUBLE SURFACTANTS

Lipid excipients those form film at oil-in-water interface and have HLB value between 8–12 can be used as water-insoluble surfactants. These substances form micelles, but they don't self-emulsify because of lack of their hydrophilic character. Most commonly used water-insoluble surfactants include polyoxyethylene (20) glyceryl trioleate (Tagot-TO), with its average HLB values between 11–11.5 and a blend of tween-80 with span-80 with HLB value 11.

5.3.5 WATER SOLUBLE SURFACTANTS

These are the excipients with HLB value greater than 12 and these possess ability to self-emulsify and thus used in self-emulsifying drug delivery systems. These synthesized as a mixture of polyethylene glycols (PEG) with hydrolyzed vegetable oils. Also, the reactions of ethylene oxide with alcohol and sorbitan esters give cetostearyl alcohol ethoxylates (cetomacragol) and ether ethoxylates. Cremophor, is another class of compound widely used as a water-soluble surfactant. It includes ethoxylated hydrogenated castor oil (Cremophor RH 40 and RH 60) and non-hydrogenated castor oil, e.g., ethoxylated castor oil (Cremophor EL).

5.3.6 CO-SURFACTANTS

Co-solvents are commonly used to enhance the solubilization, solvent capacity and aid in the self-emulsification of the system. They also aid the dissolution of hydrophilic surfactants. Commonly used co-surfactants are ethanol, glycerol, propylene glycol, polyethylene glycol, transcutol, etc.

5.3.7   OTHER ADDITIVES

Various lipid soluble anti-oxidants such as α-tocopherol, β-carotene, propyl gallate, butylated hydroxyl toluene (BHT) or butylated hydroxyl anisole (BHA) can be used to prevent the oxidation.

5.4   SOLID LIPID NANOPARTICLES (SLNS)

SLNs are solid matrix and can be described as parenteral emulsions or aqueous colloidal dispersions in which liquid oil is replaced by a solid-lipid. Pros and Cons of this system are enlisted in Table 5.2.

5.4.1   FORMULATION ASPECTS [14–19]

5.4.1.1   SOLID LIPIDS

These are biocompatible, biodegradable lipids which are solid at room and body temperature. The lipids can be used within the range of 5–40%. Examples include:

- Triglycerides (trimyristin, tripalmitin, tristearin).
- Polymers like soya phosphatidyl choline, PLGA, sodium alginate, chitosan, etc.
- Glyceryl monostearate (GMS) (Imwitor® 900).
- Glyceryl behenate (Compritol® 888 ATO, Geleol™).
- Glyceryl palmitostearate (Precirol® ATO 5).
- Wax (cetyl palmitate).

5.4.1.2   SURFACTANTS

Different surfactants are used within the range of 0.5–5%. Tween 80, poloxamer 188, lecithin, polyglycerol methylglucose distearate (Tego-Care® 450), tyloxapol, sodium cocoamphoacetate Miranol® Ultra C32) or saccharose fatty acid esters, etc. are few examples employed for SLNs. Methods of preparation for SLN are shown in Table 5.3.

TABLE 5.2   Pros and Cons of Solid Lipid Nanoparticles

Sr. No.	Pros	Cons
1.	Use of biodegradable polymers and avoidance of organic solvents decreases the risk of acute and chronic toxicity.	Poor drug loading capacity due to crystalline structure of solid lipids.
2.	Protection of chemically labile actives in the GI tract.	Polymorphic transitions and formation of perfect crystals* occur on storage which expels the drug from SLNs.
3.	Improved solubility and bioavailability of poorly water-soluble drugs. Improved penetrability into skin and localization in different skin layers.	Relatively possess high water content in the system.
4.	The manufacturing process like high shear homogenization is easy to scale-up with low cost.	Unpredictable gelling tendency.
5.	Better stability as compared to liposomes with possible lyophilization.	Initial burst release is observed.
6.	Site-specific drug delivery with injectable route.	

* Explained in Limitation section.

5.4.1.3   DRUG INCORPORATION INTO SLN$_S$

Various drug molecules can be incorporated into SLNs. Three different models are mentioned in the literature for drug incorporation into SLNs as follows

1. *Solid solution model:* In the cold homogenization method, drugs having strong interaction with lipids get dispersed in lipid matrix.
2. *Drug enriched shell model:* In the hot homogenization method, lipid core is formed in the recrystallization temperature of lipid. After the cooling step, drug gets repartitioned into the lipid phase in the surrounding membrane.
3. *Drug enriched core model:* After dispersion cooling, supersaturation of the drug occurs which is dissolved in the lipid. Drug further gets precipitated in melted lipid and finally, continued cooling leads to recrystallization of the lipid core enriched with drug [18].

TABLE 5.3   Methods of Preparation for SLNs

Nanocarrier System	Method of Preparation	References
Solid Lipid Nanoparticles (SLNs)	High shear homogenization i. Hot homogenization ii. Cold homogenization	[20–22]
	Ultrasonication/High-speed homogenization by probe sonication or Bath sonication	[23]
	Solvent emulsification/evaporation	[14]
	Solvent emulsion/diffusion	[24]
	Ultrasound-Assisted Emulsion/Evaporation Method	[21]
	Solvent diffusion/evaporation technique	[15]
	Microemulsion method	[25, 26]
	Supercritical fluid extraction method	[27]
	Spray drying method	[20]
	Double emulsion method	[28]
	Hot homogenization and ultrasonication	[16, 17]
	Film formation method following hot homogenization and ultrasonication	[17]
	Membrane contactor	[29]
	Electrohydrodynamic Spraying	[30]
	Solvent injection	[20]
	Emulsion evaporation and low temperature solidification	[31]

The important prerequisite for drug loading is that the drug should have higher solubility than expected solubility in lipid melt because it decreases in cooling melt and further gets decreased in solid lipids. The SLNs are used for oral delivery of highly lipophilic drugs to enhance their solubility and bioavailability [32].

5.4.1.4 CASE STUDIES

Olmesartan medoxomil (OLM), an antihypertensive moiety exhibits low oral bioavailability (28%) because of poor aqueous solubility, presystemic metabolism and P-glycoprotein mediated efflux. Efforts were invested by the researchers to improve its oral bioavailability using lipid nanocarriers.

SLNs were manufactured using the solvent emulsion evaporation method and effects of improved factors were analyzed using regression analysis and pareto charts. Optimized SLNs consisted glyceryl monostearate as lipid, soya phosphatidylcholine as co-emulsifier and tween 80 as a surfactant. 2.32 fold increment in relative bioavailability of the drug from SLN was observed as compared OLM plain drug [14].

Rifabutin is a major drug in the treatment of tuberculosis infections and in Mycobacterium avium complex (MAC) infection in immunocompromised individuals and HIV patients. In order to develop an oral sustained release formulation, rifabutin was incorporated in glyceryl monostearte to formulate SLN (RFB-SLNs). RFB-SLNs were found to be stable in gastric medium. Sustained release was observed up to 48 hrs and 7 days in simulated intestinal fluid (6.8 pH) and physiological buffer (PBS pH 7.4). Effective therapeutic drug concentration was maintained for 4 and 7 days in plasma and in tissues, e.g., lungs, liver, and spleen. SLNs showed fivefold enhanced bioavailability as compared to plain drug [15]. Another research showing enhancement of bioavailability of carvedilol employing coated SLNs. As the drug is highly lipid soluble, is prone to hepatic first-pass. Though, SLNs are most promising delivery systems for the enhancement of bioavailability, but gives burst release of drug in acidic environment which prevents its usage as oral delivery system. The alternative method is intraduodenal administration of SLN of the drug, but is clinically an inappropriate route for repeated administration of drugs to patients. In order to prevent the burst effect from SLNs, coating of Carvedilol SLNs was carried out using positively charged N-carboxymethyl chitosan (NCC). Monoglyceride was used as lipid, poloxamer 188 and soya lecithin as surfactants and stearyl amine to impart positive charge. Problem of burst release from SLNs in the gastric acidic environment is overcome by NCC coating. In addition, NCC coating was proved to have improved bioavailability of carvedilol compared to the uncoated SLNs [16].

In another recent study by Janga and Dudhipala, oral bioavailability of poorly water-soluble drug zaleplon was enhanced by 2.66-fold by implementing Box-Behnken design to develop solid lipid nanoparticles [17].

5.4.2 LIMITATIONS

Immediately after the formation of SLNs, the lipids get transformed into highly energetic α and β' crystalline form. But on storage with time, they may get transformed into low energetic forms like βi and β. This is responsible in the formation of perfect crystalline lattice which allow very small space for drug molecules. This results in expulsion of the encapsulated drug during storage. To overcome these limitations associated with SLNs, alternative nanostructured lipid carriers (NLCs) were emerged [33].

5.5 NANOSTRUCTURED LIPID CARRIERS (NLCS)

NLCs are the second generation SLNs in which solid lipids are replaced with a mixture of solid and liquid lipids.

5.5.1 DIFFERENT STRUCTURES OF NLCS AND THEIR COMPOSITIONS

NLCs formed in three different structures viz. imperfect type, formless type, and multiple type [19, 34].

1. **Imperfect type:** Provides imperfectly structured solid matrix. These types of NLCs can be formed by mixing spatially different lipids, for example, glycerides consisting of different fatty acids. These various fatty acid chains provide large distances and imperfections to accommodate drug molecules.

2. **Amorphous type:** This provides structureless solid amorphous matrix. These types of NLCs can formulated with a mixture of solid lipids with special lipids, for example, isopropylmyristate, hydroxyl octacosanyl hydroxystearate, or medium chain triglycerides; e.g., miglyol 812. Thus, this special structure of lipid blend forms amorphous matrix instead of crystalline one and avoids the drug expulsion.

3. **Multiple type:** This class formulates the NLCs similar to w/o/w emulsions. Drugs those have solubility higher in liquid lipids as compared to solid lipids are incorporated in this type of NLCs. The system also avoids the decomposition of drug from solid lipids.

Das and co-workers done the comparative evaluation of SLNs and NLCs of clotrimazole to check whether NLCs are better than SLNs. In their findings, they have reported better stability for NLCs with high drug loading, especially at 25°C as compared to SLNs [33]. Table 5.4 shows methods of preparation of NLCs.

Though NLCs and SLNs are extensively studied for topical drug delivery, efforts also have invested in recent days to employ NLCs and SLNs to enhance the oral solubility and bioavailability of highly lipophilic molecules. Some case studies are discussed further.

5.5.2   CASE STUDIES

Vimpocetine (VIN) is a poorly water-soluble drug, short half-life ~2 hrs and undergo extensive first-pass metabolism exhibits poor bioavailability. Pan and co-workers developed vimpocetine (VIN-NLCs). They reported the sustained *in vitro* drug release with no initial burst release. Oral bioavailability of drug in nanocarrier system was checked in Wistar rats. NLCs exhibited 332% relative bioavailability as compared to conventional VIN

TABLE 5.4   Methods of Preparation for NLCs

Nanocarrier System	Method of Preparation	References
	Hot high-pressure homogenization method	[35–45]
	Hot high-pressure homogenization Followed by ultrasonication method OR vice versa	[46, 47]
	Melt and ultrasonication method	[48, 54]
	Melt emulsion method	[47, 55, 57]
	Melt-emulsification and homogenization method	[58, 59]
	Film ultrasound method	[60, 61]
Nanostructured lipid carriers (NLCs)	Ultrasonication	[48, 62]
	Double emulsification and melt dispersion method	[63]
	Emulsion evaporation and low temperature solidification	[64–67]
	Solvent diffusion	[68–70]
	Solvent evaporation	[71]
	Microemulsion method	[72]
	Emulsion solvent diffusion and evaporation method	[73]
	Emulsion evaporation at a high temperature and solidification at a low temperature	[74]
	Nanoprecipitation/solvent diffusion method	[75]

suspension. Thus, a significant enhancement in the oral bioavailability of poorly water-soluble drug was noticed from the NLC system [42].

Oral bioavailability of sirolimus (SRL), a poorly water-soluble immunosuppressant, was improved by encapsulating it into lipids-based nanostructured lipid carriers (NLCs) Sirolimus loaded nanostructured lipid carriers (SRL-NLCs) were manufactured by a modified high-pressure homogenization method by using glycerol distearate (PRECIROL ATO–5) as the solid lipid, oleic acid as the liquid lipids, and Tween 80 as an emulsifier. 1.82 fold improved oral bioavailability of SRL-NLCs was observed as compared to commercial SRL tablets (Rapamune) in beagle dogs [76].

Trans-Ferulic acid (TFA) loaded nanostructured lipid carriers (NLCs) were prepared by microemulsion (ME)-based method using ethyl oleate as the liquid lipid and glyceryl behenate as the solid lipid. In *in vivo* pharmacokinetic studies, significant improvement in the oral bioavailability in terms of Cmax and AUC was observed in TFA loaded NLCs [77].

In other studies, Nirmal Shah and co-workers developed NLCs for raloxifene (BCS class II drug with extensive first-pass metabolism and oral bioavailability ~2%). In their findings they reported 3.75 fold increment in the oral bioavailability of the raloxifene NLCs as compared to plain drug suspension [78].

5.6 LIPOSOMES

These are lipid vesicles composed of a bilayer and/or concentric series of multiple bilayers of lipids separated by aqueous compartments formed by amphiphilic molecules such as phospholipids that encloses a a central aqueous compartments.

Possible mechanisms behind the enhanced absorption and, in turn, the bioavailability of liposomes [79].

- Increases the membrane fluidity thereby increasing the transcellular absorption.
- Increases the paracellular transport by opening the tight junctions
- Possibly inhibit P-glycoprotein and cytochrome P450 to increase intracellular concentration of API. Pros and cons of liposomes are mentioned in Table 5.5.

TABLE 5.5   Pros and Cons of Liposomes [80, 81]

Sr. No.	Pros	Cons
1.	Enhanced stability and therapeutic index by encapsulating the drug.	Low solubility and stability in GI tract
2.	Biocompatible and complete biodegradable, non-toxic, flexible and nonimmunogenic for systemic and non-systemic administrations.	Short circulating half-life*
3.	Used for active targeting by coupling with site-specific ligands	Leakage from vesicles may occur
4.	Reduce the toxicity and exposure of sensitive tissue to the drugs; e.g., amphotericin B	High production cost

* Explained in the following point.

5.6.1   FORMULATION ASPECTS

5.6.1.1   LIPIDS AND/OR PHOSPHOLIPIDS

Lipids simply consist of fatty acid molecules with differences in head moieties. Triglycerides (explained above) are the lipids with three fatty acids and a glycerol molecule. Phospholipids are similar to triglycerides except polar phosphate containing group at the first hydroxyl group of glycerol in place of the fatty acid group in triglycerides. Phospholipids are amphiphilic moieties ex. Phosphatidyl choline (PC) (lecithin), phosphatidyl ethanolamine (cephalin), phosphatidyl serine (PS), phosphatidyl inositol (PI), phosphatidyl Glycerol (PG), etc. Phosphatidylcholine from natural sources like egg or soya phosphatidylcholine are less stable and more permeable as compared to synthetic PCs; e.g., dipalmitoylphosphatidylcholine [80, 82].

5.6.1.2   STEROLS

Cholesterol is most widely used sterol able to form get incorporated into the phospholipid membrane in molar ratio 1:1 or even 2:1 of cholesterol to phospholipids. The primary function of cholesterol is to provide stability by modulating the fluidity of lipid bilayer. They impart steric hindrance to the movement of phospholipids and prevent their crystallization [83]. Methods of preparation for liposomes are given in Table 5.6.

TABLE 5.6   Methods of Preparation for Liposomes

Methods	References
1. Active loading	
i. Imparting pH gradient	[84]
2. Passive loading	
a. Mechanical dispersion method	
I. Sonication	[83]
II. French Pressure cell extrusion	[85]
III. Lipid film hydration	[80]
III. Micro-emulsification	[80]
IV. Membrane extrusion	[80]
b. Solvent dispersion method	
I. Ether injection	[86]
II. Ethanol injection	[87]
III. Reverse phase evaporation	[88]
c. Detergent removal method	
I. Dialysis	[80, 81]
II. Dilution	[81]

5.6.1.3   STEALTH LIPOSOMES

Liposomes are engulfed by macrophages when the target is beyond mono-nuclear phagocyte system (MPS), thus consequently removed from circulation. This is a major limitation of liposomal carriers. Thus, to improvise the blood circulation time of liposomes, the concept of stealth liposomes came into existence. Poly (ethylene glycol) (PEG) coated liposomes are referred as stealth liposomes. PEG is widely used as a polymeric steric stabilizer. PEG can be incorporated into liposomes via a cross-linked lipid (i.e., PEG distearoyl phosphatidyl ethanolamine (DSPE) [89]. The efficacy of these stealth liposomes is further explained in case studies.

5.6.1.4   CASE STUDIES

The PEGylated lyophilized liposomal formulation was developed for paclitaxel. This formulation has shown enhanced solubility of drug and physicochemical stability of the drug loaded liposomes as compared to marketed formulation Taxol®. A significant increase in the solubility and entrapment efficiency was observed by adding 5% v/v PEG 400 in hydration medium, whereas the use of sucrose as a lycoprotectant during lyophilization enhanced the physicochemical stability of paclitaxel [90].

PEGylated lyophilized paclitaxel containing liposomes were formulated. The use of 3% v/v of tween 80 in hydration medium depicted enhanced solubility of the drug whereas sucrose served the purpose of lycoprotectant in lyophilization process which enhanced the stability of formulation. Formulation showed the biological half-life of 17.8 hrs and increased uptake in tumor tissues. The formulation was found equipotent compared to taxol after 72 hrs due to slower drug release [91].

5.6.2   PROLIPOSOMES

To overcome the stability related issues of liposomes, proliposomes came into existence. Proliposomes are advantageous over liposomes as they are dry, free-flowing fine granules which are more stable during sterilization and storage. These systems consist of water-soluble carriers coated with phospholipids which forms liposomes upon reconstitution.

5.6.2.1   MECHANISM OF SOLUBILITY AND BIOAVAILABILITY ENHANCEMENT

The improvement of oral solubility, absorption and thus the bioavailability by proliposomes could be due to presence small their small size, presence of bile salts in GI tract interact with phospholipids to form mixed micelles and enhance the solubility of poorly soluble drugs.

5.6.2.2   METHODS OF PREPARATION OF PROLIPOSOMES

Different methods are used for the fabrication of proliposomes, to name of few:

1. Film deposition method using rotary flask evaporator [79, 92].
2. Film dispersion-freeze-drying [93].
3. Surface adsorption and solvent evaporation.

Surface adsorption on to the carriers like mannitol [94], microcrystalline cellulose (Avicel PH 102) [95] followed by evaporation of solvent.

5.6.2.3   CASE STUDIES

Proliposomes of zaleplon were prepared for improved oral delivery. Hydrogenated soyphosphatidylcholine (HSPC) and cholesterol (CHOL) were employed in varying ratios. To obtain positively and negatively charged proliposomal particles, final formulation was customized with stearylamine and dicetyl phosphate, respectively and evaluated. In order to check release rate and stability of drug, *in vitro* release test was performed. Results of solid-state characterization revealed that the drug was transformed to amorphous or molecular state from the native crystalline form. *In vivo* study of *in situ* single-pass perfusion and bioavailability in rats showed substantial enhancement of the effective permeability coefficient (Peff) and rate and extent of absorption from positively charged proliposomal formulation specifies the importance of surface charge for effective uptake across the gastrointestinal tract. Compared to control, 2 to 5 folds enhancement in bioavailability suggest the potential of proliposomes as suitable carriers for improved oral delivery of poorly soluble actives [92].

Polyphase dispersed proliposomes were prepared to improve oral bioavailability of dehydrosilymarin – an oxidized form of herbal drug silymarin. They were formulated by film dispersion-freeze-drying method and were consisting soybean phospholipids, cholesterol, isopropyl myristate, and sodium cholate. Proliposomes were characterized to endorse size and encapsulation efficiency, which were 7 to 50 nm and $81.59\% \pm 0.24\%$, respectively. The *in vitro* release of drug-loaded proliposomes was slow in acidic pH, but was continually increasing in pH 6.8, and finally reached 86.41% at 12 h. The relative bioavailability of proliposomes versus conventional suspension in rabbits was 228.85%, proving proliposomes as a useful vehicle for oral delivery of dehydrosilymarin, a drug poorly soluble in water [93]. Proliposomes for bioavailability enhancement were successfully prepared for valsartan, silymarin, nisoldipine, progesterone, and *Ginkgo biloba* [79, 94–97].

5.7   NIOSOMES

Unlike liposomes, they are made up of non-ionic surface active agents and are bilayered. With the input of appropriate energy like heat, ultrasound, hand-shaking the amphiphilic molecules get self-assembled into closed bilayers. The presence of a mixture of surfactants and charge inducing agents responsible for the formation of thermodynamically stable vesicles.

5.7.1   FORMULATION COMPONENTS

5.7.1.1.   NON-IONIC SURFACTANTS

Being an amphiphilic molecule, one portion is hydrophilic and the other is hydrophobic. These two portions can be linked by ether, amide or ester bond. Materials like amino acids, fatty acids, amides, alkyl esters, and alkyl ether surfactants can be used to prepare non-ionic surfactant vesicles.

5.7.1.2   CHOLESTEROL

The physical properties of the niosomal structures largely depend on cholesterol. It is given in the literature that, up to 30–50 mol % addition of cholesterol is required in order to form the stable niosomal structure. It is advisable to increase the cholesterol concentration as the HLB values of surfactants increase above 10 in order to compensate the effect of larger head groups on the critical packing parameter (CPP).

5.7.1.3   CHARGED MOLECULES

In addition to the abovementioned two components, charged molecules may be used to prevent aggregation of niosomes. Dicetyl phosphate (DCP) and phosphatidic acid are examples of negatively charged molecules, whereas stearylamine (SA) and cetylpyridinium chloride are positively charged molecules. 2.5–5 mole % concentration of charged molecules is suggested (for more information on niosomes refer Carafa, 2014 [98] and case studies discussed below).

5.7.1.4   CASE STUDIES

With the purpose of improving oral bioavailability of acyclovir niosomes were prepared with conventional thin film hydration method. The molar ratio of 65:60:5 cholesterol, span 60, and dicetyl phosphate, respectively was used for the preparation of niosomes. The average vesicle size was 0.95 μm with percentage entrapment of ~11% of drug used in the hydration process was estimated. Photomicrographs of niosomes were found to be unilamellar spherical shape. *In vitro* drug release profile was found to be Higuchian for free and niosomal drug with significant retarded release compared with free drug from niosomes. The average relative bioavailability of the drug from the niosomal dispersion in relation to the free solution in rabbits after single dose of 40 mg/kg was 2.55 indicating more than 2-fold increase in drug bioavailability. The mean residence time (MRT) of acyclovir reflected sustained release characteristics. The study was concluded saying, niosomal formulation could be a favorable delivery system for acyclovir with enhanced relative bioavailability and prolonged drug release profiles [99].

Similar results were reported for carvediol, clarithromycin, and levofloxacin where researchers have employed biocompatible materials such as bile salts, creatinine, and sugar, respectively for preparation of niosomes. Griseofulvin and tenofovir disoproxil fumarate niosomes were also studied and shown similar results mention in above case studies of pro/niosomes [100–104].

5.8 PRONIOSOMES

These are dry free-flowing granular product which upon addition of water, disperse or dissolves to form a multilamellar niosome suspension suitable for administration by oral or other routes. Proniosomes are developed to overcome the problems related to physical stability of niosomes like leaking, fusion, aggregation, etc. In addition, niosomes also enhance the solubility of poorly water-soluble drugs, control the drug release, and prolongs the circulation of entrapped drugs.

5.8.1 METHODS OF PREPARATION

1. Direct mixing of surfactant, drug and water-soluble carriers into absolute ethanol to form a slurry followed by evaporation of solvent [105].
2. Film deposition on carrier method [106].

5.8.2   CASE STUDIES

Free-flowing and stable proniosome formulation was developed for poorly water-soluble drugs vinpocetine; to improve its oral bioavailability using cholesterol, Span 60, and sorbitol by employing less noxious and more effective slurry method. Proniosomes of vinpocetine and its suspension after reconstitution were characterized and compared. Excellent flowability was exhibited by the proliposome. The reconstituted niosomes showed spherical morphology with smooth surface under transmission electron microscope (TEM) with high drug entrapment efficiency (89.67±3.28%). X-ray diffraction (XRD) revealed that the drug was in an amorphous or a molecular state in proniosome powder. *In vitro* release study in phosphate buffer pH 7.2 showed an enhanced release rate compared to vinpocetine suspension. Proniosome derived niosomes were found to keep their integrity and stability at different pH conditions of the GI tract. The in situ single-pass intestinal perfusion studies indicated that encapsulation of vinpocetine into niosomes largely improved the absorption of vinpocetine. From the study it was concluded that the proniosomes enhances the oral bioavailability of vinpocetine and are stable precursors for the immediate preparation of niosomes [105].

Efforts were invested by scientists to improve oral bioavailability of isradipine by converting it proniosomes and to evaluate the influence of oral bioavailability of the drug in albino Wistar rats. Various molar ratios of the different nonionic surfactants like Span20, Span40, Span60, and Span80 with a membrane stabilizing agent cholesterol were used to prepare proniosomes by film deposition on carrier technique. Dicetylphosphate was employed as charge inducer. Proniosomes were characterized using different techniques such as scanning electron microscopy (SEM), for surface topography, Fourier transform infrared spectroscopy, differential scanning calorimetry, and X-ray diffractometry to understand the solid state properties. Results revealed that prepared proliposomes have a higher dissolution rate compared to pure drug. *Ex vivo* permeation studies of these proliposomes were assessed from flux and permeation coefficient. It was stated by the researchers that enhancement ratios were significantly higher for proniosomes compared with control. The pharmacokinetic parameters were also evaluated in male albino Wistar rats. The results vindicated their stand with substantial improvement in the bioavailability (2.3 fold) from optimized proniosome formulation compared with oral suspension. The proniosome formulations were stable when stored at 4°C [106].

5.9 EMULSION BASED DRUG DELIVERY: NANOEMULSIONS, SELF-NANOEMULSIFYING DRUG DELIVERY SYSTEMS (SNEEDS) AND SOLID SELF-NANOEMULSIFYING DRUG DELIVERY SYSTEMS (S-SNEEDS)

Emulsions are considered as thermodynamically unstable formulations containing at least two immiscible liquid phases where one is dispersed in another by mechanical means. There is tremendous increase in a free energy of the system due to increase in interfacial surface area because of nanonization of a droplet. In order to stabilize such systems in its existing droplet size, surfactant/co-surfactants must be selected judicially. To plan the formulation strategy for development of emulsions, microemulsions and nanoemulsions, the relation between the work required (W) in terms of mechanical energy, interfacial area (ΔA) and surface tension (γ) must be understood. This can be explained by the equation given below,

$$W = \Delta A \times \gamma \qquad \text{(i)}$$

Thus, to generate smaller droplet size (i.e., higher interfacial area) more amount of mechanical energy is required [107].

5.10 NANOEMULSIONS

These are initially referred as mini-emulsions, sub-micron emulsions, ultrafine emulsion, and unstable microemulsions, and can be defined as fine oil/water or water/oil dispersion stabilized by an interfacial film of suitable surfactant molecules, having a droplet size range 20–200 nm.

Unfortunately, nanoemulsions are confused with microemulsions and are wrongly used as misnomers. To understand the concept of nanoemulsions, many reviews and articles are available in the literature. Research articles further help the readers to thoroughly understand the concept. The important aspects regarding nanoemulsions are explained in the following subsections.

5.10.1   THREE BASIC DIFFERENCES BETWEEN MICROEMULSIONS AND NANOEMULSIONS

- The term nanoemulsions specifically indicates the nanoscale range of the droplets which is distinctly differ from term microemulsions.
- Microemulsions are considered as thermodynamically stable preparations, at the same time though nanoemulsions are optically transparent, they are considered as thermodynamically unstable systems [11, 108].
- Microemulsions possess a high concentration of surfactants (~20% or higher), whereas nanoemulsions can be fabricated with low concentration of surfactants (3–10%) [11, 108]. Pros and cons of nanoemulsions are mentioned in Table 5.7.

TABLE 5.7   Pros and Cons of Nanoemulsions

Sr. No.	Pros	Cons
1.	Long-term colloidal stability, which in-turn gives long shelf life to the nano-carrier system [11]	Not suitable to formulate as a controlled drug delivery system.
2.	Easy manufacturing and scale-up [11]	To deliver by oral route, palatability of nanoemulsion components and compatibility of the drug with other excipients are the major concerns.
3.	Nanoemulsions demonstrated improved oral bioavailability [109–111].	
4.	Improved topical (dermal, transdermal and mucosal) drug delivery [112].	
5.	Greater esthetic and cosmetic compliance for nanoemulsions based gels [11].	
6.	Nanoemulsions used as templates to fabricate nanoparticulate systems like SLNs, polymeric nanoparticles etc. [113].	

5.10.2   FORMULATION ASPECTS OF NANOEMULSIONS

Before selecting the components to develop the nanoemulsions, it is important to know that all components must be pharmaceutically acceptable for oral drug delivery and must fall under generally regarded as safe (GRAS) category. The major components involved in the formulation of nanoemulsions are oils, water, surfactants and co-surfactants.

5.10.3 OIL PHASE

Judicious selection of the oil phase is required based on its intended use and nature and dose of active pharmaceutical ingredient. Molecular weight of oil is large enough to avoid Ostwald ripening. The examples of commonly used oils are capryol, sefsol, triacetin, castor oil, coconut oil, linseed oil, mineral oil, olive oil, peanut oil, and PUFA rich oils like safflower oil and soybean oil [114, 115].

5.10.4 SURFACTANTS AND CO-SURFACTANTS

Purpose of surfactants is to provide good stability against coalescence.

Ideal requirements of surfactants are:

– They must be able to lower interfacial surface tension to a very low value (below 10 dynes/cm).
– They must form the complete deformable flexible film which is of sufficient lipophilic character to avoid coalescence of the globules.
– They must impart appropriate zeta potential and viscosity to the carrier system for development of optimally stable formulations.
– They must be nontoxic and their organoleptic properties should be compatible with the product.

These surfactants will either form monomolecular multimolecular or particulate film [114]. Examples of commonly used surfactants are spans and tweens, hydrophilic colloids; e.g., acacia and finely divided solids like bentonite and veegum, labrasol, labrafil, labrafac, cremophor, peceol, etc.

To develop the o/w emulsion, the HLB value of nanoemulsions should be greater than 10. Here, the right blend of surfactants or surfactants and cosurfactants is needed. Also, the complete intact and flexible film cannot be formed by use of single surfactant. Propylene glycol, PEG 200, PEG 400, ethanol, etc. can be used as co-surfactants.

5.10.5   MECHANISM OF FORMATION OF NANOEMULSIONS

The mechanism of formation of nanoemulsions, can be understood better by two terms viz. Laplace pressure (π_L) and shear stress (τ) and their relationship with radius (r) of the globules. When we mix two immiscible liquids, there exists an interfacial tension (σ). This interfacial tension always acts to minimize the interfacial surface area. When we add surfactants, they exerts short range repulsion, e.g., disjoining pressure, which is able to prevent droplet coalescence [116]. In low volume of dispersed phase, when surfactants with high solubility in dispersed phase added, tend to form spherical globules of radius 'r.'

$$\text{Laplace pressure } (\pi_L) = 2\sigma/r \qquad \text{(ii)}$$

$$r \, \alpha \, \sigma/\tau \qquad \text{(iii)}$$

$$\text{i.e., } r = \sigma \, \eta_c \, \gamma'$$

where η_c is viscosity and γ' is shear rate.

According to equation ii, Laplace pressure (π_L), e.g., pressure exerted on the molecules inside the droplets is inversely proportional to the radius of the droplet. Thus droplets with smaller radius exerts higher Laplace pressure. The relationship between radius of the droplets and shear stress (τ) is given in equation (iii), which depicts that larger shear stress needs to be applied to develop the droplets with smaller size [115]. Methods of preparation of nanoemulsions are enlisted in Table 5.8.

5.10.6   CASE STUDIES

Kinetically stable nanoemulsions of artemether were prepared by ultrasonication technique with improved solubility, stability and oral bioavailability. In the preparation of nanoemulsions, drug was dissolved in internal oil phase (consisted coconut oil and span 80) and in water as an external phase, tween 80 and ethanol were dissolved. *In vitro* drug release of the drug from the nanoemulsion formulation was found significantly higher as compared to the plain drug. From pharmacokinetic studies, 2.6 fold increase in oral bioavailability of the nanoemulsion formulation was observed as compared to plain drug [109].

TABLE 5.8   Method of Preparation of Nanoemulsions

Nanocarrier system	Sr. No.	Method of preparation	References
	1.	High energy emulsification methods	
	i.	Ultrasonication	[109, 111]
	ii.	High-speed stirring followed by probe sonication	[110]
	iii.	High-pressure homogenization	[115, 117, 118]
Nanoemulsions		Using Microfluidics	
		Using high-pressure homogenizers	
	2.	Low energy emulsification methods	
	i.	Phase inversion temperature method	[114, 115, 119, 120]
	ii.	Solvent displacement method	[115, 121]
	iii.	Phase inversion composition method	[115]

Stable Itraconazole (ITR) animation was developed to enhance its oral bioavailability. ITR nanoemulsion was formulated using Capmul MCM C8 as oil, Cremophore EL as surfactant and Pluronic F68 as co-surfactant with high-speed stirring, followed by probe sonication. More than 2 folds increase in pharmacokinetic parameters like Cmax and AUC of the nanoemulsion was found as compared to the plain drug suspension confirming enhanced bioavailability of itraconazole. The developed nanoemulsion was found to be stable at both, refrigerated and room temperature conditions. [110] Similar kind of efforts was made to enhance the bioavailability of the drugs like paclitaxel [111], breviscapine [122], baicalin [123], danazol [124], resveratrol [125, 126], curcumin [127], megestrol acetate [128], colchicine [129] by developing nanoemulsion based drug delivery system.

5.11   SELF-NANOEMULSIFYING DRUG DELIVERY SYSTEMS (SNEDDS)

SNEDDS can be defined as the physically stable isotropic mixtures of natural and synthetic oil with solid or liquid surfactants, co-surfactants and solubilized drug which possesses a unique property of forming oil-in-water (o/w) emulsion on mild agitation, followed by dilution with aqueous media (e.g., GI fluid).

The required agitation is provided by digestive motility in the GI tract. In addition to aforementioned mechanisms of bioavailability enhancement of lipid-based carriers, the SNEDDS, because of their small droplet size, enhances the activity of pancreatic lipase to hydrolyze triglycerides (oils) and, in turn, responsible for formation of mixed micelles of bile salts containing drug and enhances the drug release [130]. Pros and cons of SNEDDS are highlighted in Table 5.9.

TABLE 5.9   Pros and Cons of Self-Nanoemulsifying Drug Delivery Systems (SNEDDS)

Sr. No.	Pros	Cons
1.	Nearly 100% entrapment of drug is achieved.	In case of hard and soft gelatin capsules, the cross-linking of gelatin with unsaturated lipids may occur, resulting in the aldehyde formation [131]
2.	Enhanced physical stability of the formulation, patient compliance by filling it in unit dosage forms like hard or soft gelatin capsules, hydroxypropyl methyl cellulose (HPMC) capsules.	Brittleness or softening of the capsules may occur due migration of solvents from fill to shell [131]
3.	Because of anhydrous nature, provide enhanced stability for hydrolytically susceptible drugs.	Being a liquid formulation, degradation kinetics are higher for hydrophobic drugs compared to solid formulations [11, 131].
4.	No palatability related issues as filled into capsules.	
5.	Dissolution step is bypassed and due to nano droplet size, the surface area for absorption is enhanced leading to increased absorption and bioavailability.	
6.	Formulation is stable against GI degradation.	
7.	The system shows consistent temporal effect and distributes the drug in lymphatic system further reduces the dose and dosing frequency of drug.	
8.	Ease in manufacturing and scale-up procedures.	

5.11.1 METHOD OF PREPARATION OF SNEDDS

The solubility studies of the drug are carried out in different lipids, surfactants and co-surfactants by adding excess amount of drug. The Pseudo ternary phase diagram is constructed to identify the self-emulsifying region and to select the optimum percentage of oils, surfactants and co-surfactants in the formulation of SNEDDS [132]. For constructing pseudoternary phase diagram, water titration method is widely used [133–136]. The ternary phase diagram can be drawn by different softwares like Origin Pro 8.6 [133], PCP Disso software V3.0 [137, 138] CHEMIX® ternary plot software [139, 140], and Tri plot v1–4 software [141].

5.11.2 CASE STUDIES

Efficient use of lipid formulation classification system (LFCS) type IIIA (35% hydrophilic surfactant and 65% lipid), is depicted by formulating SNEDDS for poorly water-soluble drug fenofibrate. Formulations containing hydrophilic surfactants (HCO30, cremophor EL, cremophor RH 40) demonstrated to have lower droplet size as compared to lipophilic surfactants TO–106V. Formulation found to have high solubility for fenofibrate due to the presence of a mixture of mono-, di-, and tri-glycerides in the formulation. SNEDDS significantly enhanced bioavailability of fenofibrate about 1.7 folds as compared to pure drug when it was maintained almost 90% in its solubilized form during digestion time of about 4 hrs [130].

With the objective of enhanced dissolution rate and oral bioavailability SNEDDS of grapefruit flavonoid naringenin (NRG) SNEDDS were developed NRG is an aglycone flavonoid from grapefruits, possesses anti-inflammatory, anti-carcinogenic, anti-lipid peroxidation and hepatoprotective activities. Its poor water solubility and the slow dissolution upon oral administration, restricts its use as therapeutic agent. The formulation was optimized based on the enhanced dissolution rate, optimal globule size and Polydispersity index (PDI). Improved solubility and bioavailability was noted when triacetin was used as oily phase, tween 80 as surfactant and transcutaol HP as co-surfactant [132].

Similar kind of efforts were done to enhance the bioavailability by using SNEEDS for poorly water-soluble drugs like lacidipine [142], lurasidone [133], Atazanavir [137], valsartan [139, 143], rosurvastatin [140, 144], Efavirenz [134], indirubin derivative (E 804) [145], Cyclovirobuxine D

[146], glyburide [147], irbesartan [148], avanafil [149], quercetin [150], Coenzyme (Q10) [151], and persimmon leaf extract [152].

5.12 SOLID-SNEDDS

To overcome the limitations of liquid SNEDDS, the concept of solid-SNEDDS was emerged. Solid SNEDDS are the dry, free-flowing powder system with enhanced stability as compared liquid SNDDS.

5.12.1 METHODS OF PREPARATION

The selection of the solidification method will depend upon the oily excipients used in the formulation, and physicochemical properties of therapeutic active and its compatibility with other excipients.

1. Spray drying using suitable carrier, for example, Aerosil 200, silicon dioxide, magnesium stearate [153–155].
2. Physical adsorption by using an inert carrier or adsorbent; e.g., dibasic calcium phosphate, lactose, microcrystalline cellulose, colloidal silicon dioxide and Neusilin [11, 141].
3. Direct trituration by using aeroperl [156].
4. Lyophilization [10].
5. Melt Extrusion/Extrusion Spheronization [10].
6. Melt granulation [10]

5.12.1.1 LIMITATIONS [10]

- Solidifying agent and its amount may affect the drug release.
- Degradation of drug may occur during solidification.
- Decrease in drug loading and assurance of content uniformity.

5.12.1.2 CASE STUDIES

To inhibit the HIV infection by lopinavir and to enhance its systemic bioavailability by targeting to the sanctuary site, e.g., lymphatic system,

QBD-optimized S-SNEDDS were developed by Bhoop et al. By defining QTTP, CQAs were earmarked. S-SNEDDS were prepared using maisine as lipid, tween 80 as emulgent, and transcutol HP as co-solvent. The system was adsorbed on aeroperl 300, a porous carrier and tablets were manufactured by direct compression. Matching *in vitro* dissolution profile, globule size and shape depicted the similarity between S-SNEDDS and L-SNEDDS. Augmented bioavailability by transporting the drug via lymphatic pathway was proved. Enhanced distribution of lopinavir to the lymphatic system showed site-specific inhibition of HIV. Increased values of AUC for S-SNEDDS was observed, which was supported by inhibition of the P-gp efflux and opening of tight junctions by surfactant tween 80 [138].

To enhance solubility and dissolution rate S-SNEDDS for olmesartan (OLM) was prepared using capryol 90 as oil, cremophor RH 40 as surfactant and transcutol HP as co-surfactant and system were further solidified using spray-drying technique with aerosol 200 as solid carrier. Optimized formulation showed an enhancement in solubility and *in vitro* dissolution of OLM. Drug plasma concentration from S-SNEDDS was found to be higher at all-time points compared to marketed product and plain drug. This revealed enhanced bioavailability of the drug [157]. Similar kind of efforts were done to enhance the solubility and bioavailability by using S-SNEDDS for drugs like ezetimibe and ezetimibe–simvastatin combination [11], darunavir [141], flurbiprofen [154], tacrolimus [155], avanafil [156], rosuvastatin calcium [158], dabigatran etexilate [159], etc.

5.12.2 *NANOCRYSTALS AND NANOSUSPENSIONS*

Nanocrystals are carrier-free colloidal delivery system having an interesting approach for solubility enhancement of poorly soluble drugs. These are typically stabilized with surfactants or polymeric steric stabilizers. A dispersion of drug nanocrystals in an outer liquid medium and stabilized by surface-active agents is so-called nanosuspensions. Though they are called as nanocrystals and are typically indeed crystalline, but depending on the production technique may also contain amorphous drug in part or in whole. Liquid-atomization based bottom-up techniques can often precipitate some amorphous material. Sometimes with other techniques, polymorphic changes may also take place, during nanocrystallization.

An important characteristic of API nanocrystal is that it is totally composed of API only and no other carrier material is required. Nanocrystals may be crystalline or in an amorphous state. In the latter case they are referred as "nanocrystals in amorphous state." When dispersed in liquid media, surfactants or polymeric stabilizers are needed to stabilize the nanocrystals, obtaining so-called nanosuspensions. [160, 161]

5.12.2.1   CASE STUDIES

Oridonin (Odn)–2 hydroxypropyl-β-cyclodextrin inclusion complexes (Odn-CICs) containing nanosuspensions were developed by solvent evaporation followed by wet media milling technique. Significant improvement in dissolution of oridonin through Odn-CICs was observed. Marked improvement in the intestinal effective permeability of drug was noticed in the presence of 2-hydroxypropyl-β-cyclodextrin (HP-β-CD) and poloxamer. Odn-nanosuspension demonstrated significantly increased in oral absorption with a relative bioavailability of 213.99% [162].

Nanosuspension was developed by precipitation–ultrasonication method for drug efavirenz using a soya lecithin and poloxamer 407 in combination, to control the particle size within a specific range. The optimized formulation was lyophilized to further improve the solubility and stability of nanosuspension. Nanosuspensions with a diameter of about 182.4 nm (±18.7 nm) were obtained. Significant improvement in dissolution rate and saturation solubility was achieved by nanosuspensions as compared to plain drug suspension [163].

Fenofibrate, BCS class II drug, with low aqueous solubility and associated oral bioavailability was formulated as nanocrystals. The nanosuspension of fenofibrate was formulated using poloxamer 188 as surfactant using probe sonication method. Lyophilization of nanosuspension was carried out with freeze-drying method by using mannitol as a cryoprotectant. The saturation solubility of optimized nanocrystals and plain drug at 0.5% and 1% SLS media was found to be 67.51 ± 1.5 µg/mL and 107 ± 1.9 µg/mL and 6.02 ± 1.51 µg/mL and 23.54 ± 1.54 µg/mL, respectively. The optimized nanocrystals and the pure drug displayed 73.89% and 8.53% *in vitro* drug release in 0.5% and 1% SLS media. Nanocrystals exhibited 4.73-fold enhancement in the relative bioavailability as compared to pure drug [164].

Zhonggui He and co-workers developed nanocrystals of nimodipine by microprecipitation and high-pressure homogenization followed by freeze-drying. Poloxamer F127 was used as surfactant and mannitol as a cryoprotectant. Nanocrystals with three different mean diameters were developed, e.g., 159, 503 and 833.3 nm by varying the centrifugation. The developed nanocrystal formulation was compared with marketed preparation Nimotop®. Significant improvement in aqueous solubility was noted in a size-dependent manner. There was low *in vitro* drug release was observed for nanocrystals as compared to marketed formulation. On the other hand, 2.6-fold increment in the oral bioavailability was observed for the nanocrystals with mean diameters 159 and 833.3 nm [165].

5.13   POLYMERIC NANOCARRIERS

These are stable colloidal structure and may be nanospheres or nanocapsules.

Polymeric nanospheres are matrix systems with adsorbed, dissolved, or dispersed active pharmaceutical ingredient (API) throughout the matrix. Polymeric nanocapsules are reservoir systems where the API core is surrounded by polymeric wall.

Polymeric nanocarriers are extensively used for oral drug delivery because they enhance the bioavailability of active molecules, possibly by improving the residence time in the GI tract by mucoadhesion and improve the absorption. They also protect the drugs from enzymatic and hydrolytic degradation. In the development of polymeric nanoparticles, drug release from the nanoparticles and biodegradation of the polymers are two important factors. The drug release from the nanoparticle depend either on:

- Solubility of drug in particulate system.
- Detachment of the adsorbed or surface bound drug.
- Drug diffusion from polymeric matrix or polymer erosion.
- Combination of diffusion or erosion process [166].

5.13.1   BIODEGRADABLE AND NON-BIODEGRADABLE POLYMERS

5.13.1.1   BIODEGRADABLE POLYMERS

Natural Biodegradable polymers: Different polysaccharides such as chitosan, alginate pectin, dextran.

Semisynthetic biodegradable polymers: Synthetic derivatives of chitosan such as thiolated chitosan and trimethyled chitosan have been developed [167].

Other synthetic biodegradable polymers include poly (lactide-co-glycolide) (PLGA), poly(lactide) (PLA), Poly(ε-caprolactone) (PCL), poly(lactide-co-caprolactone).

5.13.1.2 NONBIODEGRADABLE POLYMERS

Synthetic polymers such as polystyrene, poly (cyanoacrylates) like poly(isobutylcyanoacrylate) (PICBA), poly (isohexyl cyanoacrylates) (PIHCA), poly(n-butylcyanoacrylate) (PBCA), poly(acrylate) and poly(methacrylate) (Eudragit), poly (methyl methacrylate) (PMMA), and polyethyleneimine can be used for oral drug delivery.

5.13.1.3 CO-POLYMERS

Co-polymers with one part of poly(ethylene glycol) or polysaccharides were used to provide tunable surface properties and stabilize the nanoparticles without the need of additional surfactants. Various examples include poly(lactide)-poly(ethylene glycol), poly(lactide-co-glycolide)-poly(ethylene glycol), poly(epsilon-caprolactone)-poly(ethylene glycol), poly(hexadecylcyanoacrylate-co-polyethylene glycol cyanoacrylate).

5.13.1.4 STABILIZERS

Different stabilizers are employed in the fabrication of polymeric nanoparticles such as dextran, pluronic F68, poly(vinyl alcohol), various copolymers and tween® 20 or tween® 80 [168].

Methods of preparation of polymeric nanoparticles are enlisted in Table 5.10.

5.13.1.5 CASE STUDIES

Polymeric nanoparticles containing curcumin in polylactic-co-glycolic acid (PLGA) and PLGA–polyethylene glycol (PEG) (PLGA–PEG) blend were developed by a single-emulsion solvent-evaporation technique. Though more sustained release is observed with PLGA nanoparticles alone in comparison to PLGA-PEG nanoparticles better efficacy was observed with later. The mean half-life of curcumin was enhanced by approximately 4 and 6 h for PLGA and PLGA–PEG nanoparticles, respectively. There was decreased distribution and metabolism of curcumin observed, more specifically for PLGA–PEG nanoparticles. Cmax of curcumin was found to be increased by 2.9- and 7.4-fold for PLGA and PLGA–PEG nanoparticles, respectively. In comparison to curcumin aqueous suspension, curcumin-loaded PLGA and PLGA–PEG nanoparticles showed 15.6- and 55.4-fold enhancement in curcumin bioavailability. In comparison to curcumin PLGA nanoparticles, PLGA–PEG nanoparticles showed 3.5-fold enhanced bioavailability [187].

Paclitaxel, a BCS class IV drug, incorporated into folic acid-functionalized poly(D, L-lactide-co-glycolide) (PLGA) nanoparticles by the interfacial activity assisted surface functionalization technique. To study the oral absorption in terms of apparent permeability (P_{app}) of the PLGA and functionalized PLGA nanoparticles, *in vitro* model of Caco–2 cell monolayers was used. 5-fold and 8-fold enhancement in the oral P_{app} was observed for PLGA and folic acid functionalized PLGA nanoparticles. This proved the role of folic acid functionalized nanoparticles in enhancement in the oral absorption of drugs with poor bioavailability [188].

Trimethyl chitosan-cysteine conjugates were developed by Yin and co-workers for oral delivery of insulin. They reported enhanced intestinal mucoadhesion and permeability enhancement of insulin form nanocarrier system [189]. Similar kind of efforts was made to enhance the bioavailability and efficacy of insulin by incorporating it into polylactic acid-pluronic F_{127}-polylactic acid vesicles (PLA- F_{127}-PLA). The nanovesicle maintained blood glucose level for 18.5 hrs [190].

5.13.2 POLYMERIC MICELLES

These contain self-assembled amphiphilic block copolymers to form nanoscopic supramolecular core-shell structures in aqueous medium. They are versatile and can be tailored to solubilize API and release hydrophobic drugs in a controlled release manner. These systems consist of hydrophobic core which incorporates lipophilic drugs and hydrophilic corona which provides aqueous solubility and steric stability to the system [191]. Polymeric micelles were employed to deliver the hydrophobic molecules orally and to enhance their solubility, stability and bioavailability. (Explained in case studies) The co-polymers must be nontoxic, biodegradable, in contact with GI tract must be stable and able to self-assemble spontaneously [167]. Leroux et al. explained the micelle preparation, its stability and interaction with intestinal mucosa and different types of polymeric micelles.

5.13.2.1 STRATEGIES TO DEVELOP POLYMERIC MICELLES

When moderately hydrophobic poloxamers such as poly(ethylene oxide)—β-poly(propylene oxide)—β-poly(ethylene oxide), PEO-β-PPO-β-PEO are selected then the direct solubilization process is to be chosen. It involves the solubilization of polymer and drug in aqueous medium. When drug and polymers are not readily soluble in water, the organic solvent is used. Both drug and polymer are dissolved in organic solvent which later removed by an appropriate method. Now to remove the organic solvent, its miscibility with water must be taken into account. Dialysis is advised to replace the organic solvent with water when organic solvent is miscible with water. Whereas, in case of water immiscible solvents, drug entrapment takes place to form oil-in-water emulsion. Another solution casting method can be used where drug and polymer dissolved in organic solvent and solvent is allowed to evaporate to form a polymeric film-containing drug.

Phospholipid micelles are sterically stabilized nanocarriers, composed of safe biocompatible materials such as polyethylene glycol (PEGylated), phospholipids, for delivery of poorly water-soluble drugs [192].

5.13.2.2 CASE STUDIES

Self-assembling mixed polymeric micelles (MPMs) encapsulating the drug curcumin was developed. Lecithin was used which played a role to enlarge hydrophobic core of MPMs and greater solubilization of curcumin. Sodium deoxycholate and pluronic P123 were used as an amphiphilic polymers. *In vitro* and *in vivo* drug release was found to be delayed. Enhanced absolute bioavailability of curcumin was observed after oral and IV administration in rats. [193]

Novel nimodipine loaded pluronic-phosphatidylcholine polysorbate 80 mixed micelles (PPPMM) were developed to treat subarachnoid hemorrhage. PPPMM provided larger hydrophobic core volume to solubilize nimodipine. Enhanced solubility, drug absorption and long circulation time of the nimodipine was reported. The enhanced bioavailability was reported for drug, 232% in plasma and 208% in brain [194].

5.13.3 POLYMERIC DENDRIMERS

Highly branched, monodisperse, uniformly distributed polymeric macromolecules with unique host-guest entrapment properties are dendrimers. The core of dendrimers is hydrophobic and a surface is hydrophilic. Their highly asymmetric shape, interior branching and multivalent tree-like surface makes them easy to conjugate with targeting and imaging agents.

Therapeutically active agents either get encapsulated into dendritic channels (void spaces) or get attached to the functional groups available on the surface (nano-scaffolding) or combination of both. Small organic molecules preferably get encapsulated in the core and large biomolecules preferably get attached to the surface groups. Svenson and Chauhan mentioned that H-bonding, van der Waals interaction and electrostatic attraction between opposite charges on drug and polymer are responsible for the interaction. Jain and Gupta mentioned the two types of complexation viz. covalent and non-covalent complexation. It is assumed that, non-covalent complexation is mainly responsible for solubilization of drugs because more hydrophilic functional groups on the surface are available for solubilization. On the other hand, covalent bonding is preferred for targeting imaging agents, dyes and genetic materials [167, 195, 196]. In addition to the electrostatic interactions and non-covalent complex formation, pH and protonation behavior of the dendrimers also affect the solubility and drug carrying capacity of the dendrimers. Milhem et al. and Asthana et al., proved pH dependent solubility for drugs ibuprofen and flurbiprofen, respectively [197, 198].

5.13.3.1 FORMULATION ASPECTS

In drug delivery systems, polyamidoamine (PAMAM) and polypropylene imine (PPI) dendrimers are most extensively studied. Though they prepared with same ethylene diamine core, they differ in their branching units. PPI can also alternatively synthesized using diaminobutane as core material. In addition to these two, dendrimer classes such as polylysine, polyester, polyether, etc. have made significant contributions in drug delivery and solubilization [196]. Devarakonda et al. proved that the dendrimer-based carriers possess greater solubilization capacity as compared to cyclodextrin-based complexes [199].

5.13.3.2   CASE STUDIES

PAMAM dendrimers-silybin complex along with the plain silybin drug were investigated for solubility, drug release and bioavailability. The complex showed significantly higher solubility for different generations of dendrimers at different pH and concentrations, whereas slower drug release was observed as compared to drug alone. Significantly improved relative bioavailability of silybin-dendrimer complex was observed of about 178% as compared to silybin [200].

Puerarin-dendrimer complex was investigated for solubility, hemolytic toxicity and bioavailability of puerarin. Significantly higher solubility of puerarin was observed with full generation G2 and G3 dendrimers. No obvious hemolysis on erythrocytes was observed for Puerarin-dendrimer complex. Improved oral *in vivo* bioavailability was observed for the puerarin-dendrimer complex [201].

Similar kind of efforts were done to enhance the solubility and bioavailability of drugs like doxorubicin [202], NSAID drugs [203] and many more. PEGylated PAMAM dendrimers also used to enhance the solubility and bioavailability of drugs such as fluorouracil, artemether, etc. [204, 205].

5.13.4   POLYMERIC NANOGELS OR HYDROGEL NANOPARTICLES

The swollen, chemically cross-linked networks of cationic and neutral polymers of nanoscale range are called polymeric nanogels [206, 207]. These nanocarrier systems possess both properties of hydrogels and nanoparticles such as high water absorption, hydrophilic, versatile and flexible structure. Polymeric nanogels have a 3D- network structure with a diameter of 10–100 nm [167].

5.13.4.1   METHODS OF PREPARATION

Nanogels can be fabricated by one of the following methods

1. Formation of physical self-assembly of interactive polymers.
2. Polymerization of monomers in either homogenous or heterogeneous micro or nano environment.
3. Covalent cross-linking of preformed polymer chains.
4. "PRINT" (Particle Replication In Non-wetting Templates).

Active biological molecules can be incorporated in polymeric hydrogels by physical entrapment, covalent conjugation or controlled self-assembly. For loading of the drug, electrostatic, van-der Waals and/or hydrophobic interactions between the drug and the polymer matrix are mainly responsible. Poly(ethylnene glycol) (PEG) as a dispersing hydrophilic polymer, can be used to prevent the aggregation of nanogels. The loading capacity of the nanogels is higher as compared to other nanomeric systems such as liposomes, polymeric nanoparticles or polymeric micelles. This is because the low molecular mass drugs or biomacromolecules can be incorporated in the large cargo space provided by swollen polymeric gels containing water.

5.13.4.2   CASE STUDIES

Chitosan (CS) containing myricetin (Myr)/HP-β-CD inclusion complex was incorporated into nanogel. 1.73 fold enhancement in the maximum plasma concentration (Cmax) of Myr-loaded nanogels group as compared to that of the plain Myr group. The area under curve (AUC) value for nanogel formulation was found to be higher as compared to plain drug, also the relative bioavailability of Myr-loaded nanogels was found to be 220.66%, which demonstrated an improved performance of myricetin after an oral nanogel delivery system. This nanogel system mainly governed by erosion with Fickian mechanism, which provided sustained release system for drug delivery [207].

Paclitaxel containing Polyasparthydrazide (PAHy) and polyethylene glycol (PEG) nanogels were developed with the help of PAHy and the cross-linkers CPEG 1000 and CPEG 6000. In both the PTX-loaded nanogels a high entrapment efficiency (EE%) (−90%) and high drug loading (approaching −20%) was observed. Initial burst release for 2 hrs followed by sustained release of PTX-loaded nanogels were observed. Higher *in vivo* absorption of PTX-loaded PAHy-based nanogels was observed as compared to commercial formulation of Taxol® at 10 mg/kg. This was attributed to longer retention and superior adhesion and permeability of PTX-loaded PAHy-based nanogels. There was approximately 9 and 14 fold enhancement in the Cmax and AUC, respectively, observed for PTX loaded PAHy-based nanogels. The longer half-life of the PTX-loaded PAHy-based nanogels were observed as compared PTX solution. The 14-fold enhanced oral bioavailability of PTX-loaded nanogels were observed to that of Taxol® formulation [208]. PLGA nanoparticles loaded with active drug silymarin, which was further encapsulated in alginate-based hydrogel microparticles. This carrier system showed enhanced solubility, dissolution and bioavailability of silymarin with oral sustained release [209].

5.14   CYCLODEXTRIN INCLUSION COMPLEX

A class of cyclic glucopyranose oligomers, synthesized by enzymatic action on hydrolyzed starch is referred as cyclodextrins (CDs). Most native cyclodextrins are α, β and γ comprises of six, seven and eight glucopyranose units, respectively. δ-CD also available with 9 glucopyranose units. Along with conventional cyclodextrins, advanced cyclodextrins such as 2-hydroxypropyl-β-cyclodextrin, randomly methylated β-cyclodextrin, sulphobutylether β-cyclodextrin, 2-hydroxypropyl-γ-cyclodextrins got market approval [210]. Cyclodextrins with their toroidal shape have outer hydrophilic and inner hydrophobic orientation. Cyclodextrins are able to form dynamic, non-covalent and water-soluble complex. In addition to the improved solubility and bioavailability of poorly water-soluble drugs, the cyclodextrin-based complexes are able to enhance the stability of light and oxygen sensitive substances, mask the component with ill odor and taste and fix the volatile components [210–212].

5.14.1   MECHANISM OF COMPLEX FORMATION

The release of enthalpy-rich water molecules from the cavity and their displacement by more hydrophobic guest molecules present in the solution is the major driving force for complex formation. More stable lower energy state is achieved by an apolar–apolar association and decrease of cyclodextrin ring strain [211]. The parameters affecting complex formation with cyclodextrins and methods for complex formation are discussed below in Tables 5.11 and 5.12.

TABLE 5.11   Parameters Affecting Complex Formation

Sr. No.	Parameters	Comments
1.	State of complex mixture	More cyclodextrin molecules are available to form a complex in its solution state as compared to its crystalline state. Also, the guest molecule is able to form complex more rapidly in its soluble form or in dispersed form.
2.	Temperature	The cyclodextrins able to form complex more rapidly at higher temperatures but at the same time higher temperature may destabilize the complex. So, balance between these two needs to be established.
3.	Solvent Effect	Solvent should not form the complex with cyclodextrin, should be easily removed by evaporation. Water, ethanol, diethyl ether are ideal ones. In case of water, it is advisable to keep the quantity of water sufficiently low in order to form a complex at faster rate.
4.	Volatile guest molecules	In order to avoid the evaporation of volatile guest molecules, use of sealed reactors or refluxing back in mixing vessel is suggested.

TABLE 5.12   Methods of Preparation

Nanocarrier system	Method of preparation	References
β-cyclodextrin Inclusion complex	Coprecipitation	[213–217]
	Kneading	[23, 213–216, 218–222]
	Co-evaporation	[23, 220]
	Freeze-drying	[23, 222–224].
	Solvent evaporation	[215, 219, 224].
	Saturated solution method	[225]
	Physical mixture method	[215, 216, 220]
	Microwave irradiation method	[220, 222, 226]
	Ethanol water system method	[214]
	pH shift method	[224]

5.14.2   LIMITATIONS OF CYCLODEXTRIN COMPLEXES

- In case of α-CD, the cavity size is insufficient for many drugs.
- γ-CDs are expensive as compared to others.
- δ-CD has weaker complex forming ability than conventional CDs.
- β-CD complex cannot be injected intravenously as it bound with cholesterol leads to nephrotoxicity.
- CDs are unable to form a complex with hydrophilic or high molecular weight drugs.
- Most commonly used β-cyclodextrins have poor aqueous solubility as compared to others.

5.14.3   CASE STUDIES

Many scientists have done extensive work with cyclodextrins to enhance the solubility, bioavailability and stability of the large number of drugs. Supramolecular complex of a potent anticancer phytoconstituent Fisetin (FST) with hydroxypropyl β-cyclodextrin (FHIC) was made and further encapsulated in PLGA polymeric nanoparticles. Nanoparticulate formulation showed nearly 79% encapsulation of drug as inclusion complex with HP βCD. Polymeric nanoparticles showed superior pharmacokinetics and bioavailability, proved by total drug absorbed and increase in peak plasma concentration. FHIC-PNP improved the anticancer activity and apoptosis of FST against MCF–7 cells [227].

Epichlorohydrin β-cyclodextrin is used to make inclusion complex with drug altretamine (ALT-Epi- βCD) to treat ovarian cancer. Complex made with kneading method showed maximum solubility as compared to other methods. SLNs were prepared with ALT-Epi- βCD complex and with the drug alone by using modified emulsification-ultrasonication method. 2.47 fold enhanced peak plasma concentration and enhanced area under the curve of altretamine showed its increased bioavailability after oral administration [23]. Similar kind of enhancement in the oral bioavailability was noticed for the drug acyclovir [228].

5.15   CYCLODEXTRIN-BASED NANOSPONGES

Nanosponges are highly porous, new generation cyclodextrin based hyper-branched polymeric systems having unique characteristic to form nanosuspension upon dispersion in water which appears as sponges microscopically.

Hydroxyl group is primarily involved in the formation of nanoporous cross-linked networks which is specifically detected by in-depth Fourier transform infrared spectroscopy (FTIR), solid state-nuclear magnetic resonance (NMR), and Raman studies. Nanosponges typically form opalescent colloidal dispersion with size below 1 µm and possess narrow size distribution [229].

5.15.1   ADVANTAGES OVER CYCLODEXTRINS [230, 231]

- Nanosponges can be administered by parenteral route by restricting the particle size ~200–300 nm with narrow size distribution.
- More interacting site available as compared to cyclodextrins for incorporation of the drugs.
- Carbonate nanosponges possess the zeta potential ~25 mV, which form a stable suspension in aqueous phase, which do not aggregate over a period of time and have thermal stability up to temperature 300°C.

5.15.2   FORMULATION COMPONENTS

5.15.2.1. CYCLODEXTRINS

Different types of cyclodextrins α-cyclodextrin, β-cyclodextrin, γ-cyclodextrin, hydroxypropyl cyclodextrin, methyl β-cyclodextrin, and alkoxycarbonyl cyclodextrins can be used to formulate nanosponges.

5.15.2.2   CROSS LINKERS

Different cross linkers can be employed, for example, cross linkers with active carbonyl compound, such as, carbonyldimiidazole, triphosgene, diphenyl carbonate, or organic dianhydrides [232–234].

5.15.2.3   SOLVENTS

Different solvents selected based on the type of cyclodextrins; e.g., ethanol, dimethyl form amide, dimethyl acetamide, water, etc. Methods for preparation of nanosponges are given in Table 5.13.

TABLE 5.13   Methods of Preparation of Nanosponges

Cyclodextrin-based nanosponges	Solvent evaporation technique	[232–234].
	Polymer condensation method	[235]
Drug-loaded nanosponges	Freeze-drying	[232, 236]
	Solvent evaporation method	[237]

5.15.3   MECHANISM OF BIOAVAILABILITY ENHANCEMENT

Nanosponges improve the poor aqueous solubility of drugs by wetting phenomenon.

By allowing molecular dispersion of the drugs in their structures and thus avoid the dissolution step, which, in turn, enhance the apparent solubility of the drug.

5.15.4   CASE STUDIES

β-cyclodextrin nanosponges were prepared by changing the molar ratios viz. 1:2, 1:4, and 1:8 of β-cyclodextrinto carbonyldiimidazole as a cross-linker and further Tamoxifen β-cyclodextrin nanosponge complex (TNC) was developed by freeze-drying technique. The pharmacokinetic parameters like AUC and Cmax of TNC formulation after gastric intubation were found to be higher viz. 1.44 and 1.38 fold as compared to plain drug [232].

β-cyclodextrin and meloxicam complexes were prepared by physical mixing (P) and kneading method (K). Pyromellitic dianhydride (PMDA) cross-linked β-cyclodextrin complex was synthesized using polymer condensation methods and finally PMDA cross-linked β-cyclodextrin-based nanosponges were prepared by sonication (N). The solubility of formulation N was found to be higher (36.61±0.34 µg/mL). Controlled drug release patterns were noted in the *in vitro* drug release studies. The formulation was found to be stable at accelerated stability testing conditions

[235]. Similar kinds of efforts were made to enhance the bioavailability of the drugs like paclitaxel [236], Efavirenz [234], telmisartan [237], omeprazole, and nelfinavir mesylate [230].

5.16 METALS AND INORGANIC NANOCARRIERS

Various inorganic materials have been, nowadays, employed for the preparation of components to be used in industrial, biomedical and/or pharmacological applications. Silica (SiO_2), titanium dioxide (TiO_2), and zinc oxide (ZnO) nanoparticles are common among them. Many researchers have tried to develop delivery carriers for API from them. Few case studies related to these inorganic materials nanoparticles are discussed in the following subsections.

5.16.1 GOLD NANOPARTICLES

Diacerein, a BCS class II drug, was incorporated in lecithin-gold (LD-Au NPs) hybrid nanocarriers and its anti-inflammatory activity is proved against carrageenan-induced inflammation. The drug was loaded in lecithin NPs followed by coating with Au NPs, either by in-situ production or by employing pre-synthesized Au NPs. All LD-Au NPs depicted 2.5-fold enhancement in oral bioavailability by specifically releasing the drug at physiological pH of 7.4. The improved pharmacological efficacy of LD-Au NPs in terms of anti-inflammatory activity was demonstrated by suppressing inflammation at first phase. In case of acute toxicity studies, LD-Au NPs showed no hepatic damage, but the renal toxicity parameters were found to be close to the upper safety limits. It was hypothesized that along with the improved oral bioavailability and anti-inflammatory activity of hydrophobic drug, these LD-Au NPs might have potential applications in gold-based photothermal therapy as well as in tracing of inflammation at atherosclerotic and arthritic site [238].

The low aqueous solubility and bioavailability of curcumin (Ccm) can be overcome by developing the succinate linked (SA), stabilized by low molecular weight water-soluble polymer (P1) gold nanoparticles (AuNPs), e.g., (Ccm-SA-P1-AuNPs). The surface conjugation of curcumin with gold nanoparticles, which was aided by the water-soluble polymer, responsible for augmenting the solubility of curcumin in water. Ccm-SA-P1-AuNPs showed the pH dependent release. Very low amount of drug was released from nanocarrier system at physiological pH, whereas system showed ~90% release over the period of 6 hrs at pH 5.3. The PEG backbone present on the polymer surface imparts the negative charge to nanocarrier system. This negative charge is responsible for longer systemic circulation and enhanced permeation retention (EPR) effect [239].

5.16.2   AMINO FUNCTIONALIZED MESOPOROUS SILICA NANOPARTICLES

Amino group functionalized anionic surfactant templated mesoporous silica (Amino AMS) and carboxyl group functionalized cationic surfactant templated mesoporous silica (carboxyl-CMS) were synthesized for poorly water-soluble drugs by using co-condensation method via two types of silane coupling agents. Spherical nanoparticles of amino-AMS for loading and release of indomethacin (IMC) and spherical nanospheres carboxyl-CMS for loading and release of femotidine (FMT). Drug loading capacity was enhanced for both the drugs. Faster drug release of IMC loaded amino AMS was observed as compared to IMC because complex had the capacity to transform crystalline phase to amorphous phase. But in case of carboxyl-CMS complex, carboxyl strongly interact with the amino group of FMT and showed sustained drug release [240].

Composite system of curcumin-amine functionalized mesoporous silica nanoparticles (MSN) and curcumin-amine functionalized mesoporous silica microparticles (MSM) were formulated for oral delivery. The effect of particle size on oral drug release, solubility and oral bioavailability of curcumin in mice was investigated. Superior release profile, high solubility, high bioavailability was observed for MSN-A-Cur as compared to MSM-A-Cur and plain drug. Thus amino functionalized MSN carrier was demonstrated to deliver low solubility drug with improved bioavailability [241].

Dodecylamine template based hexagonal mesoporous silica (HMS) was prepared as a carrier for poorly water-soluble drug (fenofibrate). HMS material's characteristics such as easy synthesis, high surface area and wormhole pores makes it useful as a carrier in drug delivery system. HMS was prepared by pH and temperature-independent process. Fenofibrate was loaded into the HMS by a solvent immersion method using organic solvents. Characterization of HMS and drug-loaded HMS was done by differential scanning calorimetry (DSC), X-ray powder diffraction (XRPD), Fourier transform infrared spectroscopy (FTIR), scanning electron microscopy (SEM), transmission electron microscopy (TEM) and contact angle study. These were also evaluated and compared with plain drug for *in vitro* and *in vivo* study. The DSC and XRD studies revealed the amorphous nature of HMS. SEM and TEM study revealed wormhole porous structure. The contact angle study showed improvement in aqueous wetting property of the drug within the HMS (contact angle 46°). The *in vitro* drug release study showed a remarkable dissolution enhancement in HMS-based system as compared to plain drug. All these studies confirmed that HMS-based formulation significantly improves bioavailability of fenofibrate and has a potential to use as a carrier for delivery system of poorly water-soluble drugs [242].

5.16.3 NANOCOMPOSITES

5.16.3.1 AMINOCLAY LIPID HYBRID COMPOSITE

Drug fenofibrate was incorporated was incorporated into novel aminoclay-lipid hybrid composite (ALC) system by implementing antisolvent precipitation coupled with an immediate freeze-drying method. Optimized formulation possessed composition of aminoclay to krill oil; krill oil to fenofibrate and antisolvent to solvent in the ratio 3:1 w/w, 2:1 w/w and 6:4 v/v, respectively. Optimized formulation shown complete drug release within 30 min as compared to less than 15% drug release for untreated powder and physical mixture. ALC exhibited 13 and seven-fold enhancement of peak plasma concentration and area under curve of fenofibrate indicating enhanced bioavailability [243].

5.16.3.2 MESOPOROUS CARBON/LIPID BILAYER NANOCOMPOSITES

Nimodipine loaded mesoporous carbon/lipid bilayer (NIM-MCNL) nanocomposites were developed to enhance the oral solubility and bioavailability of drug. About sixfold enhanced aqueous solubility of NIM-MCNL was noticed as compared to coarse drug, also NIM-MCNL showed higher plasma drug concentration from 3 to 24 hrs in comparison with commercial Nimotop® tablets. This boosted effect in both solubility and plasma concentrations of drug is because of conversion of drug to nanometer size and nanocrystalline form. A biphasic release pattern was noticed in NIM-MCNL formulation with the initial burst release followed by sustained release. The slow release from mesoporous carbon (MC) is reasoned due to the presence of micropore and mesopore channels of MC. The lyophilization of NIM-MCLN formulation showed no dramatic initial burst release. Approximately 2.14 fold enhancement in the oral bioavailability of NIM-MCLN formulation was compared to a commercial formulation [244].

5.16.3.3   CYCLODEXTRIN-DRUG-SOLID LIPID NANOPARTICLES

Novel hydrochlorothiazide (HCT)-cyclodextrin complex was incorporated into solid lipid nanoparticles (SLNs). The Plain precirol®ATO5-based SLNs containing drug were prepared by using two surfactants Pluronic®F68 and Tween®80 and compared with the SLNs with a binary system containing hydroxypropyl-beta-cyclodextrin (HPβCD) and sulfobutyl-ether-betacyclodextrin (SBEβCD) both as physical mixture (P.M.) or co-ground product (GR). SLNs were prepared by hot high-shear homogenization followed by ultrasonication method. SBEβCD tested as complexing agent was discarded because loading of its complex with HCT into SLNs did not achieve nanometric homogeneous dispersions. On the other hand, HPβCD not only enhanced the wetting and solubility of drug, in turn, enhanced the release from complex loaded SLNs, but also seemed to play a role in stabilizing the SLN formulations. These results were found superior in case of co-ground product (GR) as compared to physical mixtures (P.M.). More than 75% drug release in a sustained fashion was observed in formulations containing Pluronic®F68 as surfactants. These Pluronic®F68-based SLN formulation loaded with HCT as co-ground with HPβCD were found to be stable over a period of three months. Formulation shown more intense and prolonged HCT pharmacological effect as compared to simple drug suspension [245].

5.16.3.4 CYCLODEXTRIN NANOCOMPOSITES

Telmisartan (TEL) is a molecule with pH dependent and poor solubility profile, whose limitations were overcome by developing novel amorphous ternary cyclodextrin nanocomposites. Two different techniques viz. complexation with β-cyclodextrin and top-down nanonization was employed to achieve the goal. Significantly enhanced *in vitro* dissolution in multimedia and biorelevant media was noticed in comparison to plain drug and marketed formulation. Uniform pharmacokinetic results were obtained for TEL nanocomposites, which showed 321% and 301% fold increase in relative bioavailability as compared to marketed formulation and 315% and 346% fold increase in relative bioavailability as compared to pure TEL in fasted and fed states, respectively [246].

5.16.4 CHARACTERIZATION OF NANOPARTICLES [33, 247–249]

Nanoparticles may be formulated as solids, as suspensions or aerosolized into gaseous forms. The evaluation and characterization of nanoparticles in each of these phases require various techniques to measure parameters like size, crystallography, composition, etc. The techniques presented below provide an overview of common techniques used for characterization of nanoparticles. Figures 5.1 & 5.2 speaks about different tests to be carried out for solid, liquid and gaseous nanoparticles. These are the basic test that can be used for characterization of the nanosystems, explained in this section. Study-specific test may be required that may vary from system to systems. Such tests are described in the literatures. Few may be traced in the references of this chapter.

5.16.5 CHARACTERIZATION OF PHARMACEUTICAL NANOPARTICLES

5.16.5.1 SOLID PHASE NANOPARTICLES

These Nanoparticles exist in a powder form or can be encapsulated into a solid medium. The powders may be prepared as 'non-compressed powders' or wet or dry 'compressed cakes' for ease of handling. Analytical tests must be done taking into consideration the use of the particles since it will affect their properties like agglomeration state. Different evaluation tests for solid and liquid nanocarriers are depicted in Figure 5.7.

Structural Characterization: Structural characterization is a factor which plays a significant role in determining various properties like shape, size, structural arrangement, morphology, spatial distribution, geometric feature, density etc. of a nanosystem. Advancement of electron microscopy helps in increasing ease of access and feasibility to determine these properties at the nanometer scale. Scanning electron microscopy (SEM) shows the image of even as small as 10 nm sizes particles and provides important information about surface morphology, spatial distribution as well as structural arrangement of nanoparticles. Even more dominant imaging tools are Transmission electron microscopy (TEM) and high resolution TEM which give meticulous geometrical characteristics and information like crystal structure and orientation of nanoparticles. Also, tools

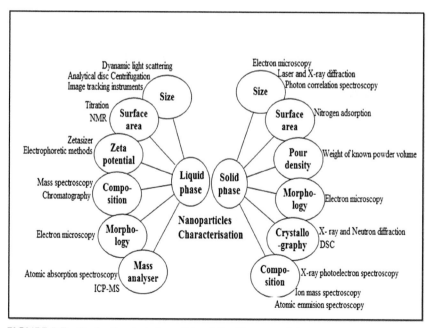

FIGURE 5.7   Evaluation tests for solid and liquid nanoparticles.

like scanning tunneling microscope (STM), scanning thermal microscopy and electrical field gradient microscopy (EFM), in combination with atomic force microscopy (AFM) can be employed to exemplify structural, magnetic, electronic and thermal characteristics along with topographical characteristics of nanosystems. Laser diffraction is the general technique used to measure particle sizes for bulk samples along with powder X-ray diffraction. Surface area can be determined by using techniques based on BET isotherm like the nitrogen adsorption technique. The morphology of nanoparticles can be analyzed by its shape and aspect ratio that is evaluated by analysis of images produced through electron micrographs.

Particle Size Distribution: Also known as polydispersity index is an important feature of the nanosystems formulations. The low value of the polydispersity index is preferred. Some techniques to determine the particle size distribution are dynamic light scattering and laser diffraction, which are used to determine the polydispersity index by detecting possible aggregates of drug nanoparticles or microparticles.

Pour Density: Can be calculated by knowing the weight of a known volume of freshly prepared nanopaticulate powder.

Composition: Surface techniques like secondary ion mass spectrometry and X-ray photoelectron spectroscopy. To analyze bulk samples, commonly they are digested after which wet chemical analyzes like atomic emission spectroscopy, mass spectrometry and ion chromatography are implemented.

Dustiness: It is the tendency of a powder to get aerosolized by mechanical agitation. The samples dustiness is very reliant on its static electrical properties and moisture content.

Particle Charge/Zeta Potential: Zeta potential help to determine the charge on the surface of the particle. Zeta potential is measured not just to optimize parameters of the formulation, but also to estimate stability during long-term storage of the colloidal dispersion. Presently the prime technique used in determination of zeta potential is laser Doppler anemometry.

Crystallography: Any possible fluctuations in the physical form of the drug brought about during nanoparticle processing are determined by either X-ray diffraction, Differential scanning calorimetry or other analytical methods. Neutron diffraction is a technique used to carry out the crystallography studies of simple lattice structures. Its use can also be extended to crystalline organic solids, except that the data analysis of such samples is much more challenging.

5.16.5.2 LIQUID PHASE NANOPARTICLES

Liquid Phase Nanoparticles are formulated as a suspension often with surfactants to prevent the agglomeration of the particles. Presence of other chemicals or even biological species as a contaminant may affect the results obtained, especially in samples taken from the environment. Therefore, to prevent unwanted matrix effects or changes to the sample care must be taken to prepare the sample.

Size: Dynamic light scattering also known as photon correlation spectroscopy and centrifugation, are the generally employed techniques. Other image-tracking instruments may also be used.

Surface Area: Simple titrations are used for estimation of surface area, but are quite tedious. NMR analysis along with other recently developed dedicated instrumentation may also be used.

Morphology: Measuring of morphology of freely moving particles in a fluid can be. Hence its deposition onto a solid surface is recommended for electron microscopy.

Zeta Potential: It assists in determining the stability of nanoparticulate dispersions. Although its direct estimation is not possible, but can be indirectly measured using an electrophoretic method.

Composition: Chemical digestion is done of the particles followed by a range of mass spectroscopy and chromatography methods. Management of the matrix interference is needed.

5.16.5.3 GASEOUS PHASE

Relatively low cost, robust equipments are commercially available to monitor aerosol nanoparticles. Generally, these instruments can be used for prolonged periods with little attention. They are also generally resistant to matrix effects. However, parameters that affect characteristics of gas such as relative humidity and volatile organic species can sometimes affect measurements. Unlike the condensed phases, aerosols cannot be stored for later analysis and so reproducible sampling is very important. Different evaluation tests for gaseous phase nanocarrier systems are shown in Figure 5.8.

5.16.5.4 STRUCTURAL CHARACTERIZATION

Size can be measured by many methods, but comparative polydispersability index determination between them is a problem. Optical and aerodynamic methods can be used, but they do not inform on morphological variations. The morphology of particles can be determined by capturing particles either by filtration or electrostatically for further imaging with the use of electron microscopy. The surface area can be calculated by charging the aerosol with a corona discharge and then measuring its charge concentration.

Concentration: It is usually expressed in the form of mass concentration. It strongly weighs the distribution curve in favor of larger particles – a single

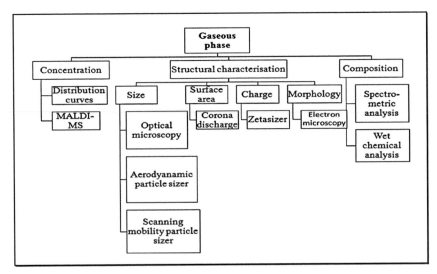

FIGURE 5.8   Evaluation tests for gaseous phase nanocarrier systems.

10 μm diameter particle weighs the same as 1 million 100 nm particles! Also MALDI-MS technique can be used for concentration determination.

Charge: A steady-state distribution of charge is a result of collisions of air ions with particles. Smaller particles carry a much lesser charge than larger particles. This results in Boltzmann-like distribution that can be determined in laboratory.

Composition: measuring aerosolized nanoparticle composition is very difficult due to the small concentration of the nanoparticulate matter present. Most techniques require the collection of particles and then subjecting them to various wet chemical or spectrometric analysis.

5.16.6   ENTRAPMENT EFFICIENCY AND LOADING CAPACITY

Entrapment efficiency is the efficacy of the nanoparticles to encapsulate drugs within its structure. And loading capacity is a function of the lipid or encapsulating material used for formulating the nanoparticles. The amount of drug not entrapped by the nanoparticles is to be determined either by filtration of the dissolved drug remaining in the dispersion or the estimation of dissolved drug in the solution. The following formulae can be employed:

$$\% \ Entrapment \ Efficiency = \frac{Amount \ of \ drug \ infiltered \ formulation - Unencapsulated \ soluble \ drug}{Actual \ amount \ of \ drug \ added} \times 100$$

$$\% Loading Capacity = \frac{Amount \ of \ encapsulated \ drug}{Amount \ of \ lipid \ used \ in \ the \ formulation} \times 100$$

5.16.6.1   TOXICITY OF NANOPARTICLES

In the above sections, we have studied that Nanotechnology is judiciously being used for almost all classes of drugs to get benefit. The drugs that failed to show therapeutic effects with conventional delivery system, when converted to nanosystems showed improved efficacy. Many researchers have modified basic systems in order to fit the requirement of either disease or the drug for the so the mankind will be advantage of the system. Before one and half decade, when the nanomedicines were introduced, people were so excited and saw the only positive side of this technology. Nowadays awareness regarding risk in using medicines is increased. Researchers are studying potential drawbacks of this 'Boon' technology. The use of occupants in nano research is one of the important considerations. The most important is the size range of the drug as well as occupants. Nano means in the range of 1 to 999 nanometers, is a general concept. One must know that particles below 100 nanometers size can directly enters cellular contents. In such cases, if the occupant employed is non-biodegradable then the chances of toxicity increase many folds. Nanoparticulate systems may cause some acute toxicities like endocytosis (caused because of inflammation or formation of granuloma); oxidative stress (caused due to cell death by generation of free radicals) and modification in protein/gene structure resulting in various immune responses culminating in an autoimmune effect. Long-term toxicities like bioaccumulation, defective distribution that halter the ultimate fate of nanosystem in the body may take place in the evaluation of their toxicities can be carried out in cell lines and MTT assay can be used to establish the cell viability. Various animal models are employed for *In vivo* determination of acute and chronic toxicities.

Few case studies are discussed below where researchers identified and reported the toxicities of the nanocarriers/particles.

5.16.6.2   SILVER NANOPARTICLES

The oral bioavailability of commercially available colloidal silver nanoparticles Collargol was studied using the rat animal model. Collargol has an average diameter nearly 15 nm and consists ~70% silver and ~30% of protein matrix. The silver content is monitored in the liver, kidney, urine and feces. With the help of asymmetric flow field-flow fractionation (AsFlFFF) coupled with UV-Vis analysis, scientist proved that intact silver was present in the feces and using Laser ablation–ICP MS imaging, scientist proved silver was able to penetrate and accumulate in the liver. This is supported by the hypothesis of presence of higher amounts of sulfur in the liver which form a complex with silver. On the other hand, in the kidneys, the silver is retained in the cortex and excreted in very minute amounts through urine over a period of 30 days. Researchers also demonstrated that oxidation of silver nanoparticles in the biological environment producing Ag(I) which complexes by a number of proteins, the major of which is metallothionein in the kidney [250].

Male Sprague-Dawley rats were administered with silver nanoparticles (AgNP) and Ag+ in 2 mg/kg as low dose and 20 mg/kg as a higher dose. The animals were kept on fasting one night before administration and 12 hrs after administration. The area under curve values after 24 hrs of administration (AUC 24 hrs) for Ag+ and AgNPs were found to be 3.81 ± 0.57 $\mu g/d/mL$ and 1.58 ± 0.25 $\mu g/d/mL$, respectively. The intact Ag+ particles showed significant tissue distribution in the liver, lungs and kidneys as compared to AgNP. Low bioavailability of AgNPs were stated due their predominant excretion through feces [251].

5.16.6.3 SILICA (SIO_2), TITANIUM DIOXIDE (TIO_2), AND ZINC OXIDE (ZNO) NANOPARTICLES

Bioavailability studies of three different inorganic components viz. silica (SiO_2), titanium dioxide (TiO_2), and zinc oxide (ZnO) in nanoparticulate and bulk size was evaluated after oral and intravenous administration. It was demonstrated that XRD patterns showed high crystallinity for bulk materials. On the other hand, the dissolution pattern of all inorganic materials was not affected by particle size. SiO_2 and TiO_2 showed no significant difference in bioavailability with respect to particle size, but bulk ZnO were more bioavailable than nano ZnO. In comparison to ZnO particles, SiO_2 and TiO_2 showed an extremely low amount entered the bloodstream, regardless of particle size. Therefore, it was concluded that the oral or intravenous absorption efficiency and bioavailability of materials were dependent on the type of material rather than its particle size [252].

5.16.7 ALUMINA

Park et al. proved that all aluminum oxide nanoparticles (AlONPs) viz. γ-AlONPs, α-AlONPs, and γ-AlOHNPs possess low stability within biological systems. Within these three, γ-AlOHNPs have higher toxicity and accumulation in comparison with γ-AlONPs and α-AlONPs. Their study concluded that the presence of hydroxyl group play an important factor in the determination of the distribution and toxicity of spherical AlONPs. In accordance with acute oral toxicity studies (OECD 2001 guideline), Balasubramanyam et al. (2009) studied aluminum oxide nano-material (Al_2O_3.NM) toxicity after oral administration of doses higher than possible exposure. They had administered 500, 1000, and 2000 mg/kg doses of Al_2O_3.NMs of particle size 30 nm, 40 nm and bulk material in Wistar rats. The similar level of significant oxidative stress was noticed for both nanomaterials in a dose dependent manner in comparison with bulk Al_2O_3. They concluded this oxidative stress was due to alteration in the antioxidant status of the cell [253].

The above studies depict that it is very much important to study the toxicity of nanoparticles. Based on the size and excipients used, Cornelia and Müller proposed a Nanotoxicological classification system (NCS) as a guide for the risk-benefit assessment of nanoparticulate drug delivery systems as shown in Figure 5.9.

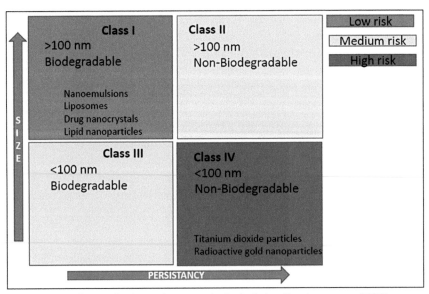

FIGURE 5.9   (See color insert.) Nanotoxicological classification system (NCS).

Though we say Green is of low risk, we need to know the degradation products of the biodegradable excipients. Essentially, they must be nontoxic to prove as a low or no risk. The same may be the case and may convert Yellow into Red or Green into Yellow or red.

5.17 CONCLUSION

Since last two decades, nanocarriers, and nanotechnology have been studied extensively for oral drug delivery. The chapter describes the application of nanoparticulate drug and/or delivery system for oral administration. Through the literature, it has been proved that this new technology is a boon to health care system. Many drugs with poor solubility and permeability have been successfully demonstrated for performance improvement for oral delivery, using nanotechnology. Some tailor-made systems are also available in the literature which are drug specific. To bypass drawback of certain systems, such as liquid nanocrystals, researchers have developed solid lipid nanocarriers. All the reforms in the nanocarrier system shows that, if one has a thorough knowledge of basic pharmacy, can produce a safer, more stable and efficacious drug products. However, in-depth characterization of the systems for their bio-fate is essentially required. Nanosystems also call for rigorous and extensive toxicity studies associated with their size.

KEYWORDS

- **cyclodextrin**
- **nanocarriers**
- **nanosponges**
- **self-nanoemulsifying drug delivery systems (SNEDDS)**

REFERENCES

1. Radtke, M., (2001). Pure drug nanoparticles for the formulation of poorly soluble drugs. *New Drugs, 3*, 62–68.
2. Lipinski, C. A. L. F., (2002). Poor aqueous solubility-an industry wide problem in drug discovery. *Am. Pharm. Rev., 5*(3), 82–85.
3. Amidon, G. L., Lennernäs, H., Shah, V. P., & Crison, J. R., (1995). A theoretical basis for a biopharmaceutic drug classification: The correlation of *in vitro* drug product dissolution and *in vivo* bioavailability. *Pharmaceutical Research, 12*(3), 413–420.
4. Radtke, M., (2001). Pure drug nanoparticles for the formulation of poorly soluble drugs. *New Drugs, 3*, 62–68.
5. Möschwitzer, J., & Müller, R. H., (2007). Drug nanocrystals-The universal formulation approach for poorly soluble drugs. In: *Nanoparticulate Drug Delivery Systems* (pp. 71–88). CRC Press.
6. Gao, L., Zhang, D., & Chen, M., (2008). Drug nanocrystals for the formulation of poorly soluble drugs and its application as a potential drug delivery system. *Journal of Nanoparticle Research, 10*(5), 845–862.
7. Junghanns, J. U., & Müller, R. H., (2008). Nanocrystal technology, drug delivery and clinical applications. *Int. J. Nanomedicine, 3*(3), 295–309.
8. Chen, H., Khemtong, C., Yang, X., Chang, X., & Gao, J., (2011). Nanonization strategies for poorly water-soluble drugs. *Drug Discovery Today, 16*(7), 354–360.
9. Kalepu, S., Manthina, M., & Padavala, V., (2013). Oral lipid-based drug delivery systems-an overview. *Acta. Pharmaceutica Sinica B., 3*(6), 361–372.
10. Gupta, S., Kesarla, R., & Omri, A., (2013). Formulation strategies to improve the bioavailability of poorly absorbed drugs with special emphasis on self-emulsifying systems. *ISRN Pharmaceutics*, 1–16.
11. Date, A. A., Desai, N., Dixit, R., & Nagarsenker, M., (2010). Self-nanoemulsifying drug delivery systems: Formulation insights, applications and advances. *Nanomedicine, 5*(10), 1595–1616.
12. Constantinides, P. P., Tustian, A., & Kessler, D. R., (2004). Tocol emulsions for drug solubilization and parenteral delivery. *Advanced Drug Delivery Reviews, 56*(9), 1243–1255.
13. Makadia, H. A., Bhatt, A. Y., Parmar, R. B., Paun, J. S., & Tank, H. M., (2013). Self-nano emulsifying drug delivery system (SNEDDS): Future aspects. A*sian Journal of Pharmaceutical Research, 3*(1), 21–27.
14. Nooli, M., Chella, N., Kulhari, H., Shastri, N. R., & Sistla, R., (2016). Solid lipid nanoparticles as vesicles for oral delivery of olmesartan medoxomil: Formulation, optimization and *in vivo* evaluation. *Drug Development and Industrial Pharmacy, 43*(4), 611–617.
15. Nirbhavane, P., Vemuri, N., Kumar, N., & Khuller, G. K., (2016). Lipid nanocarrier-mediated drug delivery system to enhance the oral bioavailability of rifabutin. *AAPS Pharm. Sci. Tech.*, 1–9.
16. Venishetty, V. K., Chede, R., Komuravelli, R., Adepu, L., Sistla, R., & Diwan, P. V., (2012). Design and evaluation of polymer coated carvedilol loaded solid lipid

nanoparticles to improve the oral bioavailability: A novel strategy to avoid intraduodenal administration. *Colloids and Surfaces B, Biointerfaces, 95*, 1–9.

17. Dudhipala, N., & Janga, K. Y., (2017). Lipid nanoparticles of zaleplon for improved oral delivery by Box-Behnken design: Optimization, *in vitro* and *in vivo* evaluation. *Drug Development and Industrial Pharmacy*, 1–26.

18. Naseri, N., Valizadeh, H., & Zakeri-Milani, P., (2015). Solid lipid nanoparticles and nanostructured lipid carriers: Structure, preparation and application. *Advanced Pharmaceutical Bulletin, 5*(3), p. 305.

19. Shidhaye, S. S., Vaidya, R., Sutar, S., Patwardhan, A., & Kadam, V. J., (2008). Solid lipid nanoparticles and nanostructured lipid carriers-innovative generations of solid lipid carriers. *Current Drug Delivery, 5*(4), 324–331.

20. Garud, A., Singh, D., & Garud, N., (2012). Solid lipid nanoparticles (SLN): Method, characterization and applications. *Int Curr Pharm J, 1*(11), 384–393.

21. Polchi, A., Magini, A., Mazuryk, J., Tancini, B., Gapiński, J., Patkowski, A., et al., (2016). Rapamycin loaded solid lipid nanoparticles as a new tool to deliver mTOR inhibitors: Formulation and *in vitro* characterization. *Nanomaterials, 6*(5), p. 87.

22. Jawahar, N., Meyyanathan, S. N., Reddy, G., & Sood, S., (2013). Solid lipid nanoparticles for oral delivery of poorly soluble drugs. *Chem. Inform., 44*(27), 1848–1855.

23. Gidwani, B., & Vyas, A., (2017). Pharmacokinetic study of solid-lipid–nanoparticles of altretamine complex edepichlorohydrin-β-cyclodextrin for enhanced solubility and oral bioavailability. *Int J Biol Macromol*, 101, 24–31.

24. Trotta, M., Debernardi, F., & Caputo, O., (2003). Preparation of solid lipid nanoparticles by a solvent emulsification–diffusion technique. *International Journal of Pharmaceutics, 257*(1), 153–160.

25. Mojahedian, M. M., Daneshamouz, S., Samani, S. M., & Zargaran, A., (2013). A novel method to produce solid lipid nanoparticles using n-butanol as an additional co-surfactant according to the o/w microemulsion quenching technique. *Chemistry and Physics of Lipids, 174*, 32–38.

26. Fadda, P., Monduzzi, M., Caboi, F., Piras, S., & Lazzari, P., (2013). Solid lipid nanoparticle preparation by a warm microemulsion based process: Influence of microemulsion microstructure. *International Journal of Pharmaceutics, 446*(1), 166–175.

27. Chattopadhyay, P., Shekunov, B. Y., Yim, D., Cipolla, D., Boyd, B., & Farr, S., (2007). Production of solid lipid nanoparticle suspensions using supercritical fluid extraction of emulsions (SFEE) for pulmonary delivery using the AERx system. *Advanced Drug Delivery Reviews, 59*(6), 444–453.

28. Li, Z., Li, X. W., Zheng, L. Q., Lin, X. H., Geng, F., & Yu, L., (2010). Bovine serum albumin loaded solid lipid nanoparticles prepared by double emulsion method. *Chem. Res. Chinese Universities, 26*(1), 136–141.

29. Charcosset, C., El-Harati, A., & Fessi, H., (2005). Preparation of solid lipid nanoparticles using a membrane contactor. *Journal of Controlled Release, 108*(1), 112–120.

30. Eltayeb, M., Bakhshi, P. K., Stride, E., & Edirisinghe, M., (2013). Preparation of solid lipid nanoparticles containing active compound by electrohydrodynamic spraying. *Food Research International, 53*(1), 88–95.

31. Tao, C., Cheng, H., Zhou, K., Luo, Q., Guo, L., & Chen, W., (2012). Preparation and characterization of Biochanin A loaded solid lipid nanoparticles. *Asian Journal of Pharmaceutics, 6*(4), p. 275.
32. Muller, R. H., Mader, K., & Gohla, S., (2000). Solid lipid nanoparticles (SLN) for controlled drug delivery–a review of the state of the art. *European Journal of Pharmaceutics and Biopharmaceutics, 50*(1), 161–177.
33. Das, S., Ng, W. K., & Tan, R. B., (2012). Are nanostructured lipid carriers (NLCs) better than solid lipid nanoparticles (SLNs): development, characterizations and comparative evaluations of clotrimazole-loaded SLNs and NLCs? *European Journal of Pharmaceutical Sciences, 47*(1), 139–151.
34. Uner, M., (2006). Preparation, characterization and physico-chemical properties of solid lipid nanoparticles (SLN) and nanostructured lipid carriers (NLC): Their benefits as colloidal drug carrier systems. *Die Pharmazie-An International Journal of Pharmaceutical Sciences, 61*(5), 375–386.
35. Khan, S., Baboota, S., Ali, J., Narang, R. S., & Narang, J. K., (2016). Chlorogenic acid stabilized nanostructured lipid carriers (NLC) of atorvastatin: Formulation, design and *in vivo* evaluation. *Drug Development and Industrial Pharmacy, 42*(2), 209–220.
36. Shi, F., Yang, G., Ren, J., Guo, T., Du, Y., & Feng, N., (2013). Formulation design, preparation, and *in vitro* and *in vivo* characterizations of β-Elemene-loaded nanostructured lipid carriers. *Int. J. Nanomedicine, 8*, 2533–2541.
37. Wang, W., Chen, L., Huang, X., & Shao, A., (2016). Preparation and characterization of minoxidil loaded nanostructured lipid carriers. *AAPS Pharm. Sci. Tech.*, 1–8.
38. Han, F., Yin, R., Che, X., Yuan, J., Cui, Y., Yin, H., & Li, S., (2012). Nanostructured lipid carriers (NLC) based topical gel of flurbiprofen: Design, characterization and *in vivo* evaluation. *International Journal of Pharmaceutics, 439*(1), 349–357.
39. Chen, S., Liu, W., Wan, J., Cheng, X., Gu, C., Zhou, H., et al., (2013). Preparation of Coenzyme Q10 nanostructured lipid carriers for epidermal targeting with high-pressure microfluidics technique. *Drug Development and Industrial Pharmacy, 39*(1), 20–28.
40. Safwat, S., Ishak, R. A., Hathout, R. M., & Mortada, N. D., (2017). Nanostructured lipid carriers loaded with simvastatin: Effect of PEG/glycerides on characterization, stability, cellular uptake efficiency and *in vitro* cytotoxicity. *Drug Development and Industrial Pharmacy*, 1–14.
41. Rahman, H. S., Rasedee, A., How, C. W., Abdul, A. B., Zeenathul, N. A., Othman, H. H., et al., (2013). Zerumbone-loaded nanostructured lipid carriers: preparation, characterization, and antileukemic effect. *Int. J. Nanomedicine, 8*, 2769–2781.
42. Zhuang, C. Y., Li, N., Wang, M., Zhang, X. N., Pan, W. S., Peng, J. J., et al., (2010). Preparation and characterization of vinpocetine loaded nanostructured lipid carriers (NLC) for improved oral bioavailability. *International Journal of Pharmaceutics, 394*(1), 179–185.
43. Shi, F., Zhao, Y., Firempong, C. K., & Xu, X., (2016). Preparation, characterization and pharmacokinetic studies of linalool-loaded nanostructured lipid carriers. *Pharmaceutical Biology, 54*(10), 2320–2328.

44. Khan, S., Shaharyar, M., Fazil, M., Baboota, S., & Ali, J., (2016). Tacrolimus-loaded nanostructured lipid carriers for oral delivery–optimization of production and characterization. *European Journal of Pharmaceutics and Biopharmaceutics, 108,* 277–288.

45. Ustundag-Okur, N., Gökçe, E. H., Bozbıyık, D. I., Egrilmez, S., Ozer, O., & Ertan, G., (2014). Preparation and *in vitro–in vivo* evaluation of ofloxacin loaded ophthalmic nano structured lipid carriers modified with chitosan oligosaccharide lactate for the treatment of bacterial keratitis. *European Journal of Pharmaceutical Sciences, 63,* 204–215.

46. Gupta, B., Poudel, B. K., Tran, T. H., Pradhan, R., Cho, H. J., Jeong, J. H., et al., (2015). Modulation of pharmacokinetic and cytotoxicity profile of imatinib base by employing optimized nanostructured lipid carriers. *Pharmaceutical Research, 32*(9), 2912–2927.

47. Yang, C. R., Zhao, X. L., Hu, H. Y., Li, K. X., Sun, X., Li, L., & Chen, D. W., (2010). Preparation, optimization and characteristic of huperzine a loaded nanostructured lipid carriers. *Chemical and Pharmaceutical Bulletin, 58*(5), 656–661.

48. Cortesi, R., Valacchi, G., Muresan, X. M., Drechsler, M., Contado, C., Esposito, E., et al., (2017). Nanostructured lipid carriers (NLC) for the delivery of natural molecules with antimicrobial activity: Production, characterization and *in vitro* studies. *Journal of Microencapsulation,* 1–10.

49. Esposito, E., Ravani, L., Drechsler, M., Mariani, P., Contado, C., Ruokolainen, J., et al., (2015). Cannabinoid antagonist in nanostructured lipid carriers (NLCs): Design, characterization and *in vivo* study. *Materials Science and Engineering, C, 48,* 328–336.

50. Uprit, S., Sahu, R. K., Roy, A., & Pare, A., (2013). Preparation and characterization of minoxidil loaded nanostructured lipid carrier gel for effective treatment of alopecia. *Saudi Pharmaceutical Journal, 21*(4), 379–385.

51. Almeida, H., Lobão, P., Frigerio, C., Fonseca, J., Silva, R., Sousa Lobo, J. M., et al., (2015). Preparation, characterization and biocompatibility studies of thermoresponsive eye drops based on the combination of nanostructured lipid carriers (NLC) and the polymer Pluronic F-127 for controlled delivery of ibuprofen. *Pharmaceutical Development and Technology, 22*(3), 336–349.

52. Das, S., Ng, W. K., & Tan, R. B., (2014). Sucrose ester stabilized solid lipid nanoparticles and nanostructured lipid carriers: I. Effect of formulation variables on the physicochemical properties, drug release and stability of clotrimazole-loaded nanoparticles. *Nanotechnology, 25*(10), p. 105101.

53. Liu, K., Sun, J., Wang, Y., He, Y., Gao, K., & He, Z., (2008). Preparation and characterization of 10-hydroxycamptothecin loaded nanostructured lipid carriers. *Drug Development and Industrial Pharmacy, 34*(5), 465–471.

54. Ribeiro, L. N., Franz-Montan, M., Breitkreitz, M. C., Alcântara, A. C., Castro, S. R., Guilherme, V. A., et al., (2016). Nanostructured lipid carriers as robust systems for topical lidocaine-prilocaine release in dentistry. *European Journal of Pharmaceutical Sciences, 93,* 192–202.

55. Wang, H., Sun, M., Li, D., Yang, X., Han, C., & Pan, W., (2017). Redox sensitive PEG controlled octaarginine and targeting peptide co-modified nanostructured lipid carri-

ers for enhanced tumour penetrating and targeting *in vitro* and *in vivo*. *Artificial Cells, Nanomedicine, and Biotechnology*, 1–10.

56. Zhao, X. L., Yang, C. R., Yang, K. L., Li, K. X., Hu, H. Y., & Chen, D. W., (2010). Preparation and characterization of nanostructured lipid carriers loaded traditional Chinese medicine, zedoary turmeric oil. *Drug Development and Industrial Pharmacy, 36*(7), 773–780.

57. Yuan, H., Wang, L. L., Du, Y. Z., You, J., Hu, F. Q., & Zeng, S., (2007). Preparation and characteristics of nanostructured lipid carriers for control-releasing progesterone by melt-emulsification. *Colloids and Surfaces B: Biointerfaces, 60*(2), 174–179.

58. Zhang, X., Pan, W., Gan, L., Zhu, C., Gan, Y., & Nie, S., (2008). Preparation of a dispersible PEGylate nanostructured lipid carriers (NLC) loaded with 10-hydroxycamptothecin by spray-drying. *Chemical and Pharmaceutical Bulletin, 56*(12), 1645–1650.

59. Lacerda, S. P., Cerize, N. N. P., & Ré, M. I., (2011). Preparation and characterization of carnauba wax nanostructured lipid carriers containing benzophenone☐3. *International Journal of Cosmetic Science, 33*(4), 312–321.

60. Chu, Y., Li, D., Luo, Y. F., He, X. J., & Jiang, M. Y., (2014). Preparation and *in vitro* evaluation of glycyrrhetinic acid-modified curcumin-loaded nanostructured lipid carriers. *Molecules, 19*(2), 2445–2457.

61. Zhang, K., Lv, S., Li, X., Feng, Y., Li, X., Liu, L., et al., (2013). Preparation, characterization, and *in vivo* pharmacokinetics of nanostructured lipid carriers loaded with oleanolic acid and gentiopicrin. *International Journal of Nanomedicine, 8*, p. 3227.

62. Kamel, R., & Mostafa, D. M., (2015). Rutin nanostructured lipid cosmeceutical preparation with sun protective potential. *Journal of Photochemistry and Photobiology B: Biology, 153*, 59–66.

63. Lee, S. A., Joung, H. J., Park, H. J., & Shin, G. H., (2017). Preparation of chitosan☐ coated nanostructured lipid carriers (CH☐NLCs) to control iron delivery and their potential application to food beverage system. *J Food Sci, 82*(4), 904–912.

64. Song, J., Fan, X., & Shen, Q., (2016). Daidzein-loaded nanostructured lipid carriers-PLGA nanofibers for transdermal delivery. *International Journal of Pharmaceutics, 501*(1), 245–252.

65. Luan, J., Zhang, D., Hao, L., Qi, L., Liu, X., Guo, H., et al., (2014). Preparation, characterization and pharmacokinetics of Amoitone B-loaded long circulating nanostructured lipid carriers. *Colloids and Surfaces B. Biointerfaces, 114*, 255–260.

66. Dai, W., Zhang, D., Duan, C., Jia, L., Wang, Y., Feng, F., et al., (2010). Preparation and characteristics of oridonin-loaded nanostructured lipid carriers as a controlled-release delivery system. *Journal of Microencapsulation, 27*(3), 234–241.

67. Patel, D., Dasgupta, S., Dey, S., Roja Ramani, Y., Ray, S., & Mazumder, B., (2012). Nanostructured lipid carriers (NLC)-based gel for the topical delivery of aceclofenac: Preparation, characterization, and *in vivo* evaluation. *Scientia Pharmaceutica, 80*(3), 749–764.

68. Hu, F. Q., Jiang, S. P., Du, Y. Z., Yuan, H., Ye, Y. Q., & Zeng, S., (2005). Preparation and characterization of stearic acid nanostructured lipid carriers by solvent diffusion method in an aqueous system. *Colloids and Surfaces B. Biointerfaces, 45*(3), 167–173.

69. Hejri, A., Khosravi, A., Gharanjig, K., & Hejazi, M., (2013). Optimisation of the formulation of β-carotene loaded nanostructured lipid carriers prepared by solvent diffusion method. *Food Chemistry, 141*(1), 117–123.

70. Hu, F. Q., Jiang, S. P., Du, Y. Z., Yuan, H., Ye, Y. Q., & Zeng, S., (2006). Preparation and characteristics of monostearin nanostructured lipid carriers. *International Journal of Pharmaceutics, 314*(1), 83–89.

71. Zhou, L., Chen, Y., Zhang, Z., He, J., Du, M., & Wu, Q., (2012). Preparation of tripterine nanostructured lipid carriers and their absorption in rat intestine. *Die Pharmazie-An International Journal of Pharmaceutical Sciences, 67*(4), 304–310.

72. Shao, Z., Shao, J., Tan, B., Guan, S., Liu, Z., Zhao, Z., He, F., & Zhao, J., (2015). Targeted lung cancer therapy: Preparation and optimization of transferrin-decorated nanostructured lipid carriers as novel nanomedicine for co-delivery of anticancer drugs and DNA. *International Journal of Nanomedicine, 10*, p. 1223.

73. Shamma, R. N., & Aburahma, M. H., (2014). Follicular delivery of spironolactone via nanostructured lipid carriers for management of alopecia. *International Journal of Nanomedicine, 9*, p. 5449.

74. Jia, L. J., Zhang, D. R., Li, Z. Y., Feng, F. F., Wang, Y. C., Dai, W. T., et al., (2010). Preparation and characterization of silybin-loaded nanostructured lipid carriers. *Drug Delivery, 17*(1), 11–18.

75. Gu, X., Zhang, W., Liu, J., Shaw, J. P., Shen, Y., Xu, Y., et al., (2011). Preparation and characterization of a lovastatin-loaded protein-free nanostructured lipid carrier resembling high-density lipoprotein and evaluation of its targeting to foam cells. *AAPS Pharm. Sci. Tech., 12*(4), 1200–1208.

76. Yu, Q., Hu, X., Ma, Y., Xie, Y., Lu, Y., Qi, J., et al., (2016). Lipids-based nanostructured lipid carriers (NLCs) for improved oral bioavailability of sirolimus. *Drug Delivery, 23*(4), 1469–1475.

77. Zhang, Y., Li, Z., Zhang, K., Yang, G., Wang, Z., Zhao, J., et al., (2016). Ethyl oleate-containing nanostructured lipid carriers improve oral bioavailability of trans-ferulic acid ascompared with conventional solid lipid nanoparticles. *International Journal of Pharmaceutics, 511*(1), 57–64.

78. Shah, N. V., Seth, A. K., Balaraman, R., Aundhia, C. J., Maheshwari, R. A., & Parmar, G. R., (2016). Nanostructured lipid carriers for oral bioavailability enhancement of raloxifene: Design and *in vivo* study. *Journal of Advanced Research, 7*(3), 423–434.

79. Nekkanti, V., Venkatesan, N., Wang, Z., & Betageri, G. V., (2015). Improved oral bioavailability of valsartan using proliposomes: Design, characterization and *in vivo* pharmacokinetics. *Drug Development and Industrial Pharmacy, 41*(12), 2077–2088.

80. Akbarzadeh, A., Rezaei-Sadabady, R., Davaran, S., Joo, S. W., Zarghami, N., Hanifehpour, Y., et al., (2013). Liposome: Classification, preparation, and applications. *Nanoscale Research Letters, 8*(1), p. 102.

81. Himanshu, A., Sitasharan, P., & Singhai, A. K., (2011). Liposomes as drug carriers. *IJPLS, 2*(7), 945–951.

82. Anwekar, H., Patel, S., & Singhai, A. K., (2011). International journal of pharmacy & life sciences. *Int. J. of Pharm. & Life Sci. (IJPLS), 2*(7), 945–951.

83. Mozafari, M. R., (2010). Nanoliposomes: Preparation and analysis. Liposomes: Methods and Protocols. *Pharmaceutical Nanocarriers, 1*, pp.29–50.

84. Mayer, L. D., Madden, T. D., Bally, M. B., & Cullis, P. R., (1993). pH gradient-mediated drug entrapment in liposomes. *Liposome Technology, 2*, 27–44.

85. Riaz, M., (1996). Liposomes preparation methods. *Pakistan Journal of Pharmaceutical Sciences, 9*(1), 65–77.

86. Deamer, D., & Bangham, A. D., (1976). Large volume liposomes by an ether vaporization method. *Biochimicaet Biophysica Acta (BBA)-Biomembranes, 443*(3), 629–634.

87. Batzri, S., & Korn, E. D., (1973). Single bilayer liposomes prepared without sonication. *Biochimicaet Biophysica Acta (BBA)-Biomembranes, 298*(4), 1015–1019.

88. Szoka, F., & Papahadjopoulos, D., (1978). Procedure for preparation of liposomes with large internal aqueous space and high capture by reverse-phase evaporation. *Proceedings of the National Academy of Sciences, 75*(9), 4194–4198.

89. Immordino, M. L., Dosio, F., & Cattel, L., (2006). Stealth liposomes: Review of the basic science, rationale, and clinical applications, existing and potential. *International Journal of Nanomedicine, 1*(3), p. 297.

90. Yang, T., Cui, F. D., Choi, M. K., Lin, H., Chung, S. J., Shim, C. K., et al., (2007). Liposome formulation of paclitaxel with enhanced solubility and stability. *Drug Delivery, 14*(5), 301–308.

91. Yang, T., Cui, F. D., Choi, M. K., Cho, J. W., Chung, S. J., Shim, C. K., et al., (2007). Enhanced solubility and stability of PEGylated liposomal paclitaxel: *In vitro* and *in vivo* evaluation. *International Journal of Pharmaceutics, 338*(1), 317–326.

92. Janga, K. Y., Jukanti, R., Velpula, A., Sunkavalli, S., Bandari, S., Kandadi, P., et al., (2012). Bioavailability enhancement of zaleplon via proliposomes: Role of surface charge. *European Journal of Pharmaceutics and Biopharmaceutics, 80*(2), 347–357.

93. Chu, C., Tong, S. S., Xu, Y., Wang, L., Fu, M., Ge, Y. R., et al., (2011). Proliposomes for oral delivery of dehydrosilymarin: Preparation and evaluation *in vitro* and *in vivo*. *Acta. Pharmacologica Sinica, 32*(7), 973–980.

94. Yan-yu, X., Yun-mei, S., Zhi-peng, C., & Qi-neng, P., (2006). Preparation of silymarin proliposome: A new way to increase oral bioavailability of silymarin in beagle dogs. *International Journal of Pharmaceutics, 319*(1), 162–168.

95. Nekkanti, V., Rueda, J., Wang, Z., & Betageri, G. V., (2016). Comparative evaluation of proliposomes and self micro-emulsifying drug delivery system for improved oral bioavailability of nisoldipine. *International Journal of Pharmaceutics, 505*(1), 79–88.

96. Potluri, P., & Betageri, G. V., (2006). Mixed-micellarproliposomal systems for enhanced oral delivery of progesterone. *Drug Delivery, 13*(3), 227–232.

97. Zheng, B., Yang, S., Fan, C., Bi, Y., Du, L., Zhao, L., et al., (2016). Oleic acid derivative of polyethylenimine-functionalized proliposomes for enhancing oral bioavailability of extract of ginkgo biloba. *Drug Delivery, 23*(4), 1194–1203.

98. Marianecci, C., Di Marzio, L., Rinaldi, F., Celia, C., Paolino, D., Alhaique, F., et al., (2014). Niosomes from 80s to present: The state of the art. *Advances in Colloid and Interface Science, 205*, 187–206.

99. Attia, I. A., El-Gizawy, S. A., Fouda, M. A., & Donia, A. M., (2007). Influence of a niosomal formulation on the oral bioavailability of acyclovir in rabbits. *AAPS Pharm. Sci. Tech., 8*(4), 206–212.

100. Arzani, G., Haeri, A., Daeihamed, M., Bakhtiari-Kaboutaraki, H., & Dadashzadeh, S., (2015). Niosomal carriers enhance oral bioavailability of carvedilol: Effects of bile salt-enriched vesicles and carrier surface charge. *International Journal of Nanomedicine, 10*, 4797.

101. Ullah, S., Shah, M. R., Shoaib, M., Imran, M., Elhissi, A. M., Ahmad, F., et al., (2016). Development of a biocompatible creatinine-based niosomal delivery system for enhanced oral bioavailability of clarithromycin. *Drug Delivery, 23*(9), 3480–3491.

102. Imran, M., Shah, M. R., Ullah, F., Ullah, S., Elhissi, A. M., Nawaz, W., et al., (2016). Sugar-based novel niosomalnanocarrier system for enhanced oral bioavailability of levofloxacin. *Drug Delivery, 23*(9), 3653–3664.

103. Jadon, P. S., Gajbhiye, V., Jadon, R. S., Gajbhiye, K. R., & Ganesh, N., (2009). Enhanced oral bioavailability of griseofulvin via niosomes. *AAPS Pharm. Sci. Tech., 10*(4), p. 1186.

104. Kamboj, S., Saini, V., & Bala, S., (2014). Formulation and characterization of drug loaded nonionic surfactant vesicles (niosomes) for oral bioavailability enhancement. *The Scientific World Journal.* Article ID 959741, pp. 1–8.

105. Song, S., Tian, B., Chen, F., Zhang, W., Pan, Y., Zhang, Q., et al., (2015). Potentials of proniosomes for improving the oral bioavailability of poorly water-soluble drugs. *Drug Development and Industrial Pharmacy, 41*(1), 51–62.

106. Veerareddy, P. R., & Bobbala, S. K. R., (2013). Enhanced oral bioavailability of isradipine via proniosomal systems. *Drug Development and Industrial Pharmacy, 39*(6), 909–917.

107. Lopez-Montilla, J. C., Herrera-Morales, P. E., Pandey, S., & Shah, D. O., (2002). Spontaneous emulsification: Mechanisms, physicochemical aspects, modeling, and applications. *Journal of Dispersion Science and Technology, 23*(1–3), 219–268.

108. Gupta, A., Eral, H. B., Hatton, T. A., & Doyle, P. S., (2016). Nanoemulsions: Formation, properties and applications. *Soft Matter, 12*(11), 2826–2841.

109. Laxmi, M., Bhardwaj, A., Mehta, S., & Mehta, A., (2014). Development and characterization of nanoemulsion as carrier for the enhancement of bioavailability of artemether. *Artificial Cells, Nanomedicine, and Biotechnology, 43*(5), 334–344.

110. Thakkar, H. P., Khunt, A., Dhande, R. D., & Patel, A. A., (2015). Formulation and evaluation of Itraconazole nanoemulsion for enhanced oral bioavailability. *Journal of Microencapsulation, 32*(6), 559–569.

111. Tiwari, S. B., & Amiji, M. M., (2006). Improved oral delivery of paclitaxel following administration in nanoemulsion formulations. *Journal of Nanoscience and Nanotechnology, 6*(9–1), 3215–3221.

112. Puglia, C., Rizza, L., Drechsler, M., & Bonina, F., (2010). Nanoemulsions as vehicles for topical administration of glycyrrhetic acid: Characterization and *in vitro* and *in vivo* evaluation. *Drug Delivery, 17*(3), 123–129.

113. Anton, N., Benoit, J. P., & Saulnier, P., (2008). Design and production of nanoparticles formulated from nano-emulsion templates-a review. *Journal of Controlled Release, 128*(3), 185–199.

114. Jaiswal, M., Dudhe, R., & Sharma, P. K., (2015). Nanoemulsion: An advanced mode of drug delivery system. *3 Biotech, 5*(2), 123–127.

115. Kotta, S., Khan, A. W., Pramod, K., Ansari, S. H., Sharma, R. K., & Ali, J., (2012). Exploring oral nanoemulsions for bioavailability enhancement of poorly water-soluble drugs. *Expert Opinion on Drug Delivery, 9*(5), 585–598.

116. Mason, T. G., (1999). New fundamental concepts in emulsion rheology. *Current Opinion in Colloid & Interface Science, 4*(3), 231–238.

117. Tagne, J. B., Kakumanu, S., & Nicolosi, R. J., (2008). Nanoemulsion preparations of the anticancer drug dacarbazine significantly increase its efficacy in a xenograft mouse melanoma model. *Molecular Pharmaceutics, 5*(6), 1055–1063.

118. Wang, X., Wang, Y. W., & Huang, Q., (2009). Enhancing stability and oral bioavailability of polyphenols using nanoemulsions. *ACS Symposium Series, 1007*, 198–212.

119. Kunieda, H., Fukui, Y., Uchiyama, H., & Solans, C., (1996). Spontaneous formation of highly concentrated water-in-oil emulsions (gel-emulsions). *Langmuir, 12*(9), 2136–2140.

120. Uson, N., Garcia, M. J., & Solans, C., (2004). Formation of water-in-oil (W/O) nanoemulsions in a water/mixed non-ionic surfactant/oil systems prepared by a low-energy emulsification method. Colloids and surfaces A. *Physicochemical and Engineering Aspects, 250*(1), 415–421.

121. Bouchemal, K., Briançon, S., Perrier, E., & Fessi, H., (2004). Nano-emulsion formulation using spontaneous emulsification: Solvent, oil and surfactant optimization. *International Journal of Pharmaceutics, 280*(1), 241–251.

122. Ma, Y., Li, H., & Guan, S., (2015). Enhancement of the oral bioavailability of breviscapine by nanoemulsions drug delivery system. *Drug Development and Industrial Pharmacy, 41*(2), 177–182.

123. Zhao, L., Wei, Y., Huang, Y., He, B., Zhou, Y., & Fu, J., (2013). Nanoemulsion improves the oral bioavailability of baicalin in rats: *In vitro* and *in vivo* evaluation. *International Journal of Nanomedicine, 8*, p. 3769.

124. Devalapally, H., Silchenko, S., Zhou, F., Mc Dade, J., Goloverda, G., Owen, A., et al., (2013). Evaluation of a nanoemulsion formulation strategy for oral bioavailability enhancement of danazol in rats and dogs. *Journal of Pharmaceutical Sciences, 102*(10), 3808–3815.

125. Zhou, J., Zhou, M., Yang, F. F., Liu, C. Y., Pan, R. L., Chang, Q., et al., (2015). Involvement of the inhibition of intestinal glucuronidation in enhancing the oral bioavailability of resveratrol by labrasol containing nanoemulsions. *Molecular Pharmaceutics, 12*(4), 1084–1095.

126. Sessa, M., Balestrieri, M. L., Ferrari, G., Servillo, L., Castaldo, D., D'Onofrio, N., et al., (2014). Bioavailability of encapsulated resveratrol into nanoemulsion-based delivery systems. *Food Chemistry, 147*, 42–50.

127. Yu, H., & Huang, Q., (2012). Improving the oral bioavailability of curcumin using novel organogel-based nanoemulsions. *Journal of Agricultural and Food Chemistry, 60*(21), 5373–5379.

128. Li, Y., Song, C. K., Kim, M. K., Lim, H., Shen, Q., Lee, D. H., et al., (2015). Nanomemulsion of megestrol acetate for improved oral bioavailability and reduced food effect. *Archives of Pharmacal Research, 38*(10), 1850–1856.

129. Shen, Q., Wang, Y., & Zhang, Y., (2011). Improvement of colchicine oral bioavailability by incorporating eugenol in the nanoemulsion as an oil excipient and enhancer. *Int. J. Nanomedicine, 6*, 1237–1243.

130. Mohsin, K., Rayan, A. A., Raish, M., Alanazi, F. K., & Hussain, M. D., (2016). Development of self-nanoemulsifying drug delivery systems for the enhancement of solubility and oral bioavailability of fenofibrate, a poorly water-soluble drug. *International Journal of Nanomedicine, 11*, p. 2829.

131. Liu, R. (2018). *Water-Insoluble Drug Formulation, 3rd edn.*; CRC Press: New York.

132. Khan, A. W., Kotta, S., Ansari, S. H., Sharma, R. K., & Ali, J., (2014). Self-nanoemulsifying drug delivery system (SNEDDS) of the poorly water-soluble grapefruit flavonoid Naringenin: Design, characterization, *in vitro* and *in vivo* evaluation. *Drug Delivery, 22*(4), 552–561.

133. Miao, Y., Sun, J., Chen, G., Lili, R., & Ouyang, P., (2016). Enhanced oral bioavailability of lurasidone by self-nanoemulsifying drug delivery system in fasted state. *Drug Development and Industrial Pharmacy, 42*(8), 1234–1240.

134. Senapati, P. C., Sahoo, S. K., & Sahu, A. N., (2016). Mixed surfactant based (SNEDDS) self-nanoemulsifying drug delivery system presenting efavirenz for enhancement of oral bioavailability. *Biomedicine & Pharmacotherapy, 80*, 42–51.

135. Ke, Z., Hou, X., & Jia, X. B., (2016). Design and optimization of self-nanoemulsifying drug delivery systems for improved bioavailability of cyclovirobuxine D. *Drug Design, Development and Therapy, 10*, p. 2049.

136. Li, Z., Zhang, W., Gao, Y., Xiang, R., Liu, Y., Hu, M., et al., (2017). Development of self-nanoemulsifying drug delivery system for oral bioavailability enhancement of valsartan in beagle dogs. *Drug Delivery and Translational Research, 7*(1), 100–110.

137. Singh, G., & Pai, R. S., (2014). Optimized self-nanoemulsifying drug delivery system of atazanavir with enhanced oral bioavailability: *In vitro/in vivo* characterization. *Expert Opinion on Drug Delivery, 11*(7), 1023–1032.

138. Garg, B., Katare, O. P., Beg, S., Lohan, S., & Singh, B., (2016). Systematic development of solid self-nanoemulsifying oily formulations (S-SNEOFs) for enhancing the oral bioavailability and intestinal lymphatic uptake of lopinavir. *Colloids and Surfaces B. Biointerfaces, 141*, 611–622.

139. Nekkanti, V., Wang, Z., & Betageri, G. V., (2016). Pharmacokinetic evaluation of improved oral bioavailability of valsartan: Proliposomes versus self-nanoemulsifying drug delivery system. *AAPS Pharm. Sci. Tech., 17*(4), 851–862.

140. Abo, E. H. A., (2015). Self-nanoemulsifying drug-delivery system for improved oral bioavailability of rosuvastatin using natural oil antihyperlipdemic. *Drug Development and Industrial Pharmacy, 41*(7), 1047–1056.

141. Inugala, S., Eedara, B. B., Sunkavalli, S., Dhurke, R., Kandadi, P., Jukanti, R., et al., (2015). Solid self-nanoemulsifying drug delivery system (S-SNEDDS) of darunavir for improved dissolution and oral bioavailability: *In vitro* and *in vivo* evaluation. *European Journal of Pharmaceutical Sciences, 74*, 1–10.

142. Subramanian, N., Sharavanan, S. P., Chandrasekar, P., Balakumar, A., & Moulik, S. P., (2016). Lacidipine self-nanoemulsifying drug delivery system for the enhancement of oral bioavailability. *Archives of Pharmacal Research, 39*(4), 481–491.

143. Li, Z., Zhang, W., Gao, Y., Xiang, R., Liu, Y., Hu, M., et al., (2017). Development of self-nanoemulsifying drug delivery system for oral bioavailability enhancement of valsartan in beagle dogs. *Drug Delivery and Translational Research, 7*(1), 100–110.

144. Kamel, A. O., & Mahmoud, A. A., (2013). Enhancement of human oral bioavailability and *in vitro* antitumor activity of rosuvastatin via spray dried self-nanoemulsifying drug delivery system. *Journal of Biomedical Nanotechnology, 9*(1), 26–39.

145. Heshmati, N., Cheng, X., Eisenbrand, G., & Fricker, G., (2013). Enhancement of oral bioavailability of E804 by self nanoemulsifying drug delivery system (SNEDDS) in rats. *Journal of Pharmaceutical Sciences, 102*(10), 3792–3799.

146. Ke, Z., Hou, X., & Jia, X. B., (2016). Design and optimization of self-nanoemulsifying drug delivery systems for improved bioavailability of cyclovirobuxine D. *Drug Design, Development and Therapy, 10*, p. 2049.

147. Liu, H., Shang, K., Liu, W., Leng, D., Li, R., Kong, Y., & Zhang, T., (2014). Improved oral bioavailability of glyburide by a self-nanoemulsifying drug delivery system. *Journal of Microencapsulation, 31*(3), 277–283.

148. Patel, J., Dhingani, A., Garala, K., Raval, M., & Sheth, N., (2014). Quality by design approach for oral bioavailability enhancement of irbesartan by self-nanoemulsifying tablets. *Drug Delivery, 21*(6), 412–435.

149. Fahmy, U. A., Ahmed, O. A., & Hosny, K. M., (2015). Development and evaluation of avanafil self-nanoemulsifying drug delivery system with rapid onset of action and enhanced bioavailability. *AAPS Pharm. Sci. Tech., 16*(1), 53–58.

150. Tran, T. H., Guo, Y., Song, D., Bruno, R. S., & Lu, X., (2014). Quercetin containing self nanoemulsifying drug delivery system for improving oral bioavailability. *Journal of Pharmaceutical Sciences, 103*(3), 840–852.

151. Khattab, A., Hassanin, L., & Zaki, N., (2016). Self-nanoemulsifying drug delivery system of coenzyme (Q10) with improved dissolution, bioavailability, and protective efficiency on liver fibrosis. *AAPS Pharm. Sci. Tech.,* 1–16.

152. Li, W., Yi, S., Wang, Z., Chen, S., Xin, S., Xie, J., & Zhao, C., (2011). Self-nanoemulsifying drug delivery system of persimmon leaf extract: Optimization and bioavailability studies. *International Journal of Pharmaceutics, 420*(1), 161–171.

153. Nasr, A., Gardouh, A., & Ghorab, M., (2016). Novel solid self-nanoemulsifying drug delivery system (S-SNEDDS) for oral delivery of olmesartan medoxomil: Design, formulation, pharmacokinetic and bioavailability evaluation. *Pharmaceutics, 8*(3), p. 20.

154. Kang, J. H., Oh, D. H., Oh, Y. K., Yong, C. S., & Choi, H. G., (2012). Effects of solid carriers on the crystalline properties, dissolution and bioavailability of flurbiprofen in solid self-nanoemulsifying drug delivery system (solid SNEDDS). *European Journal of Pharmaceutics and Biopharmaceutics, 80*(2), 289–297.

155. Seo, Y. G., Kim, D. W., Yousaf, A. M., Park, J. H., Chang, P. S., Baek, H. H., et al., (2015). Solid self-nanoemulsifying drug delivery system (SNEDDS) for enhanced oral bioavailability of poorly water-soluble tacrolimus: Physicochemical characterisation and pharmacokinetics. *Journal of Microencapsulation, 32*(5), 503–510.

156. Soliman, K. A., Ibrahim, H. K., & Ghorab, M. M., (2016). Formulation of avanafil in a solid self-nanoemulsifying drug delivery system for enhanced oral delivery. *European Journal of Pharmaceutical Sciences, 93*, 447–455.

157. Nasr, A., Gardouh, A., & Ghorab, M., (2016). Novel solid self-nanoemulsifying drug delivery system (S-SNEDDS) for oral delivery of olmesartan medoxomil: Design, formulation, pharmacokinetic and bioavailability evaluation. *Pharmaceutics, 8*(3), p. 20.

158. Abo Enin, H. A., & Abdel-Bar, H. M., (2016). Solid super saturated self-nanoemulsifying drug delivery system (sat-SNEDDS) as a promising alternative to conventional SNEDDS for improvement rosuvastatin calcium oral bioavailability. *Expert Opinion on Drug Delivery, 13*(11), 1513–1521.

159. Chai, F., Sun, L., Ding, Y., Liu, X., Zhang, Y., Webster, T. J., (2016). A solid self-nanoemulsifying system of the BCS class IIb drug dabigatranetexilate to improve oral bioavailability. *Nanomedicine, 11*(14), 1801–1816.

160. Junyaprasert, V. B., & Morakul, B., (2015). Nanocrystals for enhancement of oral bioavailability of poorly water-soluble drugs. *Asian Journal of Pharmaceutical Sciences, 10*(1), 13–23.

161. Lai, F., Pini, E., Corrias, F., Perricci, J., Manconi, M., Fadda, A. M., et al., (2014). Formulation strategy and evaluation of nanocrystalpiroxicam orally disintegrating tablets manufacturing by freeze-drying. *International Journal of Pharmaceutics, 467*(1), 27–33.

162. Zhang, X., Zhang, T., Lan, Y., Wu, B., & Shi, Z., (2016). Nanosuspensions containing oridonin/HP-β-cyclodextrin inclusion complexes for oral bioavailability enhancement via improved dissolution and permeability. *AAPS Pharm. Sci. Tech., 17*(2), 400–408.

163. Taneja, S., Shilpi, S., & Khatri, K., (2016). Formulation and optimization of efavirenz nanosuspensions using the precipitation-ultrasonication technique for solubility enhancement. *Artificial cells, Nanomedicine, and Biotechnology, 44*(3), 978–984.

164. Ige, P. P., Baria, R. K., & Gattani, S. G., (2013). Fabrication of fenofibratenanocrystals by probe sonication method for enhancement of dissolution rate and oral bioavailability. *Colloids and Surfaces B. Biointerfaces, 108*, 366–373.

165. Fu, Q., Sun, J., Zhang, D., Li, M., Wang, Y., Ling, G., et al., (2013). Nimodipinenanocrystals for oral bioavailability improvement: Preparation, characterization and pharmacokinetic studies. *Colloids and Surfaces B. Biointerfaces., 109*, 161–166.

166. Mudshinge, S. R., Deore, A. B., Patil, S., & Bhalgat, C. M., (2011). Nanoparticles: Emerging carriers for drug delivery. *Saudi Pharmaceutical Journal, 19*(3), 129–141.

167. Zhang, L., Wang, S., Zhang, M., & Sun, J., (2013). Nanocarriers for oral drug delivery. *Journal of Drug Targeting, 21*(6), 515–527.

168. Vauthier, C., & Bouchemal, K., (2009). Methods for the preparation and manufacture of polymeric nanoparticles. *Pharmaceutical Research, 26*(5), 1025–1058.

169. Bennet, D., & Kim, S., (2014). Polymer nanoparticles for smart drug delivery. In: *Application of Nanotechnology in Drug Delivery.* InTechOpen: London; pp. 257–310.

170. Galindo-Rodriguez, S., Allémann, E., Fessi, H., & Doelker, E., (2004). Physicochemical parameters associated with nanoparticle formation in the salting-out, emulsification-diffusion, and nanoprecipitation methods. *Pharmaceutical Research, 21*(8), 1428–1439.

171. Zhang, X., Dong, Y., Zeng, X., Liang, X., Li, X., Tao, W., et al., (2014). The effect of autophagy inhibitors on drug delivery using biodegradable polymer nanoparticles in cancer treatment. *Biomaterials, 35*(6), 1932–1943.

172. Chaudhary, H., & Kumar, V., (2014). Taguchi design for optimization and development of antibacterial drug-loaded PLGA nanoparticles. *International Journal of Biological Macromolecules, 64*, 99–105.

173. Liu, M., Zhou, Z., Wang, X., Xu, J., Yang, K., Cui, Q., et al., (2007). Formation of poly (L, D-lactide) spheres with controlled size by direct dialysis. *Polymer, 48*(19), 5767–5779.

174. York, P., (1999). Strategies for particle design using supercritical fluid technologies. *Pharmaceutical Science & Technology Today, 2*(11), 430–440.

175. Reis, C. P., Neufeld, R. J., Ribeiro, A. J., & Veiga, F., (2006). Nanoencapsulation I. Methods for preparation of drug-loaded polymeric nanoparticles. Nanomedicine. *Nanotechnology, Biology and Medicine, 2*(1), 8–21.

176. Nagavarma, B. V. N., Yadav, H. K., Ayaz, A., Vasudha, L. S., & Shivakumar, H. G., (2012). Different techniques for preparation of polymeric nanoparticles-A review. *Asian J. Pharm. Clin. Res., 5*(3), 16–23.

177. Ham, H. T., Choi, Y. S., Chee, M. G., & Chung, I. J., (2006). Singlewall carbon nanotubes covered with polystyrene nanoparticles by in situ miniemulsion polymerization. *Journal of Polymer Science Part A. Polymer Chemistry, 44*(1), 573–584.

178. Ziegler, A., Landfester, K., & Musyanovych, A., (2009). Synthesis of phosphonate-functionalized polystyrene and poly (methyl methacrylate) particles and their kinetic behavior in miniemulsion polymerization. *Colloid and Polymer Science, 287*(11), p. 1261.

179. Yong, C. P., & Gan, L. M., (2005). Microemulsion polymerizations and reactions. In: *Polymer Particles* (pp. 257–298). Springer Berlin Heidelberg.

180. Karode, S. K., Kulkarni, S. S., Suresh, A. K., & Mashelkar, R. A., (1998). New insights into kinetics and thermodynamics of interfacial polymerization. *Chemical Engineering Science, 53*(15), 2649–2663.

181. Yuan, Z., Ye, Y., Gao, F., Yuan, H., Lan, M., Lou, K., & Wang, W., (2013). Chitosan-graft-β-cyclodextrin nanoparticles as a carrier for controlled drug release. *International Journal of Pharmaceutics, 446*(1), 191–198.

182. Alamdarnejad, G., Sharif, A., Taranejoo, S., Janmaleki, M., Kalaee, M. R., Dadgar, M., et al., (2013). Synthesis and characterization of thiolatedcarboxymethyl chitosan-graft-cyclodextrin nanoparticles as a drug delivery vehicle for albendazole. *Journal of Materials Science: Materials in Medicine, 24*(8), 1939–1949.

183. Hao, S., Wang, B., Wang, Y., & Xu, Y., (2014). Enteric-coated sustained-release nanoparticles by coaxial electrospray: Preparation, characterization, and *in vitro* evaluation. *Journal of Nanoparticle Research, 16*(2), p. 2204.

184. Gomez, A., Bingham, D., De Juan, L., & Tang, K., (1998). Production of protein nanoparticles by electrospray drying. *Journal of Aerosol Science, 29*(5), 561–574.

185. Liversidge, G. G., & Cundy, K. C., (1995). Particle size reduction for improvement of oral bioavailability of hydrophobic drugs: I. Absolute oral bioavailability of nanocrystalline danazol in beagle dogs. *International Journal of Pharmaceutics, 125*(1), 91–97.

186. Valencia, P. M., Farokhzad, O. C., Karnik, R., & Langer, R., (2012). Microfluidic technologies for accelerating the clinical translation of nanoparticles. *Nature Nanotechnology, 7*(10), 623–629.

187. Khalil, N. M., Do Nascimento, T. C. F., Casa, D. M., Dalmolin, L. F., De Mattos, A. C., Hoss, I., et al., (2013). Pharmacokinetics of curcumin-loaded PLGA and PLGA–PEG blend nanoparticles after oral administration in rats. *Colloids and Surfaces B. Biointerfaces., 101*, 353–360.

188. Roger, E., Kalscheuer, S., Kirtane, A., Guru, B. R., Grill, A. E., Whittum-Hudson, J., et al., (2012). Folic acid-functionalized nanoparticles for enhanced oral drug delivery. *Molecular Pharmaceutics, 9*(7), p. 2103.

189. Yin, L., Ding, J., He, C., Cui, L., Tang, C., & Yin, C., (2009). Drug permeability and mucoadhesion properties of thiolated trimethyl chitosan nanoparticles in oral insulin delivery. *Biomaterials, 30*(29), 5691–5700.

190. Xiong, X. Y., Li, Y. P., Li, Z. L., Zhou, C. L., Tam, K. C., Liu, Z. Y., & Xie, G. X., (2007). Vesicles from Pluronic/poly (lactic acid) block copolymers as new carriers for oral insulin delivery. *Journal of Controlled Release, 120*(1), 11–17.

191. Gaucher, G., Satturwar, P., Jones, M. C., Furtos, A., & Leroux, J. C., (2010). Polymeric micelles for oral drug delivery. *European Journal of Pharmaceutics and Biopharmaceutics, 76*(2), 147–158.

192. Fricker, G., Kromp, T., Wendel, A., Blume, A., Zirkel, J., Rebmann, H., et al., (2010). Phospholipids and lipid-based formulations in oral drug delivery. *Pharmaceutical Research, 27*(8), 1469–1486.

193. Chen, L. C., Chen, Y. C., Su, C. Y., Wong, W. P., Sheu, M. T., & Ho, H. O., (2016). Development and characterization of lecithin-based self-assembling mixed polymeric micellar (sa MPMs) drug delivery systems for curcumin. *Scientific Reports, 6*, 37–122.

194. Basalious, E. B., & Shamma, R. N., (2015). Novel self-assembled nano-tubular mixed micelles of Pluronics P123, Pluronic F127 and phosphatidylcholine for oral delivery of nimodipine: *In vitro* characterization, *ex vivo* transport and *in vivo* pharmacokinetic studies. *International Journal of Pharmaceutics, 493*(1), 347–356.

195. Svenson, S., & Chauhan, A. S., (2008). Dendrimers for enhanced drug solubilisation. *Nanomedicine, 3*(5), 679–702.

196. Jain, N. K., & Gupta, U., (2008). Application of dendrimer–drug complexation in the enhancement of drug solubility and bioavailability. *Expert Opinion on Drug Metabolism & Toxicology, 4*(8), 1035–1052.

197. Milhem, O. M., Myles, C., Mc Keown, N. B., Attwood, D., & D'Emanuele, A., (2000). Polyamidoamine Starburst® dendrimers as solubility enhancers. *International Journal of Pharmaceutics, 197*(1), 239–241.

198. Asthana, A., Chauhan, A. S., Diwan, P. V., & Jain, N. K., (2005). Poly (amidoamine) (PAMAM) dendritic nanostructures for controlled sitespecific delivery of acidic anti-inflammatory active ingredient. *AAPS Pharm. Sci. Tech., 6*(3), 536–542.

199. Devarakonda, B., Hill, R. A., Liebenberg, W., Brits, M., & De Villiers, M. M., (2005). Comparison of the aqueous solubilization of practically insoluble niclosamide by polyamidoamine (PAMAM) dendrimers and cyclodextrins. *International Journal of Pharmaceutics, 304*(1), 193–209.

200. Huang, X., Wu, Z., Gao, W., Chen, Q., & Yu, B., (2011). Polyamidoamine dendrimers as potential drug carriers for enhanced aqueous solubility and oral bioavailability of silybin. *Drug Development and Industrial Pharmacy, 37*(4), 419–427.

201. Gu, L., Wu, Z. H., Qi, X., He, H., Ma, X., Chou, X., et al., (2013). Polyamidomine dendrimers: An excellent drug carrier for improving the solubility and bioavailability of puerarin. *Pharmaceutical Development and Technology, 18*(5), 1051–1057.
202. Ke, W., Zhao, Y., Huang, R., Jiang, C., & Pei, Y., (2008). Enhanced oral bioavailability of doxorubicin in a dendrimer drug delivery system. *Journal of Pharmaceutical Sciences, 97*(6), 2208–2216.
203. Yiyun, C., & Tongwen, X., (2005). Dendrimers as potential drug carriers. Part I. Solubilization of non-steroidal anti-inflammatory drugs in the presence of polyamidoamine dendrimers. *European Journal of Medicinal Chemistry, 40*(11), 1188–1192.
204. Bhadra, D., Bhadra, S., Jain, S., & Jain, N. K., (2003). A PEGylated dendritic nanoparticulate carrier of fluorouracil. *International Journal of Pharmaceutics, 257*(1), 111–124.
205. Bhadra, D., Bhadra, S., & Jain, N. K., (2005). Pegylated lysine based copolymeric dendritic micelles for solubilization and delivery of artemether. *J. Pharm. Pharm. Sci., 8*(3), 467–82.
206. Kabanov, A. V., & Vinogradov, S. V., (2009). Nanogels as pharmaceutical carriers: Finite networks of infinite capabilities. *Angewandte Chemie International Edition, 48*(30), 5418–5429.
207. Yao, Y., Xia, M., Wang, H., Li, G., Shen, H., Ji, G., et al., (2016). Preparation and evaluation of chitosan-based nanogels/gels for oral delivery of myricetin. *European Journal of Pharmaceutical Sciences, 91*, 144–153.
208. Guo, J., Ma, M., Chang, D., Zhang, Q., Zhang, C., Yue, Y., et al., (2015). Poly-α, β-polyasparthydrazide-based nanogels for potential oral delivery of paclitaxel: *In vitro* and *in vivo* properties. *Journal of Biomedical Nanotechnology, 11*(12), 2231–2242.
209. El-Sherbiny, I. M., Abdel-Mogib, M., Dawidar, A. A. M., Elsayed, A., & Smyth, H. D., (2011). Biodegradable p H-responsive alginate-poly (lactic-co-glycolic acid) nano/micro hydrogel matrices for oral delivery of silymarin. *Carbohydrate Polymers, 83*(3), 1345–1354.
210. Davis, M. E., & Brewster, M. E., (2004). Cyclodextrin-based pharmaceutics: Past, present and future. *Nature Reviews Drug Discovery, 3*(12), 1023–1035.
211. Del Valle, E. M., (2004). Cyclodextrins and their uses: A review. *Process Biochemistry, 39*(9), 1033–1046.
212. Rasheed, A., & VVNS, S. S., (2008). Cyclodextrins as drug carrier molecule: A review. *Scientia Pharmaceutica, 76*(4), 567–598.
213. Marzouk, M. A., Kassem, A. A., Samy, A. M., & Amer, R. I., (2010). Comparative evaluation of ketoconazole-β-cyclodextrin systems prepared by coprecipitation and kneading. *Drug Discoveries & Therapeutics, 4*(5), 380–387.
214. Miclea, L. M., Vlaia, L., Vlaia, V., Hădărugă, D. I., & Mircioiu, C., (2010). Preparation and characterization of inclusion complexes of meloxicam and α-cyclodextrin and β-cyclodextrin. *Farmacia, 58*(5), 583–593.
215. Ghosh, A., Biswas, S., & Ghosh, T., (2011). Preparation and evaluation of silymarin β-cyclodextrin molecular inclusion complexes. *Journal of Young Pharmacists, 3*(3), 205–210.

216. Sapkal, N. P., Kilor, V. A., Bhursari, K. P., & Daud, A. S., (2007). Evaluation of some methods for preparing gliclazide-β-cyclodextrin inclusion complexes. *Tropical Journal of Pharmaceutical Research, 6*(4), 833–840.

217. Heydari, A., Iranmanesh, M., Doostan, F., & Sheibani, H., (2015). Preparation of inclusion complex between nifedipine and ethylenediamine-β-cyclodextrin as nanocarrier agent. *Pharmaceutical Chemistry Journal, 49*(9), 605–612.

218. Nair, A. B., Attimarad, M., Al-Dhubiab, B. E., Wadhwa, J., Harsha, S., & Ahmed, M., (2014). Enhanced oral bioavailability of acyclovir by inclusion complex using hydroxypropyl-β-cyclodextrin. *Drug Delivery, 21*(7), 540–547.

219. Semalty, M., Panchpuri, M., Singh, D., & Semalty, A., (2014). Cyclodextrin inclusion complex of racecadotril: Effect of drug-β-cyclodextrin ratio and the method of complexation. *Current Drug Discovery Technologies, 11*(2), 154–161.

220. Sapana, B. B., & Shashikant, D. N., (2015). Preparation and characterization of [beta]-cyclodextrinnebivolol inclusion complex. *International Journal of Pharmaceutical Sciences and Research, 6*(5), p. 2205.

221. Sambasevam, K. P., Mohamad, S., Sarih, N. M., & Ismail, N. A., (2013). Synthesis and characterization of the inclusion complex of β-cyclodextrin and azomethine. *International Journal of Molecular Sciences, 14*(2), 3671–3682.

222. Badr-Eldin, S. M., Ahmed, T. A., & Ismail, H. R., (2013). Aripiprazole-cyclodextrin binary systems for dissolution enhancement: Effect of preparation technique, cyclodextrin type and molar ratio. *Iranian Journal of Basic Medical Sciences, 16*(12), p. 1223.

223. Wu, J., Shen, Q., & Fang, L., (2013). Sulfobutylether-β-cyclodextrin/chitosan nanoparticles enhance the oral permeability and bioavailability of docetaxel. *Drug Development and Industrial Pharmacy, 39*(7), 1010–1019.

224. Jantarat, C., Sirathanarun, P., Ratanapongsai, S., Watcharakan, P., Sunyapong, S., & Wadu, A., (2014). Curcumin-hydroxypropyl-β-cyclodextrin inclusion complex preparation methods: Effect of common solvent evaporation, freeze drying, and pH shift on solubility and stability of curcumin. *Tropical Journal of Pharmaceutical Research, 13*(8), 1215–1223.

225. Zhou, H. Y., Jiang, L. J., Zhang, Y. P., & Li, J. B., (2012). β-Cyclodextrin inclusion complex: Preparation, characterization, and its aspirin release *in vitro. Frontiers of Materials Science,* 1–9.

226. Nacsa, A., Ambrus, R., Berkesi, O., Szabo-Revesz, P., & Aigner, Z., (2008). Water-soluble loratadine inclusion complex: Analytical control of the preparation by microwave irradiation. *Journal of Pharmaceutical and Biomedical Analysis, 48*(3), 1020–1023.

227. Kadari, A., Gudem, S., Kulhari, H., Bhandi, M. M., Borkar, R. M., Kolapalli, V. R. M., et al., (2017). Enhanced oral bioavailability and anticancer efficacy of fisetin by encapsulating as inclusion complex with HPβCD in polymeric nanoparticles. *Drug Delivery, 24*(1), 224–232.

228. Nair, A. B., Attimarad, M., Al-Dhubiab, B. E., Wadhwa, J., Harsha, S., & Ahmed, M., (2014). Enhanced oral bioavailability of acyclovir by inclusion complex using hydroxypropyl-β-cyclodextrin. *Drug Delivery, 21*(7), 540–547.

229. Castiglione, F., Crupi, V., Majolino, D., Mele, A., Panzeri, W., Rossi, B., et al., (2013). Vibrational dynamics and hydrogen bond properties of β-CD nanosponges: an FTIR-ATR, Raman and solid-state NMR spectroscopic study. *Journal of Inclusion Phenomena and Macrocyclic Chemistry, 75*(3–4), 247–254.

230. Trotta, F., Zanetti, M., & Cavalli, R., (2012). Cyclodextrin-based nanosponges as drug carriers. *Beilstein Journal of Organic Chemistry, 8*(1), 2091–2099.

231. Trotta, F., (2011). Cyclodextrin nanosponges and their applications. Cyclodextrins in pharmaceutics, cosmetics, and biomedicine. *Current and Future Industrial Applications, 323–342.*

232. Torne, S., Darandale, S., Vavia, P., Trotta, F., & Cavalli, R., (2013). Cyclodextrin-based nanosponges: Effective nanocarrier for Tamoxifen delivery. *Pharmaceutical Development and Technology, 18*(3), 619–625.

233. Chilajwar, S. V., Pednekar, P. P., Jadhav, K. R., Gupta, G. J., & Kadam, V. J., (2014). Cyclodextrin-based nanosponges: A propitious platform for enhancing drug delivery. *Expert Opinion on Drug Delivery, 11*(1), 111–120.

234. Rao, M. R., & Shirsath, C., (2016). Enhancement of bioavailability of non-nucleoside reverse transciptase inhibitor using nanosponges. *AAPS Pharm. Sci. Tech., 1–11.*

235. Shende, P. K., Gaud, R. S., Bakal, R., & Patil, D., (2015). Effect of inclusion complexation of meloxicam with β-cyclodextrin-and β-cyclodextrin-based nanosponges on solubility, *in vitro* release and stability studies. *Colloids and Surfaces B. Biointerfaces, 136,* 105–110.

236. Torne, S. J., Ansari, K. A., Vavia, P. R., Trotta, F., & Cavalli, R., (2010). Enhanced oral paclitaxel bioavailability after administration of paclitaxel-loaded nanosponges. *Drug Delivery, 17*(6), 419–425.

237. Monica, R., Amrita, B., Ishwar, K., & Ghanshyam, M., (2012). *In vitro* and *in vivo* evaluation of b-cyclodextrin-based nanosponges of telmisartan. *J. Incl. Phenom. Macrocycl., Chem. 77*(1–4), 135–145.

238. Javed, I., Hussain, S. Z., Shahzad, A., Khan, J. M., Rehman, M., Usman, F., et al., (2016). Lecithin-gold hybrid nanocarriers as efficient and pH selective vehicles for oral delivery of diacerein-*In-vitro* and *in-vivo* study. *Colloids and Surfaces B: Biointerfaces, 141,* 1–9.

239. Dey, S., & Sreenivasan, K., (2015). Conjugating curcuminto water soluble polymer stabilized gold nanoparticles via pH responsive succinate linker. *Journal of Materials Chemistry B, 3*(5), 824–833.

240. Li, J., Xu, L., Yang, B., Wang, H., Bao, Z., Pan, W., et al., (2015). Facile synthesis of functionalized ionic surfactant template mesoporous silica for incorporation of poorly water-soluble drug. *International Journal of Pharmaceutics, 492*(1), 191–198.

241. Hartono, S. B., Hadisoewignyo, L., Yang, Y., Meka, A. K., & Yu, C., (2016). Amine functionalized cubic mesoporous silica nanoparticles as an oral delivery system for curcumin bioavailability enhancement. *Nanotechnology, 27*(50), p. 505605.

242. Jadhav, N. V., & Vavia, P. R., (2017). Dodecylamine template-based hexagonal mesoporous silica (HMS) as a carrier for improved oral delivery of fenofibrate. *AAPS Pharm. Sci. Tech., 1–10.*

243. Yang, L., Shao, Y., & Han, H. K., (2016). Aminoclay–lipid hybrid composite as a novel drug carrier of fenofibrate for the enhancement of drug release and oral absorption. *International Journal of Nanomedicine, 11*, p. 1067.

244. Zhang, Y., Zhao, Q., Zhu, W., Zhang, L., Han, J., Lin, Q., & Ai, F., (2015). Synthesis and evaluation of mesoporous carbon/lipid bilayer nanocomposites for improved oral delivery of the poorly water-soluble drug, nimodipine. *Pharmaceutical Research, 32*(7), 2372–2383.

245. Cirri, M., Mennini, N., Maestrelli, F., Mura, P., Ghelardini, C., & Di Cesare Mannelli, L., (2017). Development and *in vivo* evaluation of an innovative "hydrochlorothiazide-in cyclodextrins-in solid lipid nanoparticles" formulation with sustained release and enhanced oral bioavailability for potential hypertension treatment in pediatrics. *International Journal of Pharmaceutics, 521*(1), 73–83.

246. Sangwai, M., & Vavia, P., (2013). Amorphous ternary cyclodextrin nanocomposites of telmisartan for oral drug delivery: Improved solubility and reduced pharmacokinetic variability. *International Journal of Pharmaceutics, 453*(2), 423–432.

247. Fissan, H., Ristig, S., Kaminski, H., Asbach, C., & Epple, M., (2014). Comparison of different characterization methods for nanoparticle dispersions before and after aerosolization. *Analytical Methods, 6*(18), 7324–7334.

248. Urbán-Morlán, Z., Ganem-Rondero, A., Melgoza-Contreras, L. M., Escobar-Chávez, J. J., Nava-Arzaluz, M. G., & Quintanar-Guerrero, D., (2010). Preparation and characterization of solid lipid nanoparticles containing cyclosporine by the emulsification-diffusion method. *Int. J. Nanomedicine, 5*, 611–620.

249. Michael, T., (2010). The characterization of nanoparticles. *Analytical Methods Committee Technical Briefs (AMCTB), 48*, 1–3.

250. Jiménez-Lamana, J., Laborda, F., Bolea, E., Abad-Álvaro, I., Castillo, J. R., Bianga, J., He, M., Bierla, K., Mounicou, S., Ouerdane, L., & Gaillet, S. (2014). An insight into silver nanoparticles bioavailability in rats. *Metallomics, 6*(12), 2242–2249.

251. Song, K. S., Sung, J. H., Ji, J. H., Lee, J. H., Lee, J. S., Ryu, H. R., Lee, J. K., Chung, Y. H., Park, H. M., Shin, B. S., & Chang, H. K. (2013). Recovery from silver-nanoparticle-exposure-induced lung inflammation and lung function changes in Sprague Dawley rats. *Nanotoxicology, 7*(2), 169–180.

252.   Kim, M. K., Lee, J. A., Jo, M. R., & Choi, S. J. (2016). Bioavailability of silica, titanium dioxide, and zinc oxide nanoparticles in rats. *Journal of Nanoscience and Nanotechnology, 16*(6), 6580–6586.

253. Balasubramanyam, A., Sailaja, N., Mahboob, M., Rahman, M. F., Hussain, S. M., & Grover, P. (2009). *In vivo* genotoxicity assessment of aluminum oxide nanomaterials in rat peripheral blood cells using the comet assay and micronucleus test. *Mutagenesis, 24*(3), 245–251.

CHAPTER 6

RECENT PERSPECTIVES OF CHALCONE-BASED MOLECULES AS PROTEIN TYROSINE PHOSPHATASE 1B (PTP1B) INHIBITORS

DEBARSHI KAR MAHAPATRA, SANJAY KUMAR BHARTI, and VIVEK ASATI

Department of Pharmaceutical Chemistry, Dadasaheb Balpande College of Pharmacy, Rashtrasant Tukadoji Maharaj Nagpur University, Nagpur, Maharashtra, India

Institute of Pharmaceutical Sciences, Guru Ghasidas Vishwavidyalaya (A Central University), Bilaspur – 495009, Chhattisgarh, India, E-mail: skbharti.ggu@gmail.com

ABSTRACT

Diabetes mellitus (DM) is a heterogeneous group of disorders which is characterized by increased blood sugar level, altered metabolism of lipids, carbohydrates, and proteins and increased risk of complications from vascular disease. Protein Tyrosine Phosphatase 1B (PTP1B) has gained adequate notice due to its crucial role in type 2 diabetes (t2D) and obesity as a negative regulator of the insulin and leptin-signaling pathway. PTP-1B is primarily responsible for dephosphorylation of the insulin receptor and thus down regulates insulin signaling. PTP1B inhibitors are the latest candidate for the management of diabetes, where they prevent dephosphorylation of the insulin receptor and consequently increase insulin level. Natural products have been reported to exhibit promising anti-diabetic activity. Chalcones or 1,3-diphenyl-2E-propene-1-one, the open chain intermediate in aurones synthesis of flavones containing benzylideneacetophenone scaffold, where the two aromatic nuclei are joined by a three-carbon α, β unsaturated carbonyl bridge have shown tremendous PTP1B inhibition. In this chapter, a concrete focus on pharmacology, mechanism of action, and structural aspects along with substituents required for modulating PTP1B has been discussed. Still, none of these inhibitors have gained adequate attention at present and need to be explored and evaluated properly in terms of efficacy and toxicity to develop as therapeutic agents/formulations for the management of diabetes in future.

6.1   INTRODUCTION

Diabetes Mellitus (DM) is a heterogeneous group of disorders which is characterized by increased blood sugar level, altered metabolism of lipids, carbohydrates, and proteins and increased risk of complications from vascular disease [1]. The chronic hyperglycemic conditions are associated with dysfunction and failure of major organs like heart, eyes, nerves, blood vessels and kidneys [2]. The American Diabetes Association (ADA) defines that DM is characterized by polyuria, polydipsia, polyphagia, glycosuria, unexplained weight loss and random plasma glucose concentration of greater than 200 mg/dL along with fasting plasma glucose concentration of greater than 126 mL/dL [3]. Variations in normal glucose homeostasis occur by numerous factors like impaired insulin secretion, hepatic gluconeogenesis and reduced uptake of glucose by skeletal muscle, adipose tissues and liver [4]. In the case of type I diabetes, the body does not produce enough insulin that is required to convert sugar, starches, etc. into energy. Type II diabetes (t2D) is a condition characterized by situation where cells do not properly use insulin as a result of "resistance" [5]. The most prominent features of type II diabetes is decreased sensitivity of muscle and adipose cells to insulin. T2D is often characterized by intrinsic problems like compliance, ineffectiveness and hypoglycemic episodes with insulin and the sulfonylureas. Administration of glitazones are not effective in all t2D patients, therefore, the great need for more effective orally administered agents particularly ones that normalize both glucose and insulin levels still remains a challenge [6]. Insulin is secreted in two discrete phases from pancreatic β-cells which influence the magnitude of both fasting and postprandial blood glucose concentrations. In the beginning, a rapid release of insulin occurs, when the glucose concentration increases concurrently after a meal, which is followed by a phase of sustained increase in circulating insulin concentrations [7].

Generally, those compounds which increase the sensitivity of muscle and adipose to insulin (insulin sensitizers) are foremost choice for successful treatment of DM. For diabetotherapy, several enzyme inhibitors had been developed so far, of them Dipeptidyl Peptidase-4 (DPP-4) and PTP1B inhibitors are of foremost importance. These compounds prolong the duration of insulin by preventing its degradation/inactivation [8]. Protein Tyrosine Phosphatase 1B (PTP1B) inhibitors are the latest candidate for the management of diabetes. A large number of PTP1B inhibitors having tyrosine mimetic structures, functionalized with negatively charged moieties such as phosphonates, malonates, carboxylates, or cinnamates have been developed [9]. Recently, two inhibitors ertiprotafib and trodusquemine have advanced into clinical trials for the treatment of diabetes and obesity. Although, the second phase clinical trial for ertiprotafib was discontinued due to lack of efficacy [10]. Natural and semi (synthetic) chalcones have shown significant anti-diabetic property by inhibiting PTP1B enzyme without showing major associated diabetic complications. At present, these inhibitors have not received adequate attention and need further exploration regarding efficacy and toxicological profiles to develop as formulations.

6.2 PTP1B: ROLE IN DIABETES MELLITUS

Protein kinases and phosphatases are groups of enzymes responsible for mediating various intra-cellular functions such as mediation of metabolic and cellular actions of insulin, etc. using their phosphorylation and de-phosphorylation reactions [11]. Among them, Protein Tyrosine Phosphatase 1B (PTP1B), an intercellular non-receptor Protein Tyrosine Phosphatase is localized to the cytoplasmic face of the endoplasmic reticulum and is expressed ubiquitously, including in the classical insulin-targeted tissues such as liver, muscle and fat [12]. PTP1B has gained adequate notice due to its crucial role in type 2 diabetes (t2D) and obesity as a negative regulator of the insulin and leptin-signaling pathway [13]. Metabolic insulin signal transduction occurs through activation of the insulin receptor (IR), including autophosphorylation of tyrosine (Tyr) residues in the insulin receptor activation loop. Several protein tyrosine phosphatases

(PTPs), such as receptor protein tyrosine phosphatase (rPTP-a), leuko-cyte antigen-related tyrosine phosphatase (LAR), SH2-domain-containing phosphotyrosine phosphatase (SHP2), and protein tyrosine phosphatase 1B (PTP1B) have been implicated in the dephosphorylation of the IR. PTP1B downregulates insulin signaling by dephosphorylating the insulin receptor (IR), insulin receptor substrate-1 (IRS-1) and insulin receptor substrate-2 (IRS-2) [14]. In insulin pathway, PI3K is the key enzyme that downstream metabolic signaling. PI3K specifically phosphorylates PI substrates to produce PIP2 which activates PDK1. This process, in turn, activates protein kinase AKT, an essential component for insulin-stimulated GLUT4 translocation to plasma membrane, which increases uptake of glucose (Figure 6.1). Inhibition of PTP1B leads to increased insulin level, cellular sensitivity and uptake of glucose. Various studies have shown that PTP1B-knockout mice exhibit enhanced insulin sensi-tivity, improved glucose tolerance and resistance to diet-induced obesity treatment of diabetic mice with PTP1B antisense oligonucleotides reduced the expression level of the enzyme and subsequently normalized the blood glucose and improved insulin sensitivity [15]. Clinical studies have also demonstrated that PTP-1B is primarily responsible for dephosphory-lation of the insulin receptor and thus down regulates insulin signaling [16]. Collectively, these biochemical, genetic and pharmacological studies provide strong proof-of-concept, validating the notion that inhibition of PTP1B could address both diabetes and obesity and making PTP1B an exciting target for drug development [17]. Therefore, the search for potent small molecule protein tyrosine phosphatase 1B inhibitors is a major thrust area in the management of type 2 diabetes mellitus.

PTP-mediated catalysis proceeds via two-step mechanism wherein the initial step, a nucleophilic attack by the sulfur atom of the thiolate side chain of the Cys on the substrate phosphate, coupled with protonation of the tyrosyl-leaving group of the substrate by the side chain of a conserved acidic residue (Asp181 in PTP1B) acting as a general acid. This leads to formation of a cysteinyl-phosphate catalytic intermediate. In the later step, mediated by Gln 262, which coordinates a water molecule, and Asp181, which functions as a general base, there is hydrolysis of the catalytic inter-mediate and release of phosphate (Figure 6.2) [18]. Although several types of PTP1B inhibitors have been reported, because of the low selectivity and poor pharmacokinetic properties, new types of PTP1B inhibitors with

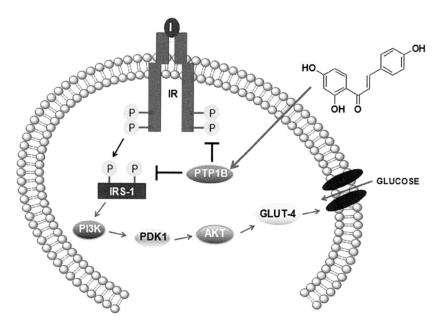

FIGURE 6.1   (See color insert.) The physiological role of PTP-1B in glucose metabolism.

improved pharmacological properties are still being sought. Chalcones may be believed to play a major role in the inhibition of PTP1B and can be used to manage diabetes and associated complications with better selectivity and improved pharmacokinetic properties.

6.3   CHALCONES

Natural products have been reported to exhibit promising anti-diabetic activity. They have been the mainstay of various biological activities, of them flavonoids class remained the principle candidate [19]. Flavonoids are a group of heterogeneous heat stable polyphenols with various health benefits. There are more than 4000 polyphenolic compounds have probably existed in the plant kingdom for over 1 billion years [20]. They are ubiquitously found in fruits, vegetables, tea, wine, and are usually subdivided into six classes including flavonols (e.g., quercetin, kaempferol),

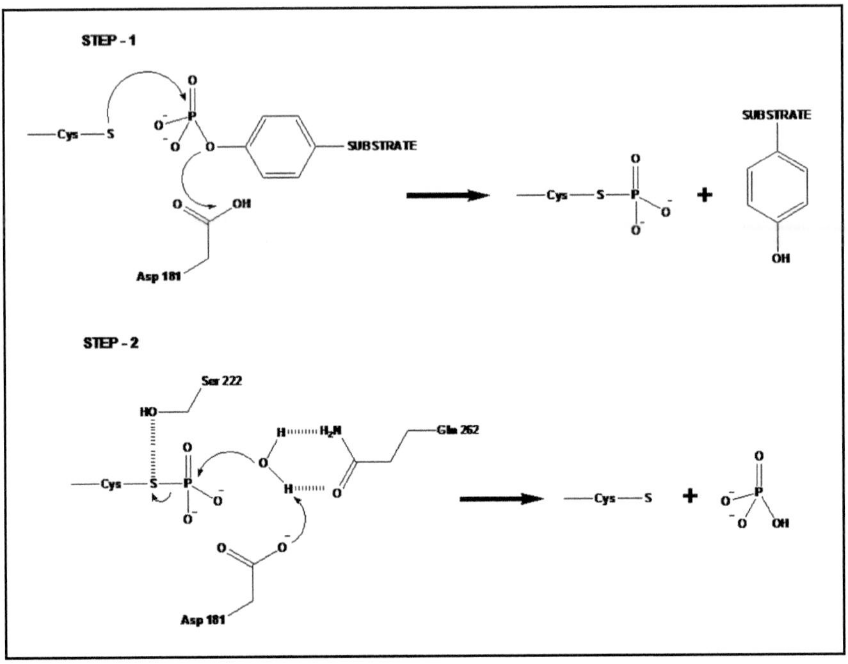

FIGURE 6.2 The process and biochemical pathway of PTP-mediated catalysis in human body.

flavones (e.g., apigenin, luteolin), flavanones (e.g., hesperidin, naringenin), flavan-3-ols (e.g., catechin, theaflavin, and gallic esters of catechin and theaflavins), anthocyanidins (e.g., pelargonidin, cyanidin) and isoflavones (e.g., genistein, daidzein) [21]. Various studies have suggested that dietary intake of natural flavonoids displayed protective, modulatory, and mimetic properties that reduce the risk of tumors formation, provide effective hypoglycemic control, etc. Chalcones or 1,3-diphenyl-2E-propene-1-one is an open chain intermediate in aurones synthesis of flavones that exists in many conjugated forms in nature. They are the precursors of flavonoids and isoflavonoids. It contains a benzylideneacetophenone scaffold where the two aromatic nuclei are joined by a three-carbon α, β unsaturated carbonyl bridge [22]. Kostanecki and Tambor, first synthesized a series of natural chromophoric products comprising of α, β unsaturated carbonyl bridge and termed them "chalcone" [23]. Chalcones gained popularity among researchers in this century as compared to other scaffolds due to its uncomplicated chemistry, simplicity in chemical synthesis, multiplicity of substitutions and multifarious pharmacological potentials such as anti-hypertensive [24], anti-arrhythmic [25], anti-platelet [26], anti-diabetic [27], anti-neoplastic [28], anti-angiogenic [29], anti-retroviral [30], anti-inflammatory [31], anti-gout [32], anti-histaminic [33], anti-oxidant [34], anti-obesity [35], hypolipidemic [36], anti-tubercular [37], anti-filarial [38], anti-invasive [39], anti-malarial [40], anti-protozoal [41], anti-bacterial [42], anti-fungal [43], anti-ulcer [44], anti-steroidal [45], immunosuppressant [46], hypnotic [47], anxiolytic [48], anti-spasmodic [49], anti-nociceptive [50], osteogenic [51], etc.

6.4   CHALCONES AS PTP1B INHIBITORS

A large number of chalcones were isolated from nature which has been reported to exhibit potential PTP1B inhibition activity (Figure 6.3). Hoang et al. isolated three methylcyclohexene substituted chalcones derived Diels-Alder type compounds from *Morus bombycis* namely, kuwanon J (1), kuwanon R (2), and kuwanon V (3). All these derivatives showed remarkable inhibition of PTP1B with IC_{50} in range of 2.7–13.8 μM in mixed–type manner. The number of hydroxyl moieties play essential role on activity. The hydroxyl groups not only provided the needed penetration into the active site but also likely to produce an effective hydrogen bonding interaction with the amide backbone of the active-site loop. These facts suggest that with an increase in number of OH groups in chalcone-derived Diels–Alder-type compounds, the potential inhibitory effects against PTP1B increases tremendously. The compound (2), having seven OH groups demonstrate strong dose-independent inhibition, compared to (3), which contains six OH groups. Compound (1) has an additional OH group at C-2, which increases the potency of (1) upto 3 times with respect to (3). The order of inhibitory activity of chalcone-derived Diels–Alder-type compounds can be summarize as 1>2>3. [52]. Broussochalcone (4), isolated from *Broussonetia papyrifa* was reported to effectively inhibit PTP1B with IC_{50} of 21.5 μM. The two-hydroxyl groups at both the rings are assumed to be responsible for effective inhibition. As the number of hydroxyl groups increases, the inhibitory activity increases concurrently [53]. A novel chalcone, abyssinone-VI-4-O-methyl ether (5) was isolated from ethyl acetate-soluble extract of the root bark of *Erythrina mildbraedii* exhibited *in vitro* PTP1B inhibitory activity with IC_{50} value 14.8 μM [54]. Licochalcone A, isolated from *Glycyrrhiza inflata* and its semi-synthetic derivatives have been reported to be inhibitor of PTP1B. The isolated chalcones isoliquiritigenin (6), echinatin (7), licochalcone A (8), licochalcone C (9), licochalcone E (10), licochalcone B (11), and licochalcone D (12). The semi-synthetic derivatives (13) and (14) were fabricated by methylation and compounds (15) and (16) were formed by acetylating the licochalcone A. Compounds (17) and (18) were prepared by THP-protection of the 4-hydroxy group in the A ring followed by methylation or acetylation of the 4'-hydroxyl group in the B ring and subsequent cleavage of the THP-ether under acidic conditions. Acetylation of compound (13) and compound (17) affords compounds (19) and (20). The semi-synthetic derivative (13) presented the highest activity [55–56]. Isoliquiritigenin (ISL) restores PTP1B activity by inhibiting PTP1B oxidation and IR/PI3K/AKT phosphorylation during the early stages of insulin-induced adipogenesis. The antioxidant capacity of ISL attenuated insulin IR/PI3K/AKT signaling through inhibition of PTP1B oxidation, and ultimately attenuated insulin-induced adipocyte differentiation of 3T3-L1 cells [58].

Chalcone scaffold is one of the privileged scaffolds across medicinal chemistry due to ease of synthesis, a large number of chalcone derivatives having potential PTP1B inhibition activity have been recently synthesized rationally. Chalcones and their derivatives are synthesized classically by Claisen-Schmidt condensation between benzaldehyde and acetophenone employing a solution of sodium hydroxide (40%) as catalyst [59]. Recently, microwave assisted synthesis of chalcone gained popularity where irradiation of above chemicals with domestic microwave is often employed [60]. Based on the natural structural template, Chen *et al.* synthesized few novel heterocyclic ring-substituted chalcone derivatives as potent inhibitor of PTP1B where derivatives (21–29) showed best inhibitory activity. SAR studies revealed that electron-withdrawing groups on ring B showed better PTP1B inhibitory activity than the compounds containing electron-donating groups. Compound (28) showed best result with 99.17% inhibition with IC_{50} value 3.12±0.18 μM [61]. A series of furan chalcone derivatives were reported to exhibit similar activity where two compounds (30) and (31) showed most effective PTP1B inhibition in competitive manner as experimentally displayed by IC_{50} values of 2.49 and 2.9, respectively. The hydroxylated derivatives containing 2,4-OH

(30), 2-OH **(32)**, and 3-OH **(33)** also showed better inhibitory activity [62]. A number of 2,'4,'6'-trihydroxy chalcone derivatives **(34–42)** have been represented as promising pharmacological candidates for *in vitro* PTP1B inhibition which was similar to reference drugs Na_3VO_4 and oleanolic acid, respectively. From above studies, authors concluded that electron-donating groups on ring B are the key factor in exhibiting significant inhibitory activity [63–64].

Macdentichalcone **(43)**, an unprecedented polycyclic dimeric chalcone featuring a unique quinonoid moiety, was isolated from *Macaranga denticulata*, together with 1-(5,7-dihydroxy-2,2,6-trimethyl-2H-1-benzopyran-8-yl)-3-phenyl-2-propen-1-one **(44)**, a known monomeric chalcone proposed as a biosynthetic precursor of **(43)**. Both compounds presented noteworthy inhibitory activity against PTP1B *in vitro* with IC_{50} values of 21.0 ± 3.4 μM and 22.0 ± 3.9 μM, respectively [65]. The well-known coumarin based chalcone compounds, (2E)-1-(5,7-dihydroxy-2,2,6-trimethyl-2H-benzopyran-8-yl)-3-(4-methoxyphenyl)-2-propen-1-one **(45)**, (2E)-1-(5,7-dihydroxy-2,2-dimethyl-2H-benzopyran-8-yl)-3-phenyl-2-propen-1-one **(46)**, and laxichalcone **(47)** showed inhibitory activities against protein tyrosine phosphatase 1B (PTP1B) *in vitro* with IC_{50} values of [66]. Among them, six chalcones, xanthoangelol K **(48)**, xanthoangelol **(49)**, xanthoangelol F **(50)**, 4-hydroxyderricin **(51)**, xanthoangelol D **(52)**, and xanthoangelol E **(53)** showed strong PTP1B inhibitory effect with IC_{50} values of 0.82, 1.97, 1.67, 2.47, 3.97, 1.43, and 2.53 μg/mL, respectively. A kinetic study revealed that compound **(48)** inhibited PTP1B with characteristics typical of a competitive inhibitor. Molecular docking simulations elucidated that ring B of 1 may anchor in a pocket of PTP1B and the molecule is stabilized by hydrogen bonds with Arg47, Asp48, and p–p interaction with Phe182 of PTP1B [67].

6.5 CONCLUSION

The structural features of anti-diabetic chalcones inhibiting protein tyrosine phosphatase 1B (PTP-1B), one of the most promising hypoglycemic target have been highlighted in this chapter. A sufficient stress have been given to the molecular pathway involved in insulin-glucose interphase with probable mechanism of chalcone modulators; and structure-activity relationships (SARs) of the 1,3-diphenyl-2E-propene-1-one based

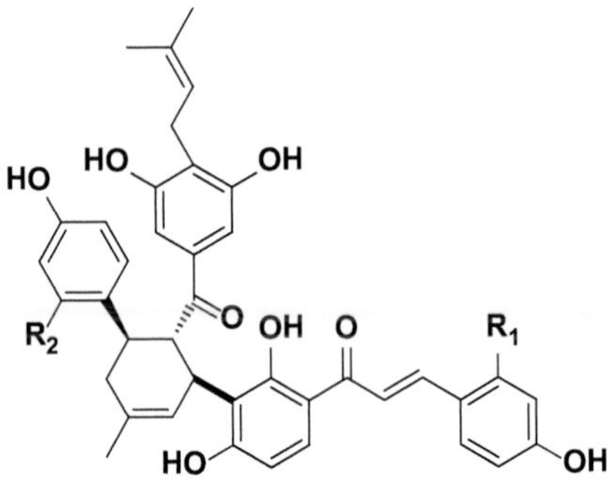

(1) R_1 = OH, R_2 = OH ; **(2)** R_1 = H, R_2 = OH ; **(3)** R_1 = H, R_2 = H

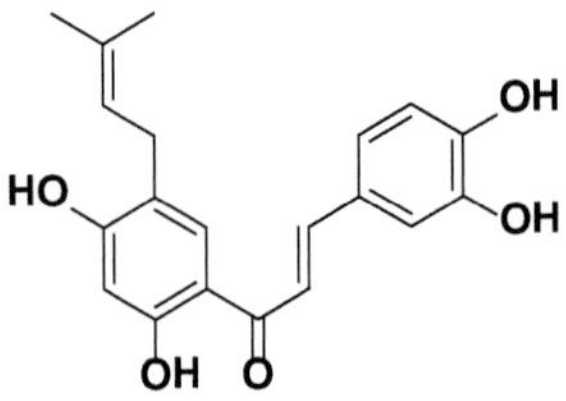

(4) 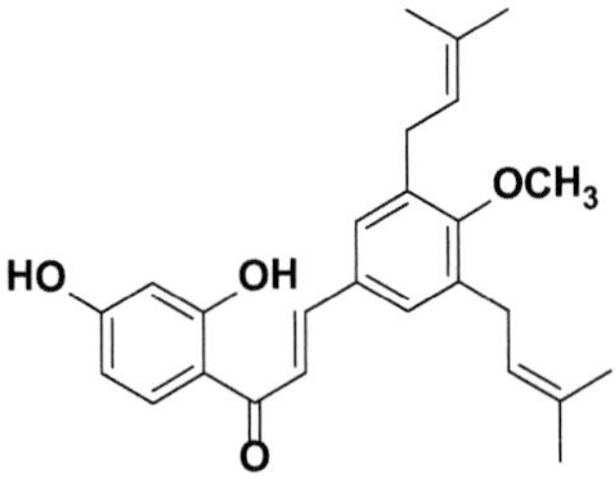   **(5)**

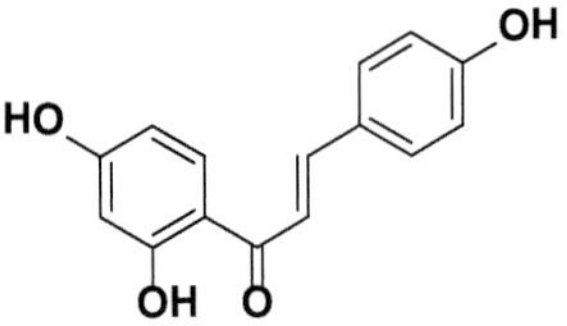

(6)

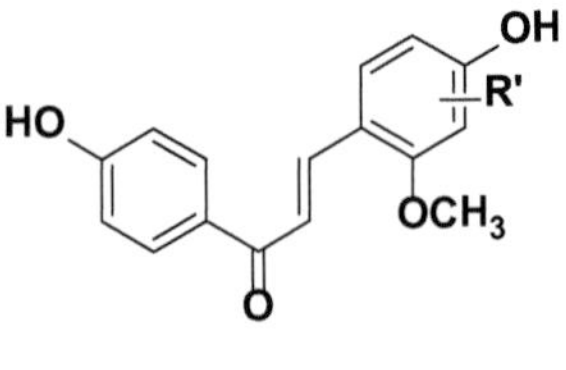

(7) R' = H ;   **(8)** R' = 5'- 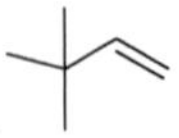   ; **(9)** R' = 3'- 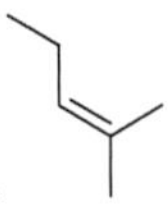   ; **(10)** R' = 5'-

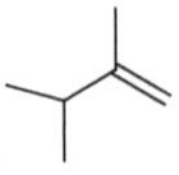

FIGURE 6.3   List of reported PTP-1B inhibitors of chalcone scaffolds.

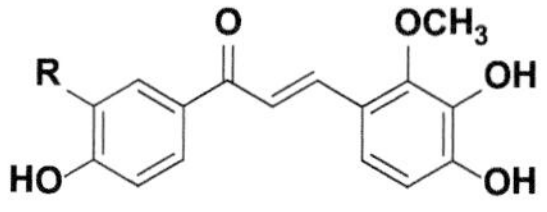

(11) R = H ; (12) R =

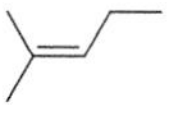

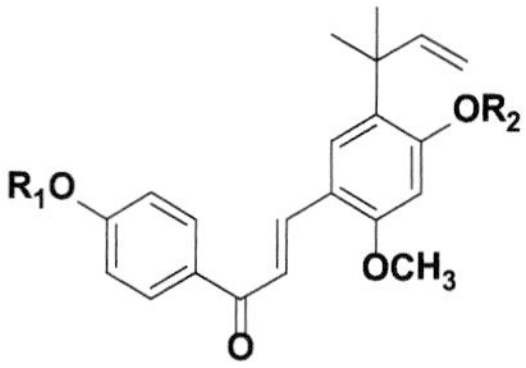

(13) R_1 = CH_3, R_2 = H ; (14) R_1 = CH_3, R_2 = CH_3 ; (15) R_1 = $COCH_3$, R_2 = H ;
(16) R_1 = $COCH_3$, R_2 = $COCH_3$; (17) R_1 = H, R_2 = CH_3 ; (18) R_1 = H, R_2 = $COCH_3$;
(19) R_1 = CH_3, R_2 = $COCH_3$; (20) R_1 = $COCH_3$, R_2 = CH_3

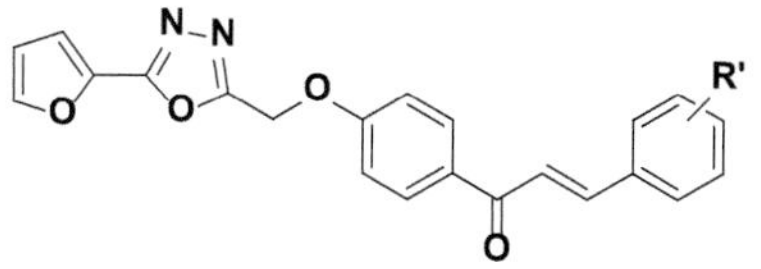

(21) R' = H ; (22) R' = 4-F ; (23) R' = 2-Cl ; (24) R' = 4-Cl ; (25) R' = 2,6-$(Cl)_2$; (26) R' = 2-Br ; (27) R'
= 4-Br ; (28) R' = 2-F ; (29) R' = 3-Cl

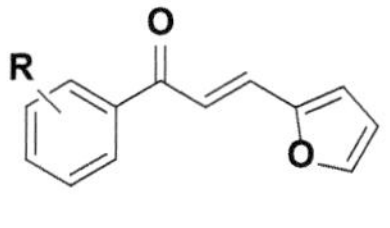

(30) R = 2, 4-OH ; (31) R = 3-Cl ; (32) R = 2-OH ; (33) R = 3-OH

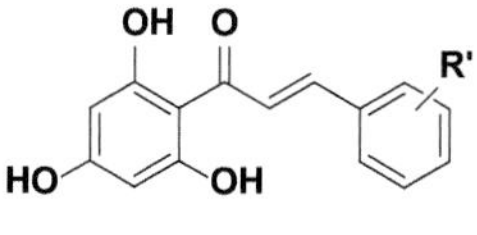

(34) R' = 3-OCH_3, 4-OH ; (35) R' = 3-OCH_3, 4-$OCH_2CH=CH_2$; (36) R' = 4-OCH_3 ;
(37) R' = 3, 4-$(OCH_3)_2$; (38) R' = 4-CH_3 ; (39) R' = H ; (40) R' = 2-F ; (41) R' = 3-Cl ;
(42) R' = 2,4-Cl

FIGURE 6.3 *(Continued)*

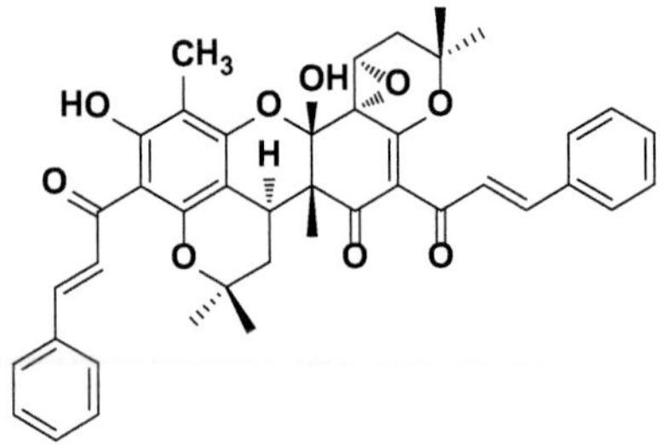

(43)

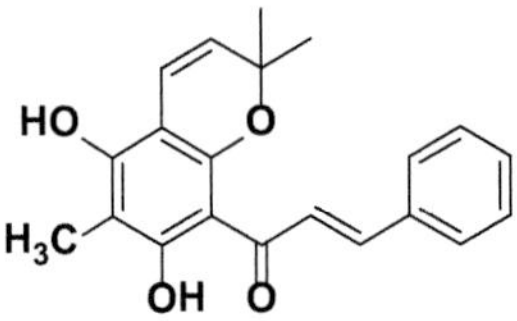

(44)

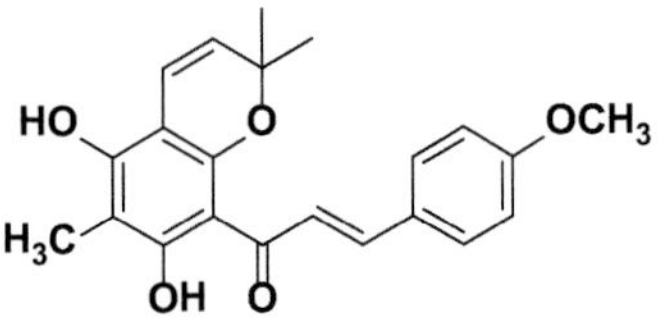

(45)

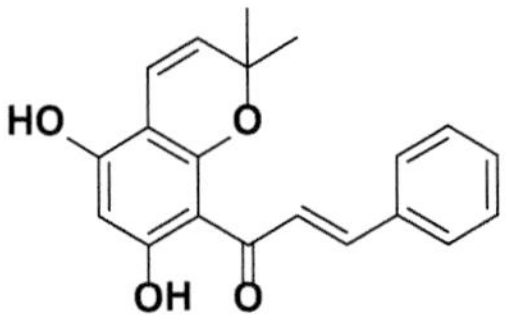

(46)

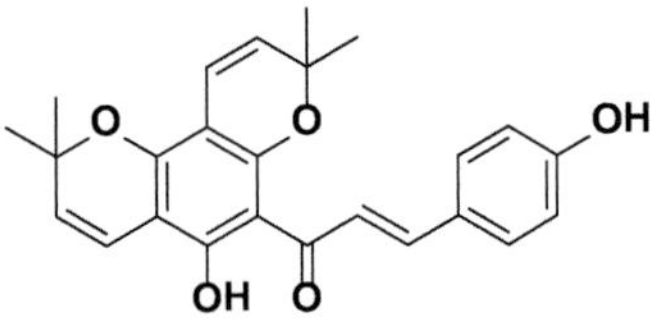

(47)

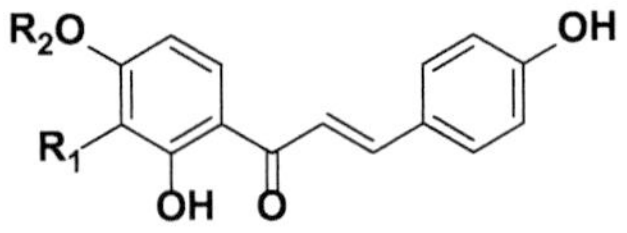

(48) $R_1 =$ 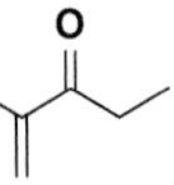 , $R_2 = CH_3$; **(49)** $R_1 =$ 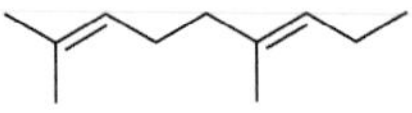 , $R_2 = H$;

(50) $R_1 =$, $R_2 = CH_3$; **(51)** $R_1 =$, $R_2 = CH_3$;

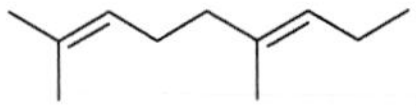

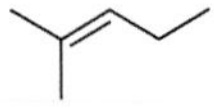

(52) $R_1 =$, $R_2 = CH_3$; **(53)** $R_1 =$ 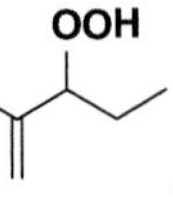 , $R_2 = CH_3$

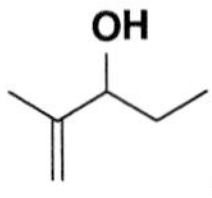

FIGURE 6.3 *(Continued)*

inhibitors where the profound role of electron withdrawing/donating groups along with importance of heteroaryl ring are comprehensively discussed. As a modulator, at present, none of them have gained proper consideration in medicinal chemistry for hypoglycemic control. Although, they have a huge future prospective to be regarded as potential players for anti-hyperglycemic activity owing to their simple chemistry, multifarious modulation potentials, etc. which will certainly open new avenues in the upcoming decade(s).

KEYWORDS

- chalcone
- diabetes
- inhibitors
- protein tyrosine phosphatase
- PTP1B
- structure-activity relationships

REFERENCES

1. Carig, C. R., & Stizel, R. E., (2004). *Modern Pharmacology with Clinical Applications*, 5th edn,; Lippincott Williams & Wilkins: Philadelphia; pp. 285–286.
2. Ritter, J. M., Lewis, L. D., Mant, T. G. K., & Ferro, A., (2008). *A Textbook of Clinical Pharmacology and Therapeutics*. London: Hodder Arnold; pp. 201–202.
3. American Diabetes Association, (2010). Diagnosis and classification of diabetes mellitus, *Diabetes Care, 33*, 62–69.
4. Mahapatra, D. K., & Bharti, S. K., (2016). *Drug Design*, 1st ed.; Lippincott Williams & Wilkins: Philadelphia; pp. 112–113.
5. Rawat, P., Kumar, M., Rahuja, N., Srivastava, D. S. L., Srivastava, A. K., & Maurya, R., (2011). Synthesis and antihyperglycemic activity of phenolic C-glycosides. *Bioorganic and Medicinal Chemistry Letters, 21*, 228–233.
6. http://oasys2.confex.com/acs/227nm/techprogram/P726617.HTM (Accessed Feb 5, 2015).
7. Brunton, L., Parker, K., Blumenthal, D., & Buxton, I., (2008). *Goodman and Gilman's Manual of Pharmacology and Therapeutics,* New York: The McGraw-Hill; pp. 1042–1045.
8. Wani, J. H., John-Kalarickal J., Fonseca, V. A., (2008). Dipeptidyl peptidase-4 as a new target of action for type 2 diabetes mellitus: A systematic review. *Cardiology Clinics, 26*, 639–648.

9. Dixit, M., Saeed, U., Kumar, A., Siddiqi, M. I., Tamrakar, A. K., Srivastava, A. K., et al., (2008). Synthesis, molecular docking and PTP1B inhibitory activity of functionalized 4,5-dihydronaphthofurans and dibenzofurans. *Medicinal Chemistry, 4*, 18–24.

10. Nguyen, L. K., Matallanas, D., Croucher, D. R., Von Kreigsheim, A., & Kholodenko, B. N., (2013). Signalling by protein phosphatases and drug development: A systems-centered view. *FEBS Journal, 280*, 751–765.

11. Zhang, W., Hong, D., Zhou, Y., Zhang, Y., Shen, Q., Li, J. Y., et al., (2006). Ursolic acid and its derivative inhibit protein tyrosine phosphatase 1B, enhancing insulin receptor phosphorylation and stimulating glucose uptake. *Biochimica et Biophysica Acta, 1760*, 1505–1512.

12. Zhang, S., & Zhang, Z. Y., (2007). PTP1B as a drug target. Recent developments in PTP1B inhibitor discovery. *Drug Discovery Today, 12*, 373–381.

13. Asante-Appiah, E., & Kennedy, B. P., (2003). Protein tyrosine phosphatases: The quest for negative regulators of insulin action. *American Journal of Physiology Endocrinology and Metabolism, 284*, 663–670.

14. Johnson, T. O., Ermolieff, J., & Jirousek, M. R., (2002). Protein tyrosine phosphatase 1B inhibitors for diabetes. *Nature Reviews, 1*, 696–709.

15. Zhang, Z., & Lee, S., (2003). PTP1B inhibitors as potential therapeutics in the treatment of Type 2 Diabetes and obesity. *Expert Opinion in Investigating Drugs, 12*, 223–233.

16. He, R., Zeng, L., He, Y., & Zhang, Z., (2012). Recent advances in PTP1B inhibitor development for the treatment of type 2 diabetes and obesity. In: Jones, R. M., (ed.), *New Therapeutic Strategies for Type 2 Diabetes: Small Molecule Approaches* (pp. 142–176). London: Royal Society of Chemistry.

17. Koren, S., (2007). Inhibition of the protein tyrosine phosphatase PTP1B: Potential therapy for obesity, insulin resistance and type-2 diabetes mellitus. *Best Practice & Research Clinical Endocrinology & Metabolism., 21*, 621–640.

18. Tonks, N. K., (2003). PTP1B: From the sidelines to the front lines! *FEBS Letters, 546*, 140–148.

19. Prakash, O., Kumar, A., & Kumar, P. A., (2013). Anticancer potential of plants and natural products: A review. *American Journal of Pharmacological Sciences, 6*, 104–115.

20. Flavonoids: Flavones, flavonols, anthocyanins, and related compounds wwwlifeillinoisedu/ib/425/lecture11html (Accessed September 5, 2015).

21. Farooqui, A. A., (2012). Phytochemicals, signal transduction, and neurological disorders. Springer Science & Business Media.

22. Dimmock, J. R., Elias, D. W., Beazely, M. A., & Kandepu, N. M., (1999). Bioactivities of chalcones. *Current Medicinal Chemistry, 6*, 1125–1149.

23. Kostanecki, S. V., & Tambor J., (1899). Ueber die sechs isomeren Monoxybenzalacetophenone (Monoxychalkone). *Chemische Berichte., 32*, 1921–1926.

24. Mahapatra, D. K., & Bharti, S. K., (2016). Therapeutic potential of chalcones as cardiovascular agents. *Life Sciences, 148*, 154–172.

25. Yarishkin, O. V., Ryu, H. W., Park, J., Yang, M. S., Hong, S., & Park, K. H., (2008). Sulfonate chalcone as new class voltage-dependent K+ channel blocker. *Bioorganic and Medicinal Chemistry Letters, 18*, 137–140.

26. Zhao, L., Jin, H., Sun, L., Piao, H., & Quan Z., (2005). Synthesis and evaluation of antiplatelet activity of trihydroxychalcone derivatives. *Bioorganic and Medicinal Chemistry Letters*, *15*, 5027–5029.

27. Mahapatra, D. K., Asati, V., & Bharti, S. K., (2015). Chalcones and their therapeutic targets for the management of diabetes: Structural and pharmacological perspectives. *European Journal of Medicinal Chemistry*, *92*, 839–865.

28. Mahapatra, D. K., Bharti, S. K., & Asati, V., (2015). Anti-cancer chalcones: Structural and molecular target perspectives, *European Journal of Medicinal Chemistry*, *98*, 69–114.

29. Lee, Y. S., Lim, S. S., Shin, K. H., Kim, Y. S., Ohuchi, K., & Jung, S. H., (2006). Anti-angiogenic and anti-tumor activities of 2'-Hydroxy-4'-methoxychalcone. *Biology Pharmaceutical Bulletin*, *29*, 1028–1031.

30. Rizvi, S. U. F., Siddiqui, H. L., Johns, M., Detorio, M., & Schinazi, R. F., (2012). Anti-HIV-1 and cytotoxicity studies of piperidyl-thienyl chalcones and their 2-pyrazoline derivatives. *Medicinal Chemistry Research*, *21*, 3741–3749.

31. Mahapatra, D. K., Bharti, S. K., & Asati V., (2017). Chalcone derivatives: Anti-inflammatory potential and molecular targets perspectives. *Current Topics in Medicinal Chemistry*, *17*, 1–24.

32. Kim, D. W., Curtis-Long, M. J., Yuk, H. J., Wang, Y., Song, Y. H., Jeong, S. H., et al., (2014). Quantitative analysis of phenolic metabolites from different parts of *Angelica keiskei* by HPLC–ESI MS/MS and their xanthine oxidase inhibition. *Food Chemistry*, *153*, 20–27.

33. Yamamoto, T., Yoshimura, M., Yamaguchi, F., Kouchi, T., Tsuji, R., Saito, M., et al., (2004). Anti-allergic activity of naringenin chalcone from a tomato skin extract. *Bioscience Biotechnology and Biochemistry*, *68*(8), 1706–1711.

34. Aoki, N., Muko, M., Ohta, E., & Ohta, S., (2008). *C*-Geranylated chalcones from the stems of *Angelica keiskei* with superoxide-scavenging activity. *Journal of Natural Products*, *71*, 1308–1310.

35. Birari, R. B., Gupta S, Mohan, C. G., & Bhutani, K. K., (2011). Antiobesity and lipid lowering effects of Glycyrrhiza chalcones: Experimental and computational studies. *Phytomedicine*, *18*, 795–801.

36. Sashidhara, K. V., Palnati, G. R., Sonkar, R., Avula, S. R., Awasthi, C., & Bhatia, G., (2013). Coumarinchalcone fibrates: A new structural class of lipid lowering Agents. *European Journal of Medicinal Chemistry*, *64*, 422–431.

37. Mahapatra, D. K., Bharti, S. K., & Asati V., (2015). Chalcone scaffolds as anti-infective agents: Structural and molecular target perspectives. *European Journal of Medicinal Chemistry*, *101*, 496–524.

38. Sashidhara, K. V., Rao, K. B., Kushwaha, V., Modukuri, R. K., Verma, R., & Murthy, P. K., (2014). Synthesis and antifilarial activity of chalcone-thiazole derivatives against a human lymphatic filarial parasite, *Brugia malayi*. *European Journal of Medicinal Chemistry*, *81*, 473–480.

39. Wang, L., Chen, G., Lu, X., Wang, S., Hans, S., Li, Y., et al., (2015). Novel chalcone derivatives as hypoxia-inducible factor (HIF)-1 inhibitor: Synthesis, anti-invasive and anti-angiogenic properties. *European Journal of Medicinal Chemistry*, *89*, 88–97.

40. Tomar, V., Bhattacharjee, G., Kamaluddin, R. S., Srivastava, K., & Puri, S. K., (2010). Synthesis of new chalcone derivatives containing acridinyl moiety with potential antimalarial activity. *European Journal of Medicinal Chemistry, 45*, 745–751.

41. Chen, M., Christensen, S. B., Blom, J., Lemmich, E., Nadelmann, L., Fich, K., et al., (1993). Licochalcone A, a novel antiparasitic agent with potent activity against human pathogenic protozoan species of Leishmania. *Antimicrobial Agents Chemotherapy, 37*, 2550–2556.

42. Abdullah, M. I., Mahmood, A., Madni, M., Masood, S., & Kashif, M., (2014). Synthesis, characterization, theoretical, anti-bacterial and molecular docking studies of quinoline based chalcones as a DNA gyrase inhibitor. *Bioorganic Chemistry, 54*, 31–37.

43. Lahtchev, K. V., Batovska, D. I., Parushev, S. P., Ubiyvovk, V. M., & Sibirny, A. A., (2008). Antifungal activities of chalcones: A mechanistic study using various yeast strains, *European Journal of Medicinal Chemistry, 43*, 2220–2228.

44. Sashidhara, K. V., Avula, S. R., Mishra, V., Palnati, G. R., Singh, L. R., Singh, N., et al., (2015). Identification of quinoline-chalcone hybrids as potential antiulcer agents. *European Journal of Medicinal Chemistry, 89*, 638–653.

45. Bail, J. L., Pouget, C., Fagnere, C., Basly, J., Chulia, A., & Habrioux, G., (2001). Chalcones are potent inhibitors of aromatase and 17 β-hydroxysteroid dehydrogenase activities. *Life Sciences, 68*, 751–761.

46. Luo, Y., Song, R., Li, Y., Zhang, S., Liu, Z. J., Fu, J., et al., (2012). Design, synthesis, and biological evaluation of chalcone oxime derivatives as potential immunosuppressive agents. *Bioorganic and Medicinal Chemistry Letters, 22*, 3039–3043.

47. Cho, S., Kim, S., Jin, Z., Yang, H., Han, D., Baek, N. I., et al., (2011). Isoliquiritigenin, a chalcone compound, is a positive allosteric modulator of GABAA receptors and shows hypnotic effects. *Biochemical and Biophysical Research Communications, 413*, 637–642.

48. Jamal, H., Ansari, W. H., & Rizvi, S. J., (2008). Evaluation of chalcones-a flavonoid subclass, for, their anxiolytic effects in rats using elevated plus maze and open field behaviour tests. *Fundamental Clinical Pharmacology, 22*, 673–681.

49. Sato, Y., He, J., Nagai, H., Tani T., & Akao, T., (2007). Isoliquiritigenin, one of the antispasmodic principles of *Glycyrrhiza ularensis* roots, acts in the lower part of intestine. *Biological Pharmaceutical Bulletin, 30*, 145–149.

50. De Campos-Buzzi, F., Padaratz, P., Meira, A. V., Correa, R., Nunes, R. J., & Cechinel-Filho, V., (2007). 4'-Acetamidochalcone derivatives as potential antinociceptive agents. *Molecules, 12*, 896–906.

51. Ortolan, X. R., Fenner, B. P., Mezadri, T. J., Tames, D. R., Correa, R., & De Campos Buzzi F., (2014). Osteogenic potential of a chalcone in a critical-size defect in rat calvaria rat. *Craniomaxillofacial Surgery, 42*, 520–524.

52. Hoang, D. M., Ngoc, T. M., Da, N. T., Ha, D. T., Kim, Y. H., Luon, H. V., et al., (2009). Protein tyrosine phosphatase 1B inhibitors isolated from Morus bombycis. *Bioorganic and Medicinal Chemistry Letters, 19*, 6759–6761.

53. Chen, R. M., Hu, L. H., An, T. Y., Li, J., & Shen, Q., (2002). Natural PTP1B Inhibitors from Broussonetia papyrifera. *Bioorganic and Medicinal Chemistry Letters, 12*, 3387–3390.

Recent Perspectives of Chalcone-Based Molecules

54. Na, M. K., Jang, J., Njamen, D., Mbafor, J. T., Fomum, Z. T., Kim, B. Y., et al., (2006). Protein tyrosine phosphatase-1b inhibitory activity of isoprenylated flavonoids isolated from erythrina mildbraedii. *Journal of Natural Products, 69*, 1572–1576.

55. Yoon, G., Lee, W., Ki, S., & Cheon, S. H., (2009). Inhibitory effect of chalcones and their derivatives from Glycyrrhiza inflate on protein tyrosine phosphatase 1B. *Bioorganic and Medicinal Chemistry Letters, 19*, 5155–5157.

56. Guo, Z., Niu, X., Xiao, T., Lu, J., Li, W., & Zhao, Y., (2015). Chemical profile and inhibition of α-glycosidase and protein tyrosine phosphatase 1B (PTP1B) activities by flavonoids from licorice (Glycyrrhiza uralensis Fisch). *Journal of Functional Foods, 14*, 324–336.

57. Kim, E., Kim, C., Kang, Y. C., Liu, Z., Kim, S. N., Kim, H. J., et al., (2016). Molecular modeling of licochalcone E as protein tyrosine phosphatase 1B inhibitor. *Bulletin of the Korean Chemical Society, 37*, 2102–2105.

58. Park, S. J., Choe, Y. G., Kim, J. H., Chang, K. T., Lee, H. S., & Lee, D. S., (2016). Isoliquiritigenin impairs insulin signaling and adipocyte differentiation through the inhibition of protein-tyrosine phosphatase 1B oxidation in 3T3-L1 preadipocytes. *Food and Chemical Toxicology, 93*, 5–12.

59. Detsi, A., Majdalani, M., Kontogiorgis, C. A., Hadjipavlou-Litin, D., & Kefalas, P., (2009). Natural and synthetic 2'-hydroxy-chalcones and aurones: Synthesis, characterization and evaluation of the antioxidant and soybean lipoxygenase inhibitory activity. *Bioorganic and Medicinal Chemistry, 17*, 8073–8085.

60. Srivastava, Y. K., (2008). Ecofriendly microwave assisted synthesis of some chalcones. *Rasayan Journal of Chemistry, 1*, 884–886.

61. Chen, Z., Sun, L., Zhang, W., Shen, Q., Gao, L., Li, J., et al., (2012). Synthesis and biological evaluation of heterocyclic ring-substituted chalcone derivatives as novel inhibitors of protein tyrosine phosphatase 1B. *Bulletin of Korean Chemical Society, 33*, 1505–1508.

62. Sun, L., Jiang, Z., Gao, L., Sheng, L., Quan, Y., Li, J., & Piao, H., (2013). Synthesis and biological evaluation of furan-chalcone derivatives as protein tyrosine phosphatase inhibitors. *Bulletin of Korean Chemical Society, 34*, 1023–1024.

63. Sun, L., Gao, L., Ma, W., Nan, F., Li, J., & Piao, H., (2012). Synthesis and biological evaluation of 2,4,6-trihydroxychalcone derivatives as novel protein tyrosine phosphatase 1B inhibitors. *Chemical Biology and Drug Design, 80*, 584–590.

64. Zhao, S. L., Peng, Z., Zhen, X. H., Jin, H. G., Han, Y., Qu, Y. L., et al., (2015). Potent CDC25B and PTP1B phosphatase inhibitors: 2',4',6'-trihydroxylchalcone derivatives. *Medicinal Chemistry Research, 24*, 2573–2579.

65. Lei, C., Zhang, L. B., Yang, J., Gao, L. X., Li, J. Y., Li, J., et al., (2016). Macdenti chalcone, a unique polycyclic dimeric chalcone from macaranga denticulata. *Tetrahedron Letters, 57*, 5475–5478.

66. Zhang, L. B., Lei, C., Gao, L. X., Li, J. Y., Li, J., & Hou, A. J., (2016). Isoprenylated flavonoids with PTP1B inhibition from macaranga denticulata. *Natural Products and Bioprospecting, 6*, 25–30.

67. Li, J. L., Gao, L. X., Meng, F. W., Tang, C. L., Zhang, R. J., Li, J. Y., & Zhao, W. M., (2015). PTP1B inhibitors from stems of angelica keiskei (Ashitaba). *Bioorganic and Medicinal Chemistry Letters, 25*, 2028–2032.

CHAPTER 7

BRIEFING THERAPEUTIC APPROACHES IN ANTICOAGULANT, THROMBOLYTIC, AND ANTIPLATELET THERAPY

KUNTAL MANNA and MANIK DAS

Department of Pharmacy, Tripura University (A Central University), Suryamaninagar–799022, Tripura, India

ABSTRACT

Abnormal blood clotting disorders (hypercoagulable states) are the leading causes of mortality and morbidity in many of developing and developed countries. Hypercoagulable states can be described as a group of inherited or acquired conditions associated with a predisposition to venous thrombosis (including upper and lower extremity deep venous thrombosis with or without pulmonary embolism, cerebral venous thrombosis, and intra-abdominal venous thrombosis), arterial thrombosis (including myocardial infarction, stroke, acute limb ischemia, and splanchnic ischemia), or both. In past decades, researcher tried to understand about the etiology of thrombosis and its risks. It has been found that people with several thrombosis associated risks never reported thrombosis in their life and the phenomena is enigmatic. Recent investigations are trying to find out this mystery, but challenging due to the complex phenomena of clotting and bleeding. The most effective itinerary for the blood clot management is prevention. For high-risk patients, mainstream prophylaxis against thrombosis and its complications often includes powerful anti-clotting medications. These require careful monitoring and inconvenient dietary restrictions. Conventional medications used to prevent blood clots, such as warfarin, increase the potential for serious bleeding as well as the risk of mortality from traumatic injuries.

Next-generation anticoagulant medications that overcome these vascular and skeletal risks are emerging, yet they still lack high-quality evidence from clinical trials to solidify them as first-line treatments. This chapter deals with the recent advances in understanding the pathophysiologic basis of the hypercoagulable states and presents a systematic clinical approach to thrombosis.

7.1 INTRODUCTION

Localized clotting of blood, which may occur in the arterial or venous circulation due to pathogenic changes in the blood vessel wall and in the blood itself, termed as thrombosis. Thrombosis is designated for a spectrum of common diseases presented in Figure 7.1. Deep vein thrombosis occurs most often in the large veins of the legs. Pulmonary embolism is a complication of deep vein thrombosis that can occur if part of the thrombus breaks away, travels to the lungs and lodges in a pulmonary artery, resulting in the disruption of blood flow. In the case of arterial thrombosis, atherosclerotic plaque rupture and embolism are the leading mechanisms responsible for acute coronary or cerebrovascular events, which lead to many stroke and heart attacks [1–3]. The treatment regimens of arterial thrombosis and venous thrombosis are different. Currently available antithrombotic drugs are effective in the treatment of arterial thrombosis and venous thrombosis in patients with cardiovascular disease, but these drugs possess bleeding complications as the main adverse site effect [4, 5]. Anticoagulation is a treatment with significant and life-threatening complications requiring that the balance of risk and benefit be individually assessed in each patient. Accordingly, there has been a massive independent and commercial research effort to identify novel antithrombotic strategies with larger therapeutic window for the treatment and prevention of these thrombotic conditions. Larger therapeutic window implies the large difference between the dose that averts thrombosis and the dose that prompts bleeding. The historical mainstay of antithrombotic therapy has been aspirin, warfarin and unfractionated heparin, and continue to dominate the current antithrombotic armamentarium, though the last two decades have seen impressive advances on antithrombotic agents (direct factor Xa and thrombin inhibitor) [6, 7].

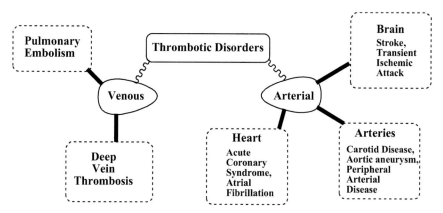

FIGURE 7.1 Thrombotic disorders.

7.2 MOLECULAR MECHANISMS OF HEMOSTASIS AND THROMBOSIS

Hemostasis can be defined as the termination of bleeding from a cut or injured vessel, a normal process of coagulation. Thrombosis occurs when the endothelial lining of blood vessels is ruptured or removed. Both hemostasis and thrombosis encompass coagulation, where involvement of blood vessels, platelets, and plasma proteins (zymogens) takes place. Subsequently, formation and dissolution of platelet aggregates occur in a downstream pathway. Platelets and coagulation factors (Table 7.1) are two major components of the coagulation system. As hemostatic plugs form and grow after injury, mechanisms exist to maintain or dissolve them as needed while allowing the normal flow in the remainder of the vasculature. Imbalances in this complex regulatory network of coagulation and anticoagulation, leads to thrombosis within the arteries and veins. In hemostasis, there is initial vasoconstriction of the damaged vessel, causing lessened blood flow distal to the injury. Then, three stages are alike for both hemostasis and thrombosis. First is the development of a loose and impermanent platelet aggregate at the site of injury. At the site of vessel wall injury, platelets and plasma clotting factors become exposed to the subendothelial collagen and the endothelial basement membrane, which releases adenosine diphosphate (ADP), a potent platelet aggregator, and tissue factor, which launches the clotting cascade. Platelets are activated by thrombin through protease activated receptor-1 (PAR-1) present on the platelets, or by ADP released from other activated platelets and bind to collagen. Thrombin is formed in the coagulation cascade at the same site. After activation by thrombin, platelets change shape and form the hemostatic plug (in hemostasis) or a thrombus (in thrombosis) in the presence of fibrinogen and aggregate. It is due to the up-regulation of glycoprotein (GP) IIb/IIIa receptors and subsequent binding of fibrinogen and von Willebrand's factor (vWF) to activated platelets via GP IIb/IIIa receptors. Second is the construction of a fibrin mesh. Mesh binds to the platelet aggregate to form a more stable hemostatic plug or thrombus. Third is the most common phase of hemostasis and thrombosis where partial or complete dissolution of the hemostatic plug or thrombus occurs by plasmin. However, unsuccessful dissolution leads to thrombotic conditions [1, 4, 5, 8, 9].

TABLE 7.1 Coagulation Factors

Factor	Common Name	Factor	Common Name
I	Fibrinogen	VIII	Antihemophilic A factor
II	Prothrombin	IX	Antihemophilic B factor or Christmas factor
III	Tissue thromboplastin or tissue factor	X	Stuart or Stuart-Prower factor
IV	Calcium (Ca^{2+})	XI	Plasma thromboplastin antecedent
V	Proaccelerin, labile factor	XII	Hageman factor, contact factor
VII	Proconvertin, stable factor	XIII	Fibrin stabilizing factor, Prekallikrein factor, High-molecular-weight kininogen

7.3 CLOTTING CASCADES

The coagulation initiates through intrinsic or extrinsic clotting cascade as presented in Figure 7.2. These cascades are not independent. Instigation of coagulation (extrinsic pathway) occurs in arteries or veins by the tissue factor (TF) which is in response to tissue injury. The extrinsic pathway is essential for the initiation of fibrin formation while the intrinsic pathway is involved in fibrin growth and maintenance.TF is a small molecular weight glycoprotein, which is expressed on the surface of macrophages. Classically, tissue factor is not present in the plasma, but only present on cell

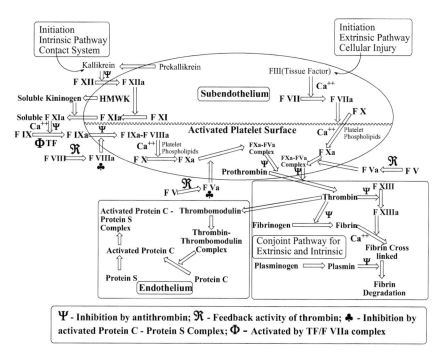

FIGURE 7.2 Pathway of coagulation and endogenous regulation.

surfaces at a wound site. As tissue factor is "extrinsic" to the circulation, the pathway was thusly named [1].

TF binds and activates factor VII to form tissue factor/VIIa complex. This complex activates factor IX in the intrinsic pathway or factor X in the extrinsic pathway. After the commencement of the pathway, the TF/VIIa activation of factor X is rapidly deregulated by TFPI (tissue factor pathway inhibitor), produced by endothelial cells. Thus, in the presence of calcium, activated factor IX binds to an activated factor VIII on the surface of the activated platelets (phospholipid surface) to form an intrinsic xase or tenase complex (IXa-VIIIa-PL), which initiates the activation of factor X. The intrinsic pathway is instigated through the activation of factor XII by the high-molecular-weight kininogen(HMWK) and kallikrein in the presence of collagen. Factor XIIa converts factor XI to XIa, which, in turn, activates factor IX to factor IXa. Finally, the activation of factor X takes place as shown in Figure 7.2. Activated factor Xa then binds with factor Va,

Ca^{2+}, and phospholipids (PLs) from the platelet membrane to form Xa–Va–PL complex (prothrombinase). Prothrombinase converts prothrombin (FII) to thrombin (FIIa). Thrombin lastly cleaves fibrinogen to release isolated fibrin monomers, which polymerize with each other to form complex fibrin. Factor XIII (activated by thrombin itself) stabilizes fibrin complexes and hereafter hemostatic plug or thrombus formation [1, 4, 8–10].

7.4   ENDOGENOUS REGULATION OF COAGULATION

The function of endothelium is to promote blood fluidity. If there is an injury, the normal response of endothelium is to endorse coagulation at the site of injury and prohibit the coagulation mechanism to propagate beyond the site of injury. Endothelium maintains both procoagulant and anticoagulant functions [11]. As a procoagulant function, it aids TAFI (thrombin activatable fibrinolytic inhibitor) activation, which leads to the removal of lysine residues from the fibrin clot. Hence, TAFI makes a fibrin clot less recognizable as a substrate of plasmin and fibrin clot becomes more persistent. As an anticoagulant function of endothelium, endogenous inhibitors maintain the balance of coagulation to heal the injury and inhibit the pathogenic clot formation. Inhibition takes place by TFPI, antithrombin, and the protein C pathways. TFPI produced by endothelial cells, inhibits the TF/FVIIa complex, and further activation of factor Xa and factor Va. TFPI also potentiates the effect of heparin. Activated protein C along with its cofactor protein S, can cleave coagulation cofactor Va and VIIIa. Thrombin when binds to the endothelial cell receptor thrombomodulin, it activates protein C. Limiting the amounts of factor Va and factor VIIIa at the site of growing clot essentially shuts down the ability to activate more thrombin and decrease the amount of blood clot. Antithrombin with cofactor heparin, is a potent inhibitor of thrombin, factor IXa and Xa and also TF/VIIa complex as shown in Figure 7.2. The activity of antithrombin is accelerated 1000 times in the presence of Heparin. The anticoagulant effects of heparin can be antagonized by strongly cationic polypeptides such as protamine (reversal agent of heparin), which bind strongly to heparin, thus inhibiting its binding to antithrombin. Individuals with inherited deficiencies of antithrombin are prone to develop venous thrombosis, providing evidence that antithrombin has a physiologic function and that the coagulation system in humans is normally in a dynamic state [12].

7.5   RISK FACTORS FOR THROMBOSIS

Thrombotic disorders may develop due to genetically inherited mutation of gene(s) encoded for antithrombin, protein C, and protein S, involved in endogenous inhibition of coagulation. Some factors are acquired as a direct or indirect result of trauma, systemic illness (acute or chronic), or an altered physiological state [13].

7.5.1   GENETIC CAUSES

Factor V Leiden, named after the city in the Netherlands (first identified) is a unique single point mutation in the gene for factor V. It is a common genetic defect associated with thrombophilia. Instead of arginine at position 506, glutamine is the outcome of mutation in the amino acid sequence of factor V and hence the cleavage site of factor V Leiden is resistant to the inactivation by activated protein C [14, 15].

Prothrombin Gene 20210G/A mutation occurs at the 20210 position in the 30 un-translated regions of the prothrombin gene. G/A implies glutamine is filled in for arginine, linked with an amplified risk of thromboembolism due to elevated levels of plasma prothrombin. Patients with this mutation reported to be positive for the factor V Leiden mutation too [16].

Hyperhomocysteinemia is also a possible risk factor for the occurrence of atherosclerosis. Homocysteine is an amino acid and a breakdown product of protein metabolism. High concentrations (>15 mmol/L in plasma) of homocysteine have been linked to an increased risk of heart attacks and strokes and also thought to contribute to plaque formation by damaging arterial walls [17, 18].

7.5.2   ACQUIRED CAUSES

Anti-phospholipid antibody such as the lupus anticoagulant and anti-cardiolipin antibody in patients are associated with a predisposition for blood clots in several ways, such as promoting platelets to clump together, reducing the production of activated protein C, and inhibiting the anti-coagulant activity of antithrombin. Antiphospholipid antibody syndrome (APS) is an autoimmune disorder, where the body produces antibodies against a phospholipid of its own cells. APS is characterized by antibodies. Anti-phospholipid antibodies were first identified in patients with systemic lupus erythematosus (SLE). Thus the name was coined as lupus anticoagulants (LA) as anti-phospholipid antibody. However, the vast majority of patients identified with lupus anticoagulants does not have SLE and hence, the term LA is considered imprecise. It is found that, 70% of the clotting events occur in the veins, and the remaining 30% events occur in the arteries in patients with a lupus anticoagulant [19, 20].

Pregnant women with a previous history of a DVT are more prone to develop DVT. Risk is due to diminished levels of natural anticoagulants such as total and free protein S, or antithrombin. Elevated levels of procoagulant proteins such as FVIII, von Willebrand factor and fibrinogen also promote the risk of thrombosis. During pregnancy, the lessened rate of blood flow increases the risk of blood pooling [21, 22].

Hormone therapy which includes estrogens and oral contraceptives increases the risk of venous thrombosis. They increase the levels of procoagulant factors (VII, X, XII, and XIII) and diminish levels of the endogenous anticoagulants (protein S and anti-thrombin). The use of hormone therapy, therefore, may be associated with an increased risk of blood clots, especially in women with abnormalities of coagulation, such as factor V Leiden [23, 24].

Malignancies such as adenocarcinoma (pancreatic, ovarian, breast, and brain) are highly associated with an increased prevalence of thrombosis. It might be due to elevated levels of procoagulant substances in advanced stage of malignancies [25, 26].

Acute and chronic infectious and inflammatory disorders cause endothelial cell damage due to secretion of interleukins, C-reactive protein, TNF-α and endotoxins. This damage may induce clotting cascade. Hence, patients suffering from such diseases have increased risk of developing thrombotic conditions [27, 28].

Major surgical procedures are highly associated with thrombosis. During post-surgical periods following surgical interventions like hip or knee replacement surgery, traumatic hip fractures and neurosurgery, the risk of developing thrombotic conditions is high. However, the risk is less in those patients who receive prophylactic anticoagulants routinely [29, 31].

Heparin-Induced Thrombocytopenia (HIT) occurs due to immune system mediated adverse drug reaction against the administration of heparin. This is the consequence of formation of antibody-platelet factor IV–heparin complex. The complex activates platelets and thrombotic conditions are subsequently propagated. Heparin is discontinued instantaneously and alternative anticoagulants like direct thrombin inhibitors are recommended for the treatment [32, 33].

7.6 ANTICOAGULANTS

7.6.1 HEPARINS

7.6.1.1 UNFRACTIONATED HEPARIN (UFH)-ANTITHROMBIN DEPENDENT INHIBITION OF THROMBIN, XA, IXA AND XIA

Endogenous heparin acts as an anticoagulant when binds to serpin antithrombin (AT), which covalently inactivates both thrombin and FXa by binding to their active sites. Pharmaceutical heparin, e.g., unfractionated heparin (UFH) was isolated from porcine intestinal mucosa. In 1916, the anticoagulant properties of unfractionated heparin were first described. In 1968, antithrombin was identified and isolated. Since then, it has become evident that heparin binds to the protein antithrombin and it can also facilitate antithrombin-mediated inhibition of FIXa and FXIa, along with the inhibition of thrombin and FXa as shown in Figure 7.3. Antithrombin alone is not a potent anticoagulant, but when bound to UFH, the anticoagulant activity of antithrombin is accelerated approximately 1000-fold. Clinical use of unfractionated heparin (UFH) was started since the 1930s for the prevention and treatment of thrombosis. It is currently used for cardiovascular surgery and for the prevention of venous thromboembolism. Though use UFH is decreased with the discovery of LMWH. A complication of administering unfractionated heparin is the syndrome of heparin induced thrombocytopenia, which is associated with high rates of both arterial thrombosis and venous thrombosis. Long-term high dose (4 months at 15,000 U or more) heparin administration can lead to severe osteopenia.

In the rare patient with hypoaldosteronism, heparin may induce hyper-kalemia. Heparin is not effective at inhibiting fibrin-bound thrombin. Thrombin binds to fibrin at its exosite 1, and heparin can form a tight bridge from the fibrin surface to exosite 2 of thrombin. This tight ternary complex does not allow the antithrombin-heparin complex to access the exosite 2 of thrombin, and therefore inhibition of thrombin cannot proceed [34–37].

7.6.1.2 LOW-MOLECULAR-WEIGHT HEPARINS (LMWHS)-ANTITHROMBIN DEPENDENT INHIBITION OF XA AND THROMBIN

Low-molecular-weight heparins, e.g., fractionated heparin was introduced two decades ago. Low-molecular-weight heparins (LMWHs) are hetero-geneous mixtures of sulfated glycosaminoglycans. LMWHs are derived from unfractionated heparin by controlled depolymerization processes,

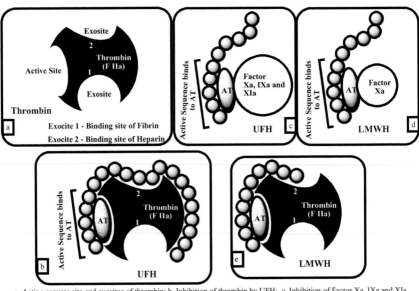

a. Active enzyme site and exosites of thrombin; b. Inhibition of thrombin by UFH; c. Inhibition of Factor Xa, IXa and XIa by UFH; d. Inhibition of Factor Xa by LMWH; e. Inhibition (Partial)of thrombin by LMWH. All inhibitions are antithrombin dependent.

FIGURE 7.3 Mechanism of action of UFH and LMWH.

fractionation methods, or both. These molecules are an indirect inhibitor of both factor Xa and thrombin since the binding is antithrombin mediated [38, 39].

The molecular weight (MW) of LMWHs is lesser than the MW of UFHs. To inactivate thrombin, heparin molecule must form a ternary complex as a bridge between antithrombin and thrombin, which requires minimum 18 saccharide units and thus smaller heparin molecules, cannot facilitate the interaction between antithrombin and thrombin. The implication of the minimum size requirement is that there must be at least 13 saccharides next to the pentasaccharide for thrombin binding. Therefore, LMWHs have greater affinity towards factor Xa, but lesser to thrombin as shown in Figure 7.3. If the MW of LMWHs is greater than 5000 Da, the anti-IIa specific activity of LMWH is drastically reduced and selectivity increased towards the inhibition of factor X. Enoxaparin, dalteparin, tinzaparin, parnaparin, nadroparin, certoparin, bemiparin, reviparin are the approved LMWHs. Tinzaparin is withdrawn from the market. These agents have improved subcutaneous bioavailability; dose-independence clearance; longer biological half-life; lower incidence of thrombocytopenia; and a reduced need for routine laboratory monitoring in comparison to UFH [38–41].

7.6.1.3 FONDAPARINUX, IDRAPARINUX AND IDRABIOTAPARINUX-ANTITHROMBIN DEPENDENT INHIBITOR OF XA

Fondaparinux (Figure 7.4) is a synthetic pentasaccharide which is antithrombin dependent factor Xa inhibitor. Within heparin and heparin sulfate this monomeric sequence is thought to form the high-affinity binding site for the anti-coagulant factor antithrombin. In comparison to heparin, fondaparinux does not inhibit thrombin. Fondaparinux has several potentially important advantages over other UFH and LMWHs. It is synthetic and has no viral or other animal contaminants. It has a well-defined, pure, homogenous molecular structure. It does not interact with Platelet factor 4 [42, 43].

Idraparinux and idrabiotaparinux (Figure 7.4) are polymethylated derivatives of fondaparinux. Their elimination half-life increases from 7 days after single administration up to 60 days after a 6–12 month treatment period. However, the development of idraparinux and idrabiotaparinux

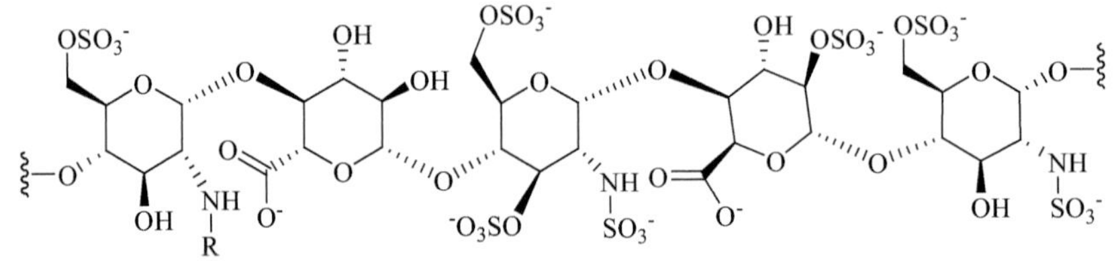

R= SO$_3^-$ OR COCH$_3$

Heparin and LMWH (Partial structure showing pentasaccharide sequence that binds to anti-thrombin)

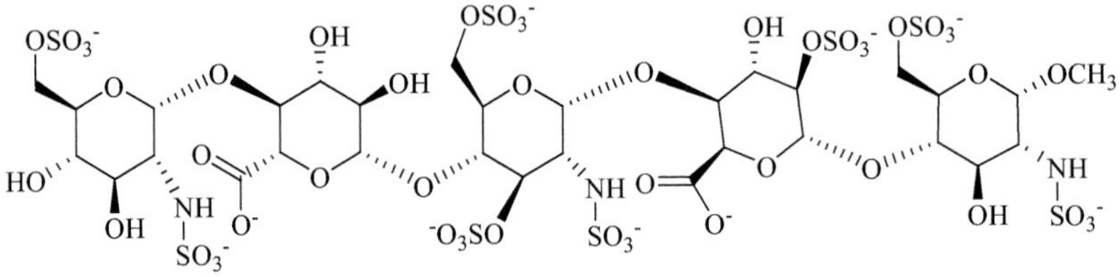

Fondaparinux

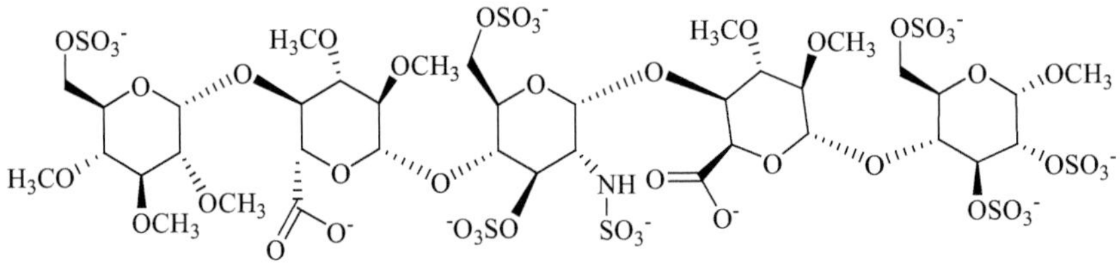

Idraparinux

FIGURE 7.4 Structure of heparins, fondaparinux and idraparinux.

is halted. Biotinylated idraparinux, named idrabiotaparinux, is structurally similar to idraparinux, with the addition of a biotin segment. It has the same anticoagulant activity as idraparinux. Avidin exposes a strong affinity to biotin and can be given intravenously to rapidly bind, neutralize and eliminate idrabiotaparinux [43, 44].

7.6.2 COUMARIN DERIVATIVES-VITAMIN K ANTAGONISTS

Vitamin K antagonists inhibit the enzyme vitamin K epoxide reductase, which uses vitamin K to modify several coagulation zymogens (factor VII, factor IX, factor X and prothrombin). Therefore, antithrombotic effect is a consequence of their inhibition of the vitamin K-dependent post-translational γ-carboxylation of glutamic acid (Gla) residues in the previously mentioned procoagulant zymogens. γ-carboxyglutamic acid residues (a specific domain) are an essential requirement for these proteins for permitting the binding of procoagulants to phospholipid surfaces and hence proper assembly into the active tenase and prothrombinase complexes. Deficiency of Gla domains turns FVIIa, FIXa, FXa, and thrombin into physiologically very poor procoagulants. These antagonists are used for long-term anticoagulant therapy. These inhibitors were introduced more than 50 years ago and still in clinical use today. Vitamin K antagonists are Warfarin and its analogs, such as phenprocoumon, acenocoumarol, and dicumarol (Figure 7.5).

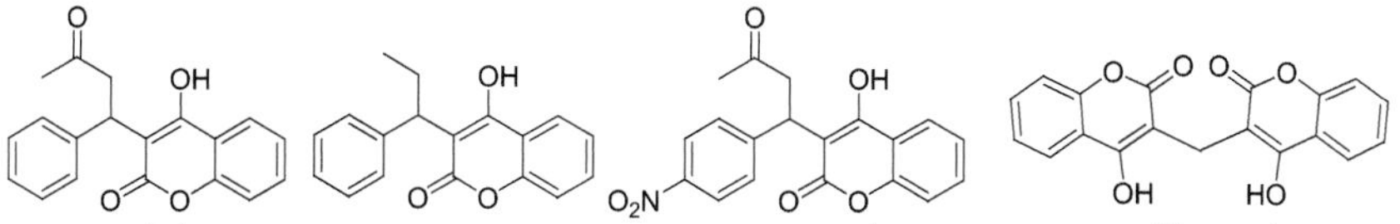

FIGURE 7.5 Vitamin K antagonists.

They inhibit the reductive recycling of vitamin K epoxide to vitamin K hydroquinone by inhibiting the enzyme vitamin K epoxide reductase as shown in Figure 7.6. Vitamin K hydroquinone is the cofactor to γ-glutamylcarboxylase, which affects the actual carboxylation of glutamic acid residues. The major effect of anticoagulation is primarily because of inhibition of thrombin generation. Warfarin is the most commonly prescribed vitamin K antagonist. Despite careful therapeutic monitoring, the incidence of major bleeding is a serious adverse effect. The clinically used preparation of warfarin is racemic. However, the enantiomers are not equipotent. In fact, (S)-warfarin is at least fourfold more potent as an anticoagulant than the (R)-warfarin. Warfarin and other coumarin derivatives undergo extensive hepatic oxidative metabolism catalyzed by CYP2C6 isozyme to give 6- and 7-hydroxyl warfarins as the major inactive metabolites. Those individuals with compromised hepatic function are at greater risk for warfarin toxicity secondary to diminished clearance.

Many of the drug-drug interactions are associated with enhanced or inhibited metabolism of warfarin via CYP2C9 induction or inhibition. As

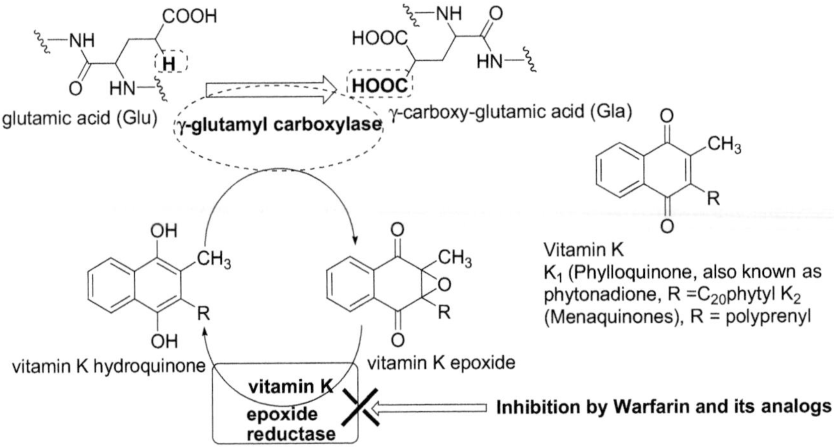

FIGURE 7.6   Mechanism of action of vitamin K antagonists.

warfarin is highly protein bound drug, any other substances that displace bound drug from protein binding sites increases the levels of free drug and, as a result, can cause warfarin toxicity, which usually is manifested by hemorrhage. The activity of warfarin is affected by diet and by genetic make-up: polymorphisms in the gene that encodes vitamin K epoxide reductase and in the cytochrome P450 gene *CYP2C9* account for up to 50% of the inter-individual variability of warfarin dosing. Dicumarol, phenprocoumon, and acenocoumarol have also been investigated as anti-coagulants in humans, although no significant advantages were found [45–47].

7.6.3   DIRECT FACTOR XA INHIBITOR

Currently available direct factor Xa inhibitors are rivaroxaban, apixaban, and edoxaban (Figure 7.7). **Rivaroxaban** is given orally. It can inhibit both free factor Xa and factor Xa bound in the prothrombinase complex. Inhibition of Factor Xa interrupts the intrinsic and extrinsic pathway of the blood coagulation cascade by inhibiting both thrombin formation and development of thrombi [48]. **Apixaban** is a highly selective; orally bioavailable and reversible direct inhibitor of free and clot-bound factor Xa. It was approved in Europe in 2012. Subsequently, it was approved in

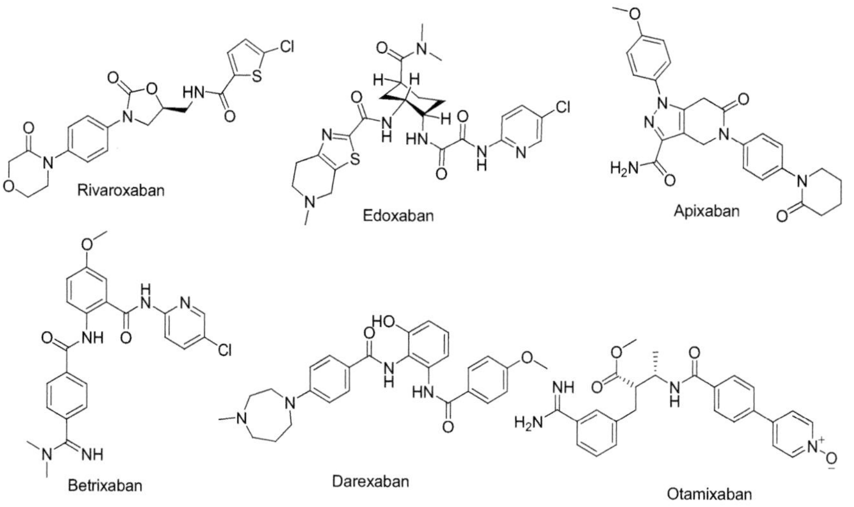

FIGURE 7.7 Factor Xa inhibitor.

the U.S. in 2014 for treatment and secondary prophylaxis of deep vein thrombosis (DVT) and pulmonary embolism (PE) [49].

Edoxaban was approved in July 2011 in Japan for the prevention of venous thromboembolisms (VTE) following lower-limb orthopedic surgery. It was also approved by the FDA in January 2015 for the prevention of stroke and non-central nervous system systemic embolism. It inhibits free factor Xa and prothrombinase activity and inhibits thrombin-induced platelet aggregation [50]. Portola submitted a New Drug Application (NDA) to the U.S. Food and Drug Administration (FDA) seeking approval to market **betrixaban** for extended-duration prophylaxis of venous thromboembolism (VTE) in acute medically ill patients with risk factors for VTE [51]. **Darexaban** (YM150) is a direct inhibitor of factor Xa created by Astellas Pharma. The development of darexaban was discontinued in September 2011 [52]. **Otamixaban** is an experimental injectable anticoagulant direct factor Xa inhibitor that was investigated for the treatment for acute coronary syndrome. In 2013, Sanofi announced that the drug candidate showed poor performance in a Phase III clinical trial [53]. The advantages of the xabans over vitamin K antagonists include no requirement for routine anticoagulation monitoring as well as a fast and reliable onset of action [54–56].

7.6.4 DIRECT THROMBIN INHIBITORS (DTIS)

DTIs act by directly inhibiting thrombinto delay clotting and do not require a cofactor such as antithrombinto exert their effect. They have been developed and investigated for their utility in prophylaxis and treatment of venous thromboembolism (VTE), prevention of thromboembolic complications in patients with HIT or at risk for HIT and undergoing percutaneous coronary intervention (PCI), acute coronary syndromes (ACS) with and without percutaneous transluminal coronary angioplasty (PTCA), secondary prevention of coronary events after ACS and nonvalvular atrial fibrillation. Thrombin has three domains: the active site and exosites 1 and 2 (Figure 7.8). Exosite 1 is the fibrin-binding site of thrombin. Exosite 2 serves as the heparin-binding domain. Both soluble thrombin and fibrin-bound thrombin are inhibited by DTIs. DTIs do not bind with plasma

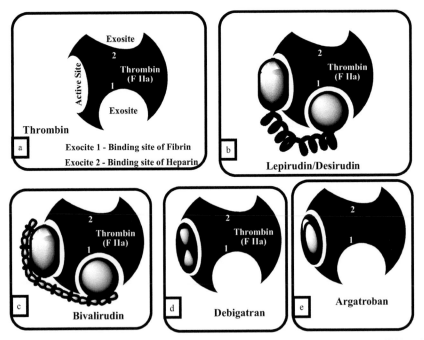

a. Active enzyme site and exosites of thrombin; b. Inhibition of thrombin by Lepirudin/ Desirudin; c. Inhibition of thrombin by Bivalirudin; d. Inhibition of thrombin by Debigatran; e. Inhibition of thrombin by Argatroban.

FIGURE 7.8 Mechanism of action of direct thrombin inhibitors.

proteins, indirectly inhibit platelet aggregation. They do not have the risk of thrombocytopenia. The bivalent DTIs include bivalirudin and lepirudin, and bind both the active site and exosite 1 of thrombin. Univalent DTIs bind only the active site and include argatroban and dabigatran [57–61].

7.6.4.1 PARENTERAL DIRECT THROMBIN INHIBITORS

Hirudin, the first parenteral DTI, was isolated in the late 1800s. It is a peptide originally isolated from the salivary glands of the medicinal leech, *Hirudo medicinalis*. However, issues related to unsuccessful purification and availability led to its abandonment following the introduction of heparin. **Lepirudin and desirudin** are derivatives of hirudin that were developed by recombinant technology in *Saccharomyces cerevisiae*. Both are 65-amino acid protein. Compared with hirudin, their affinity is 10 times weaker for thrombin. They are still considered the most potent of all the thrombin inhibitors. Both recombinant hirudins (r-hirudins) are bivalent direct thrombin inhibitors that bind simultaneously to the active site and exosite 1 domain on thrombin, an interaction that increases their specificity for thrombin as presented in Figure 7.8. They also have the highest affinity for thrombin as they rapidly form essentially irreversible, 1:1 stoichiometric complexes. Lepirudin was approved for anticoagulation in patients with HIT by the Food and Drug Administration (FDA) in 1998. It can inhibit free and clot-bound thrombin. It is not activated by platelet factor 4, making it more effective in the presence of platelet-rich thrombi [62–65].

Bivalirudin is an engineered 20-amino acid; synthetic, bivalent analog of hirudin with a thrombin inhibition activity. It has been approved for use in patients with unstable angina undergoing percutaneous coronary intervention. It is nearly 800 times weaker than that of hirudin. The amino terminus consists of the active site inhibitory sequence, D-Phe- Pro-Arg, which is connected by a flexible tetra-glycine linker to 12 amino acids from the carboxy terminus of hirudin that bind to exosite 1 as shown in Figure 7.8. The Pro-Arg peptide bond can be slowly cleaved by the catalytic site of thrombin; hence bivalirudin functions as a reversible inhibitor with a short half-life (20 to 30 min). It has decreased bleeding risk and improved safety profile when compared with r-hirudins [66, 67].

Argatroban (Figure 7.9) is a small (527 Da) synthetic piperidine carboxylic acid derivative of L-arginine. It is a univalent DTI that non-covalently and reversibly binds to the active site on thrombin as shown in Figure 7.7. Commercially available argatroban is a racemic mixture of the *R*- and *S*-diastereoisomers in a ratio of approximately 65 to 35, with the S-isomer having about twice the thrombin-inhibitory potency of the R-isomer. It is licensed in the United States for the prophylaxis or treatment of thrombosis in patients with HIT and for anticoagulation in patients with a history of HIT or at risk of HIT undergoing PCI. It is given as an intravenous infusion and does not require a bolus injection [68, 69].

7.6.4.2   ORAL DIRECT THROMBIN INHIBITORS

Ximelagatran (Figure 7.9) is the oral, double prodrug of melagatran. It was the first oral direct thrombin inhibitor developed. The approval of ximelagatran in Europe was given for prevention of venous thromboembolism in major elective orthopedic surgeries. However, ximelagatran was removed from the European market approximately 20 months later because therapy greater than 35 days was associated with a risk of hepatotoxicity [70–71].

Dabigatran etexilate (Figure 7.9) is a prodrug of dabigatran. It is a small peptidomimetic that directly inhibits thrombin by binding to its active site

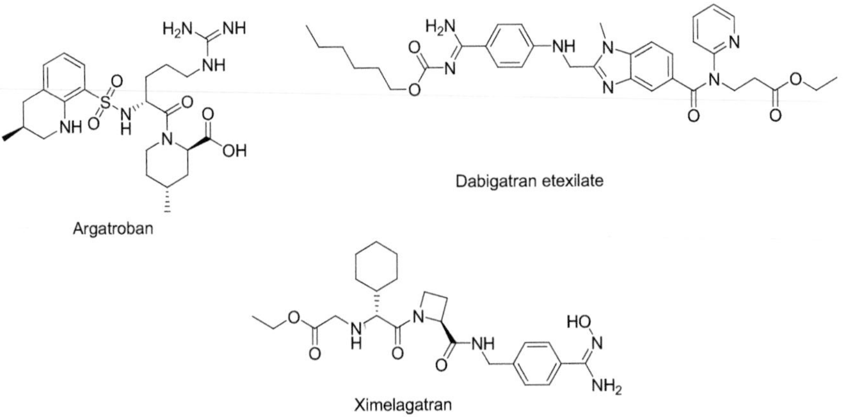

FIGURE 7.9   Parenteral and oral direct thrombin inhibitors.

via ionic interactions. It rapidly and reversibly inhibits both clot-bound and free thrombin in a concentration-dependent manner. It was approved in Canada and Europe in 2008 for the prevention of venous thromboembolism after elective total hip replacement and/or total knee replacement. In October 2010, the USFDA advisory committee approved dabigatran etexilate for stroke prevention in patients with atrial fibrillation. Significant clinical advantages of this drug include a rapid onset of action, lack of interaction with cytochrome P450 enzymes or with other food and drugs, excellent safety profile, lack of need for routine monitoring, a broad therapeutic window and a fixed-dose administration. Long-term use of dabigatran has not been associated with liver toxicity [72–74].

7.7 ANTIPLATELET DRUGS

Antiplatelet drugs stop blood cells (called platelets) from sticking together and forming a blood clot. In healthy vasculature, circulating platelets are sustained in an inactive state by nitric oxide and prostacyclin released by endothelial cells lining the blood vessels. Endothelial cells also release ADPase (adenosine diphosphatase), which vitiates ADP released from red blood cells and activated platelets, thereby preventing further activation of ADP. Whenever there is an injury, platelets are directed to the site of the injury, where they cluster together to form a blood clot. The receptors (GP Ia, IIa, and IIb/IIIa) shown in Figure 7.10 that are constitutively expressed on the platelet surface adhere to exposed collagen and von Willebrand factor (vWF). Platelets secrete ADP from their dense granules, and synthesize and release thromboxane A_2 (TXA_2). The ADP and TXA_2 act as platelet agonists by activating ambient platelets and recruiting them to the site of vascular damage. Thrombin can activate platelets by binding through PAR–1 receptors. While converting fibrinogen to fibrin; thrombin also aids as a potent platelet agonist and recruits more platelets to the site of vascular damage. Glycoprotein (GP) IIb/IIIa ($\alpha IIb\beta 3$), the most abundant receptor on the platelet surface, undergoes a conformational change after activation of platelets. This change increases the capacity of platelets to bind fibrinogen. Divalent fibrinogen molecules bridge adjacent platelets together to form platelet aggregates. Fibrin strands, generated by the action of thrombin, then weave these aggregates together to form a platelet-fibrin mesh. This stops the bleeding in normal physiological condition to repair the cut or a wound. But, sometimes, platelets will clump together inside a blood vessel that is injured, swollen (inflamed), or that has plaque build-up (atherosclerosis), e.g., forming a blood clot (thrombosis) inside the vessel which may occlude the blood vessel to supply the blood into the vital organs. They can also cause blood clots to form around stents, artificial heart valves, and other devices that are placed inside the heart or blood vessels. Anti-platelets are usually given to those patients who have a history of coronary artery disease (CAD), heart attack, angina, stroke or transient ischemic attacks (TIAs), and peripheral vascular disease (PVD). These may also be given to patients during and after angioplasty and stent procedures and after coronary artery bypass surgery. Some patients with atrial fibrillation or valve disease also prescribed anti-platelets. Mechanism of action(s) of various antiplatelet agents are presented in Figure 7.10 [75–79].

7.7.1 ASPIRIN AND TRIFLUSAL-COX INHIBITOR

Aspirin (Figure 7.11) is a well-established antiplatelet drug regarded as the cornerstone of treatment for patients with any vascular disease. Aspirin irreversibly inhibits cyclooxygenase-1 (COX-1) by its ability to acetylate the hydroxyl group of Ser–529 near the active site. It blocks the binding of COX-1 to its endogenous substrate arachidonic acid. Thus, inhibit the subsequent production of TXA_2 via the synthesis of PGH from arachadonic acid. Due to lack of a nucleus, protein synthesis does not occur in platelets. Therefore, cannot replenish the acetylated COX-1 and antithrombotic effect remains for the duration of about 7 to 10 days, which is the normal life span of the platelet. Antithrombotic effects of aspirin include the dose-dependent inhibition of platelet function, boosting the fibrinolysis, and a clampdown of blood coagulation. However, higher dosage of aspirin only adds the side effects like internal bleeding and upper gastrointestinal irritations [80–82]. **Triflusal** (Figure 7.11) is a platelet aggregation inhibitor. It is found to inhibit COX and cAMP (cyclic adenosine monophosphate) phosphodiesterase. cAMP phosphodiesterase inhibition leads to increased levels of cAMP. Elevated cAMP levels decrease platelet aggregation [83, 84].

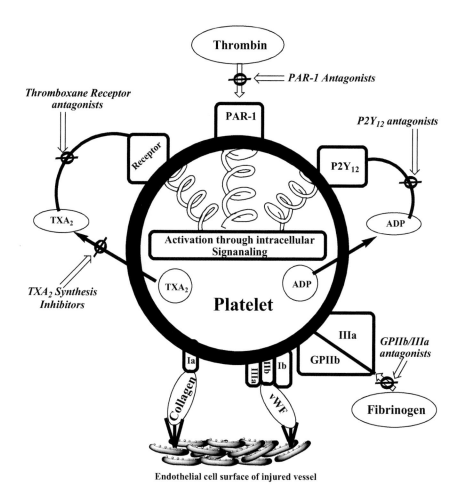

FIGURE 7.10 Mechanism of antiplatelet drugs.

7.7.2 DIPYRIDAMOLE AND CILOSTAZOL – PHOSPHODIESTERASE ENZYME INHIBITOR

Dipyridamole (Figure 7.12) is a pyrimidopyrimidine derivative inhibits phosphodiesterase enzymes that normally break down cAMP. Inhibition leads to increase in cellular cAMP levels and blocking the platelet aggregation response to ADP and/or cGMP. It also inhibits the cellular reuptake of adenosine into platelets, red blood cells, and endothelial cells, leading

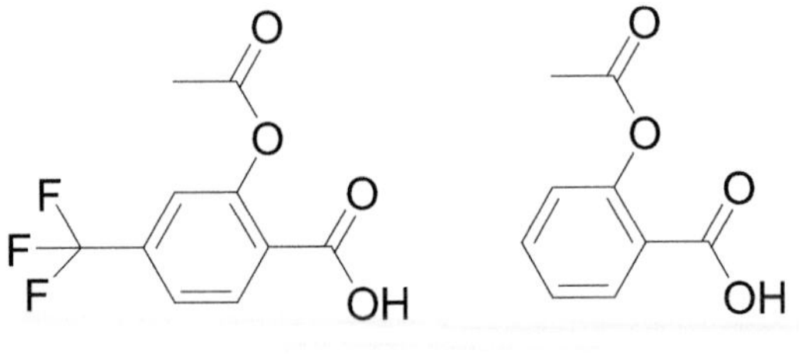

Triflusal   Aspirin

FIGURE 7.11   Antiplatelet as COX-inhibitor.

to increased extracellular concentrations of adenosine [85]. **Cilostazol** (Figure 7.12) is a quinolinone-derivative and a selective inhibitor of phosphodiesterase type 3 (PDE_3) with a therapeutic focus on increasing cAMP [86, 87].

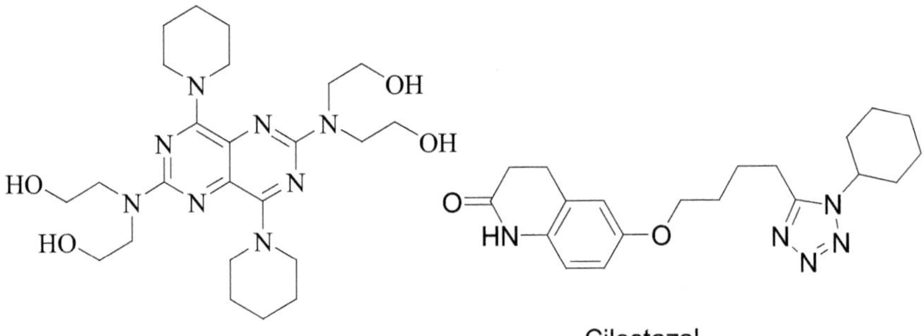

Dipyridamole   Cilostazol

FIGURE 7.12   Antiplatelet as phosphodiesterase enzyme inhibitor.

7.7.3   CLOPIDOGREL, PRASUGREL, CANGRELOR, TICAGRELOR, ELINOGREL AND TICLOPIDINE-PLATELET P2Y12 PURINERGIC RECEPTOR ANTAGONISTS

ADP induced platelet aggregation takes place through binding of ADP to G protein-coupled P2Y purinergic receptors. **Clopidogrel** (Figure 7.13) is a thienopyridine that irreversibly inhibits $P2Y_{12}$ on the platelet surface.

It must be metabolized in the liver to generate the active metabolites that inhibit the ADP receptor [88]. New $P2Y_{12}$ inhibitors include prasugrel, cangrelor, and ticagrelor (Figure 7.13). Like clopidogrel, **prasugrel** is a thienopyridine that requires hepatic metabolism to generate active metabolites. It is more efficient than that of clopidogrel. It produces more rapid, more consistent, and more potent inhibition of ADP-induced platelet aggregation. **Cangrelor and ticagrelor** are direct-acting reversible inhibitors of $P2Y_{12}$. Cangrelor, which is administered intravenously, has a rapid onset and offset of action. Ticagrelor is administered orally [89, 90]. **Elinogrel** (Figure 7.13) is an experimental antiplatelet drug acting as a $P2Y_{12}$ inhibitor. However, development was discontinued in 2012. **Ticlopidine** (Figure 7.13) is an antiplatelet drug in the thienopyridine family. However, because of its rare but serious side effects of neutropenia and thrombotic thrombocytopenic purpura, its use remained limited [91, 92].

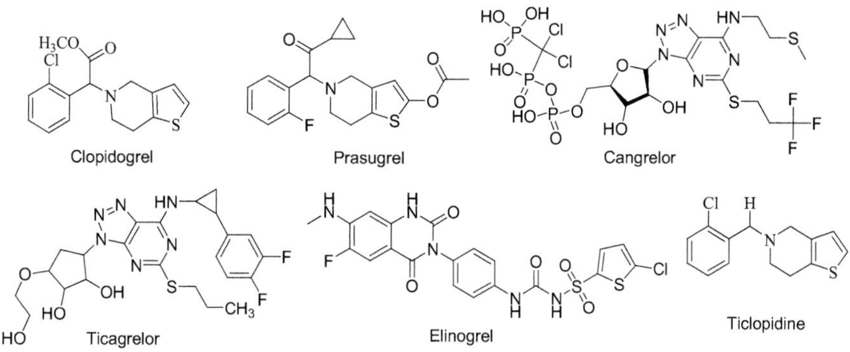

FIGURE 7.13   Antiplatelet as platelet P2Y12 purinergic receptor antagonist.

7.7.4 ABCIXIMAB, TIROFIBAN AND EPTIFIBATIDE – GLYCOPROTEIN IIB/IIIA RECEPTOR ANTAGONIST

Abciximab (Figure 7.14) is a monoclonal antibody and a glycoprotein IIb/IIIa receptor antagonist, inhibits platelet aggregation by preventing the binding of fibrinogen, von Willebrand factor, and other adhesive molecules to platelet [93]. **Tirofiban** (Figure 7.14) is also a glycoprotein IIb/IIIa antagonist. It is contraindicated in patients with a history of thrombocytopenia [94]. **Eptifibatide** (Figure 7.14) is a cyclic heptapeptide derived from a protein found in the venom of the south-eastern pygmy rattlesnake. It has a short half-life. The drug is usually given with aspirin or clopido-grel and with heparins [95, 96].

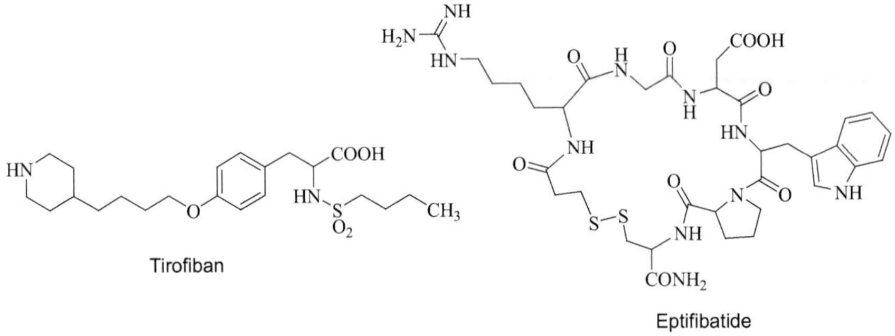

FIGURE 7.14   Antiplatelet as glycoprotein IIb/IIIa receptor antagonist.

7.7.5   VORAPAXAR – PAR–1 ANTAGONIST

Vorapaxar (Figure 7.15) is a thrombin receptor (protease-activated receptor, PAR-1) antagonist based on the natural product himbacine. It is given to patients with a history of myocardial infarction (heart attack) or persons with peripheral arterial disease [97].

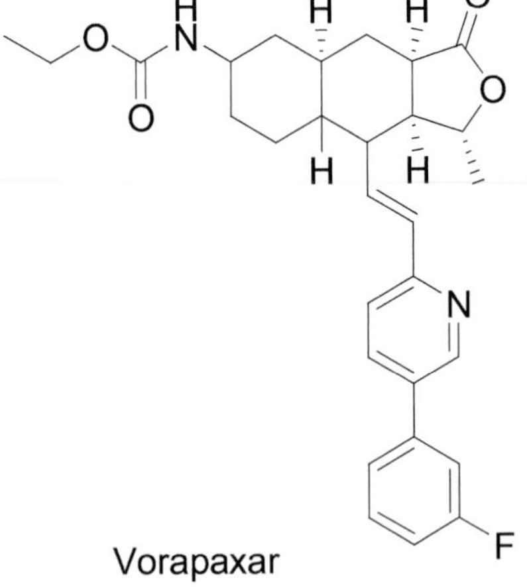

Vorapaxar

FIGURE 7.15   Antiplatelet as PAR–1 antagonist.

7.7.6 APPROACHES AS PLATELET ADHESION INHIBITION

Inhibition of collagen-platelet interaction may be an excellent way to inhibit platelet aggregation. Collagen is abundant in atherosclerotic plaques. Humanized monoclonal antibodies and aptamers against the receptors, small-molecule peptide inhibitors, and proteins derived from the medicinal leech are the research strategy going on to inhibit collagen-platelet interaction. Some are tested in humans, but none was extended to phase III clinical trial [3, 98].

7.8 THROMBOLYTIC DRUGS

Thrombolytic drugs dissolve the preexisting thrombus in both arteries and veins by generating plasmin from inactive plasminogen. Plasmin degrades fibrin to soluble peptides. The actions of thrombolytic drugs are attained by either potentiating endogenous fibrinolytic pathways or mimicking natural thrombolytic molecules. The drugs are different in their efficiency, fibrin selectivity, and side-effect profile. Presently available thrombolytic agents are derived from bacterial products or manufactured using recombinant DNA technology. Tissue plasminogen activator (t-PA) and streptokinase were introduced as a life-saving treatment for acute myocardial infarction in the mid-1980s. The disadvantages include the risk of hemorrhage and high costs. t-PA is synthesized by the vascular endothelial cells and is considered as the endogenous thrombolytic agent. The first recombinant t-PA was **alteplase**. It is manufactured by recombinant biotechnology and referred to as recombinant t-PA (rt-PA) which catalyzes the conversion of plasminogen to plasmin. It is used in the treatment of coronary artery thrombosis, pulmonary embolism, and acute stroke. It is highly expensive in compare to streptokinase. **Streptokinase** is produced by hemolytic *Streptococci* and a nonfibrinogen-specific fibrinolytic agent. It does not exhibit plasmin activity by proteolytic cleavage of plasminogen. It binds non-covalently to plasminogen in a 1:1 equimolar fashion and thereby deliberates plasmin activity. The greatest benefit of streptokinase appears to be achieved by early intravenous administration. **Urokinase** is the most familiar thrombolytic agent among interventional radiologists and it is often used for peripheral vascular thrombosis. It is approved for pulmonary embolism and for lysis of coronary thrombi, but not for mortality reduction in acute myocardial infarction. Recombinant techniques are used to produce urokinase in *Escherichia coli*. Urokinase directly cleaves plasminogen to produce plasmin [99–106]. **Prourokinase** is a new fibrinolytic agent. It is a single chain urokinase and has been produced both in glycosylated and nonglycosylated forms. It is a relatively inactive precursor that must be converted to urokinase before it becomes active *in vivo*. **Reteplase** is a recombinant nonglycosylated form of human t-PA produced in *Escherichia coli* and has a longer half-life of 13–16 min. It binds fibrin and has the ability to penetrate into thrombi. **Tenecteplase** is developed by introducing modifications to the complementary DNA for natural human t-PA. It is a 527 amino acid glycoprotein. It binds to fibrin and selectively converts thrombus bound plasminogen to plasmin, which degrades the fibrin matrix of the thrombus. It has higher fibrin specificity. **Lanoteplase** (nPA) is a variant of t-PA with greater fibrinolytic activity and slower clearance from the plasma. **Staphylokinase** is a 136 amino-acid protein produced by certain strains of *Staphylococcus aureus* and has a unique mechanism of fibrin selectivity. Recombinant staphylokinase has been demonstrated to provide beneficial effects among patients with acute myocardial infarction. Recombinant desmodus salivary plasminogen activator-1 (r DSPA-1, **desmoteplase**) is a naturally occurring enzyme in the saliva of the blood-feeding vampire bat (*Desmodus rotundus*) is genetically related to t-PA. It targets and destroys fibrin. It is more fibrin dependent and fibrin specific than t-PA. Compared to t-PA, it is non-neurotoxic, causes less fibrinogenolysis, less antiplasmin consumption, and results in faster and more sustained reperfusion as revealed by animal studies. The benefit of fibrinolytic therapy decreases as time progresses following the onset of symptoms. Pre-hospital administration of thrombolysis is found to be advantageous if administered within 70 min in terms of reducing death, stroke, serious bleed, and infarct size [107–111].

7.9   OTHER EMERGING TARGET AND EXPERIMENTAL APPROACHES

There is a continuous unmet need for the development of new lead due to lacunae of existing drugs. Ongoing inquisitiveness of researcher in anti-thrombotic arena led to the establishment of diverse targets. Targets are tissue factor/factor VIIa complex, factor IXa, Va, VIIIa, and factor XIIIa.

Inhibition of tissue factor/factor VIIa complex block the initiation of coagulation, whereas inhibition of factor IXa, factor VIIIa and factor Va, block the propagation of coagulation. Inhibition of factor XIIIa, has the potential to increase the exposure of the thrombus for lysis. Because, factor XIIIa cross-links the fibrinogen to form clots. Cross-linking stabilizes the fibrin polymer and renders it more stubborn to degradation by plasmin [112]. Parenteral tifacogin is a recombinant form of tissue factor pathway inhibitor which deregulates the factor X activation by tissue factor/factor VIIa complex [113, 114]. Factor VIIai, recombinant factor VIIa that has its active site irreversibly blocked competes with factor VIIa for tissue factor binding, thereby preventing the initiation of coagulation by the factor VIIa/tissue factor complex [115]. The parenteral agents which inhibit factor IX are factor IX-directed monoclonal antibodies [116] and pegnivacogin [117], a factor IXa-directed aptamer. Factor Va is directly inhibited by drotrecogin alfa (activated), a recombinant form of activated protein C, however, halted due to lack of efficacy and the potential to cause harm because of bleeding [118]. TB–402 is a human IgG_4 monoclonal antibody that partially inhibits factor VIIIa. It was generated by introducing a point mutation into the gene encoding MAB-LE2E9, a factor VIII-directed monoclonal antibody [119]. Tridegin is a specific inhibitor of factor XIIIa. It is a peptide isolated from the giant Amazon leech, *Haementeria ghilianii*, and enhances fibrinolysis *in vitro* when added before clotting of fibrinogen [120, 121]. Destabilase is a leech enzyme hydrolyzes cross-links and provides an alternative approach to reversing the consequences of factor XIIIa-mediated fibrin crosslinking. Tridegin and destabilase has not been tested in humans [122, 123]

Currently marketed anticoagulant, antiplatelet, and thrombolytic agents are used to treat the major classes of thrombotic conditions. The milestone of discovery and development of currently available drugs had begun a century ago as presented in Figure 7.16. The continuous effort of medicinal chemists has given the antithrombotic choices that we have today. Yet, numbers of questions remain unanswered. Risk of bleeding and hemorrhage is an unavoidable scenario. Treatment options are diverged based on the individual circumstances of the patient and outcome of ongoing clinical trials. Concomitant disease conditions along with thrombosis have a major influence in the treatment criteria selection [124]. Anticoagulants are used to treat a wide variety of conditions that involve arterial or venous thrombosis, including prevention of venous thromboembolism and long-term prevention of ischemic stroke in patients with atrial fibrillation. As long-term anticoagulant therapy for VTE without cancer, dabigatran, rivaroxaban, apixaban, or edoxaban was suggested over vitamin K antagonist (VKA) therapy and VKA therapy over low-molecular-weight heparin in the CHEST Guideline and Expert Panel Report [125]. For VTE with cancer, LMWH over VKA, dabigatran, rivaroxaban, apixaban, or edoxaban. But the recommendations for who should stop anticoagulation at 3 months or receive extended therapy were not suggested. Thrombolytic therapy was suggested for pulmonary embolism with hypotension, and systemic therapy over catheter-directed thrombolysis. For recurrent VTE on a non-LMWH anticoagulant, LMWH was suggested. For recurrent VTE on LMWH, increased LMWH dose was suggested. Some suggestions are strong, but not based on high-quality evidence and further research is needed [125]. In patients with acute ischemic stroke, IV recombinant tissue plasminogen activator (r-tPA) was recommended if treatment can be initiated within 3 h or 4.5 h of symptom onset. In patients with a history of noncardioembolic ischemic stroke or TIA, long-term treatment with aspirin (75–100 mg once daily), clopidogrel (75 mg once daily), aspirin/extended release dipyridamole (25 mg/200 mg bid), or cilostazol (100 mg bid) over no antiplatelet therapy, oral anticoagulants, the combination of clopidogrel plus aspirin, or triflusal was recommended. Of the recommended antiplatelet regimens, clopidogrel or aspirin/extended-release dipyridamole over aspirin or cilostazol was suggested. In patients with a history of stroke or TIA and atrial fibrillation, oral anticoagulation was suggested over no antithrombotic therapy, aspirin, and combination therapy with aspirin and clopidogrel. These recommendations can help clinicians make evidence-based treatment decisions with their patients who have had strokes as mentioned in the American College of Chest Physicians Evidence-Based Clinical Practice Guidelines [126]. For long-term prevention of atrial fibrillation in low-risk patients, aspirin is recommended. Either aspirin or warfarin is recommended for AF patients who have moderate risk. Warfarin is recommended for high-risk groups with AF. Dabigatran is recommended in the prevention of stroke and systemic embolism associated with nonvalvular atrial fibrillation. Apixaban is recommended to lower the risk of stroke and embolism in patients with nonvalvular atrial fibrillation and also to reduce the risk of recurring DVT and PE after initial therapy. Stable angina, unstable angina, and acute myocardial infarction are treated with aspirin and low dose warfarin. Aspirin is recommended for the treatment of stable angina. For unstable angina, long-term aspirin therapy with low dose is recommended. Hospitalized patients because of unstable angina receive heparin or LMWH treatment for at least 48 hrs. Lepirudin can be given instead of heparin for patients with a history of HIT. AMI patients are usually immediately put on aspirin and treated with a thrombolytic. AMI Patients with received streptokinase previously, alteplase, or tenecteplase can be administered according to individual circumstances of patient. Adjunctive therapy with heparin (or lepirudinin individuals with a history of HIT) is recommended [4, 127–129]. Low-molecular-weight heparin was recommended for the prevention and treatment of VTE in pregnant women instead of unfractionated heparin. For pregnant women with acute VTE, anticoagulants were suggested to be continued for at least 6 weeks postpartum (for a minimum duration of therapy of 3 months) compared with shorter durations of treatment [130]. Use of LMWH is recommended for antithrombotic therapy in neonates and children with special care. But the evidences are weak [131]. For a comprehensive detail outcome of the clinical trials and basis of treatment criteria selection regarding antithrombotic therapy, it is suggested to read the entire volume of Chest [Volume 141, Issue 2, Supplement, Pages 1S–70S, e1S-e801S (February 2012)]

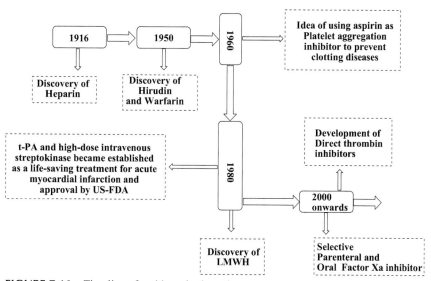

FIGURE 7.16   Timeline of antithrombotic options.

7.10 LABORATORY MONITORING OF ANTITHROMBOTIC THERAPY AND REVERSAL AGENTS

Monitoring of anticoagulant therapy is essential to avert unwanted bleeding. The prothrombin time (PT) is measured is to assess the activity of vitamin-K dependent factors, which include factor II, VII, IX and X. When calcium ions with an excess of thromboplastin added to anticoagulated plasma is a direct measure of the prothrombin amount in the plasma. In normal plasma, this clot formation takes 10 to 13 seconds. PT results are expressed in terms of international normalized ratios (INRs) due to variances in commercially available thromboplastins. Patients on warfarin therapy are optimally maintained with an INR of 2.0 to 3.0. However, this is varied depending on the individual patients with multiple circumstances. Heparin therapy is based on the assay named activated partial thromboplastin time (aPTT). In the aPTT assay, a surface activator (elegiac acid, kaolin, or silica) is used to activate the intrinsic pathway and clot formation begins. Like PT, the time taken for this clot formation is recorded. In normal plasma, the average aPTT result is 25 to 45 seconds. A therapeutic aPTT in a patient receiving heparin is 70 to 140 seconds usually. This assay mainly monitors factor II and X [5]. Direct thrombin inhibitors also monitored by aPTT assay. Other assay procedures are activated clotting time (ACT), ecarin clotting time (ECT), prothrombinase-induced clotting time (PiCT), chromogenic anti-IIa, diluted thrombin time (dTT), and anti-Xa assays. However, PT and aPTT assay are significant in monitoring [5, 132–138]. Apart from the monitoring, there is a need for a reversal agent or antidote to control the bleeding risk. Protamine sulfate is established reversal agent for heparins. For direct oral anticoagulats, most authorities recommend the use of four-factor prothrombin complex concentrates, although the evidence to support their use in terms of improving outcomes is insufficient. At the present time, there are three reversal agents. Idarucizumab is a monoclonal antibody, which is a drug-specific antidote targeted to reverse the direct thrombin inhibitor, dabigatran. Andexanet alfa is a recombinant protein and a class-specific antidote targeted to reverse the oral direct factor Xa inhibitors as well as the indirect inhibitor, enoxaparin. Ciraparantag (Figure 7.17) is a universal antidote targeted to reverse the direct thrombin and factor Xa inhibitors as well as the indirect inhibitor, enoxaparin. It consists of two L-arginine units connected with a piperazine containing linker chain [139–147].

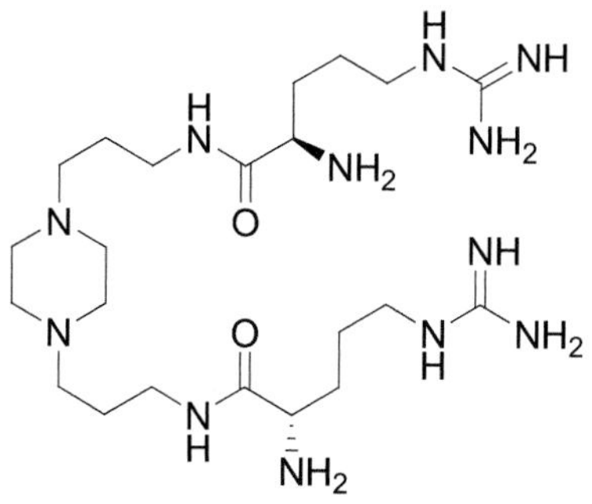

Ciraparantag

FIGURE 7.17   Antidote ciraparantag.

7.11   FUTURE DIRECTIONS

Development of a drug without endorsing bleeding in antithrombotic therapy is yet to be found. The complex bleeding and coagulation phenomena might be a reason. Reversal agents of anticoagulation are limited. Wide ranges of thrombotic condition make the treatment options more challenging. Earlier detection of thrombotic conditions by measuring the level of clotting factors could be the way to minimize the risk in patients. DTIs and direct factor Xa inhibitors have been shown useful in treating multiple thrombotic conditions. Nevertheless, continuous efforts in finding novel targets and novel drug candidate may solve the inevitable troubles of antithrombotic therapy in near future.

KEYWORDS

- **antithrombotic therapy**
- **ciraparantag**
- **thrombotic condition**

REFERENCES

1. Goodnight, S. H., & Hathaway, W. E., (2001). *Disorders of Hemostasis and Thrombosis: A Clinical Guide* (2nd edn.). McGraw-Hill Professional, New York.

2. Depta, J. P., & Bhatt, D. L., (2015). New approaches to inhibiting platelets and coagulation. *Annu. Rev. Pharmacol. Toxicol., 55*, 373–397.

3. Gross, P. L., & Weitz, J. I., (2009). New antithrombotic drugs. *Clin. Pharmacol. Ther., 86*(2), 139–146.

4. Bisacchi, G. S., (2003). *Anticoagulants, Antithrombotics, and Hemostatics.*, In: Abraham D. J., (6th ed.). *Burger's Medicinal Chemistry and Drug Discovery* (Vol. *3*, p. 283), John Wiley & Sons.

5. Lu, M. C., & Lemke, T. L., (2008). *Antithrombotics, Thrombolytics, Coagulants, and Plasma Extenders.* In: Lemke, T. L., & Williams, D. A., (6th edn.). *Foye's Principles of Medicinal Chemistry* (p. 820), Lippincott Williams & Wilkins: New York.

6. Mackman, N., (2008). Triggers, targets and treatments for thrombosis. *Nature, 451*(7181), 914–918.

7. Tsiara, S., Pappas, K., Boutsis, D., & Laffan, M., (2011). New oral anticoagulants: Should they replace heparins and warfarin? *Hellenic. J. Cardiol., 52*(1), 52–67.

8. Rand, M. L., & Murray, R. K., (2003). *Hemostasis & Thrombosis.* In: Robert, K., Murray, R. K., Granner, D. K., Mayes, P. A., & Rodwell, V. W., (26th edn.). *Harper's Illustrated Biochemistry* (p. 598), Mc Graw-Hill: New York.

9. Colman, R. W., Marder, V. J., Clowes, A. W., George, J. N., & Goldhaber, S. Z., (2006). *Hemostasis and Thrombosis: Basic Principles and Clinical Practice, 5th edn.* Philadelphia: JB Lippincott Co.

10. Manly, D. A., Boles, J., & Mackman, N., (2011). Role of tissue factor in venous thrombosis. *Annu. Rev. Physiol., 73*, 515–525.

11. Wu, K. K., & Thiagarajan, P., (1996). Role of endothelium in thrombosis and hemostasis. *Annu. Rev. Med., 47*(1), 315–331.

12. Dahlback, B., (2005). Blood coagulation and its regulation by anticoagulant pathways: Genetic pathogenesis of bleeding and thrombotic diseases. *J. Intern. Med., 257*(3), 209–223.

13. Agrawal, Y. K., Vaidya, H., Bhatt, H., Manna, K., & Brahmkshatriya, P., (2007). Recent advances in the treatment of thromboembolic diseases: Venous thromboembolism. *Med. Res. Rev., 27*(6), 891–914.

14. Segers, K., Dahlbäck, B., & Nicolaes, G. A., (2007). Coagulation factor V and thrombophilia: Background and mechanisms. *Thromb. Haemost., 98*, 530–542.

15. Van Cott, E. M., Khor, B., & Zehnder, J. L., (2016). Factor V Leiden. *Am. J. Hematol., 91*(1), 46–49.

16. Adler, G., Agnieszka, G., Valjevac, A., Czerska, E., Kiseljakovic, E., & Salkic, N. N., (2015). Prevalence of genetic prothrombotic risk factors: 1691G > A FV, 20210G > A PT and 677C > T MTHFR mutations in the Bosnian population. *Ann. Hum. Biol., 42*(6), 576–580.

17. Mascarenhas, J. V., Albayati, M. A., Shearman, C. P., & Jude, E. B., (2014). Peripheral arterial disease. *Endocrinol. Metab. Clin. North. Am., 43*(1), 49–166.

18. Lehotský, J., Tothová, B., Kovalská, M., Dobrota, D., Beňová, A., Kalenská, D., & Kaplán, P., (2016). Role of homocysteine in the ischemic stroke and development of ischemic tolerance. *Front. Neurosci., 10*, 538.

19. Gómez-Puerta, J. A., Cervera, R., (2014). Diagnosis and classification of the antiphospholipid syndrome. *J. Autoimmun., 48*, 20–25.

20. Gebhart, J., Posch, F., Koder, S., Perkmann, T., Quehenberger, P., Zoghlami, C., et al., (2015). Increased mortality in patients with the lupus anticoagulant: The Vienna lupus anticoagulant and thrombosis study (LATS). *Blood, 125*(22), 3477–3483.

21. Kourlaba, G., Relakis, J., Kontodimas, S., Holm, M. V., & Maniadakis, N., (2016). A systematic review and meta analysis of the epidemiology and burden of venous thromboembolism among pregnant women. *Int. J. Gynaecol. Obstet., 132*(1), 4–10.

22. Kamimoto, Y., Wada, H., Ikejiri, M., Nakatani, K., Sugiyama, T., Osato, K., et al., (2015). High frequency of decreased antithrombin level in pregnant women with thrombosis. *Int. J. Hematol., 102*(3), 253–258.

23. Martinelli, I., Lensing, A. W., Middeldorp, S., Levi, M., Beyer-Westendorf, J., Van Bellen, B., et al., (2015). Recurrent venous thromboembolism and abnormal uterine bleeding with anticoagulant and hormone therapy use. *Blood.* blood-2015-08-665927, doi:https://doi.org/10.1182/blood-2015-08-665927.

24. Sandset, P. M., (2013). Mechanisms of hormonal therapy related thrombosis. *Thromb. Res., 131*, 4–7.

25. Khorana, A. A., & Mc Crae, K. R., (2014). Risk stratification strategies for cancer-associated thrombosis: An update. *Thromb. Res., 133*, 35–38.

26. Smith, T. R., Nanney, A. D., Lall, R. R., Graham, R. B., Mc Clendon, J., Lall, R. R., et al., (2015). Development of venous thromboembolism (VTE) in patients undergoing surgery for brain tumors: Results from a single center over a 10-year period. *J. Clin. Neurosci., 22*(3), 519–525.

27. Riva, N., Donadini, M. P., & Ageno, W., (2015). Epidemiology and pathophysiology of venous thromboembolism: Similarities with atherothrombosis and the role of inflammation. *Thromb. Haemost., 113*(6), 1176–1183.

28. Rabinovich, A., Cohen, J. M., Cushman, M., Wells, P. S., Rodger, M. A., Kovacs, M. J., et al., (2015). Inflammation markers and their trajectories after deep vein thrombosis in relation to risk of post thrombotic syndrome. *J. Thromb. Haemost., 13*(3), 398–408.

29. Douketis, J. D., Spyropoulos, A. C., Kaatz, S., Becker, R. C., Caprini, J. A., Dunn, A. S., et al., (2015). Perioperative bridging anticoagulation in patients with atrial fibrillation. *N. Engl. J. Med., 373*(9), 823–833.

30. Kim, J. Y., Khavanin, N., Rambachan, A., Mc Carthy, R. J., Mlodinow, A. S., De Oliveria, G. S., et al., (2015). Surgical duration and risk of venous thromboembolism. *JAMA Surg., 150*(2), 110–117.

31. Cushman, M., (2007). Epidemiology and risk factors for venous thrombosis. *Sem. Hematol., 44*(2), 62–69.

32. Magnani, H. N., & Gallus, A., (2006). Heparin-induced thrombocytopenia (HIT). *Thromb. Haemost, 95*, 967–981.

33. Walenga, J. M., & Prechel, M. M., (2016). *Heparin-Induced Thrombocytopenia (HIT).* In: *Anticoagulation and Hemostasis in Neurosurgery* (p. 183), Springer.

34. Onishi, A., St Ange, K., Dordick, J. S., & Linhardt, R. J., (2016). Heparin and antico-agulation. *Front. Biosci., 21*, 372–1392.

35. Appadu, B., & Barber, K., (2016). Drugs affecting coagulation. *Anaesthesia & Intensive Care Medicine, 17*(1), 55–62.

36. Whiteley, W. N., Adams, H. P., Bath, P. M., Berge, E., Sandset, P. M., Dennis, M., et al., (2013). Targeted use of heparin, heparinoids, or low-molecular-weight heparin to improve outcome after acute ischaemic stroke: An individual patient data meta-analysis of randomised controlled trials. *Lancet. Neurol., 12*(6), 539–545.

37. Harbrecht, U., (2011). Old and new anticoagulants. *Hamostaseologie., 31*(1), 21–27.

38. Holzheimer, R. G., (2004). Low-molecular-weight heparin (LMWH). *Eur. J. Med. Res., 9*, 225–239.

39. Weitz, J. I., (1997). Low-molecular-weight heparins. *N. Engl. J. Med., 337*(10), 688–699.

40. Gray, E., Mulloy, B., & Barrowcliffe, T. W., (2008). Heparin and low-molecular-weight heparin. *Thromb. Haemost., 99*(5), 807–818.

41. Xu, Y., Cai, C., Chandarajoti, K., Hsieh, P. H., Li, L., Pham, T. Q., Sparkenbaugh, E. M., et al., (2014). Homogeneous low-molecular-weight heparins with reversible anticoagulant activity. *Nat. Chem. Biol., 10*(4), 248–250.

42. 42.ARIXTRA®-FDA. http://www.accessdata.fda.gov/drugsatfda_docs/label/2005/021345s010lbl.pdf (Accessed 26th August, 2017).

43. Brandao, G., Junqueira, D. R., Rollo, H. A., & Sobreira, M. L., (2015). Pentasaccha-rides for the treatment of deep vein thrombosis. *Cochrane. Database. Syst. Rev., 7.* Art. No.: CD011782. doi:10.1002/14651858.CD011782.

44. Perry, D., (2016). Overview of Anticoagulants. In: *Handbook of Thromboprophylaxis* (3rd edn., p. 3). Springer.

45. Ansell, J., Hirsh, J., Poller, L., Bussey, H., Jacobson, A., & Hylek, E., (2004). The pharmacology and management of the vitamin K antagonists: The seventh ACCP conference on antithrombotic and thrombolytic therapy. *Chest., 126*(3), 204–233.

46. Ansell, J., Hirsh, J., Hylek, E., Jacobson, A., Crowther, M., & Palareti, G., (2008). Pharmacology and management of the vitamin K antagonists: American college of chest physicians evidence-based clinical practice guidelines. *Chest., 133*(6), 160–198.

47. De, Caterina, R., Husted, S., Wallentin, L., Andreotti, F., Arnesen, H., Bachmann, F., et al., (2013). Vitamin K antagonists in heart disease. Current status and perspectives (Section III). *Thromb. Haemost., 110*(6), 1087–1107.

48. Janssen Pharmaceuticals. Xarelto (rivaroxaban) prescribing information, https://www.xareltohcp.com/shared/product/xarelto/prescribing-information.pdf (accessed August 13, 2017).

49. Bristol-Myers Squibb. Eliquis (apixaban) prescribing information. http://packagein-serts.bms.com/pi/pi_eliquis.pdf. (accessed August 13, 2017).

50. Daiichi Sankyo. Savaysa (edoxaban) prescribing information. http://dsi.com/pre-scribing-information-portlet/get PIContent?product Name=Savaysa&inline=true. (accessed August 13, 2017).

51. Betrixaban: FXa Inhibitor. https://www.portola.com/clinical-development/betrixa-ban-fxa-inhibitor/. (Accessed 26 August, 2017).

52. YM150, an oral direct factor Xa inhibitor astellas. https://www.astellas.com/en/corporate/news/pdf/110928_en.pdf. (accessed 26 August, 2017).

53. Sanofi provides update on phase 3 studies of two investigational compounds. http://en.sanofi.com/Images/33127_20130603_rdupdate_en.pdf. (accessed 26 August, 2017).

54. Kakkos, S. K., Kirkilesis, G. I., & Tsolakis, I. A., (2014). Choice-efficacy and safety of the new oral anticoagulants dabigatran, rivaroxaban, apixaban, and edoxaban in the treatment and secondary prevention of venous thromboembolism: A systematic review and meta-analysis of phase III trials. *Eur. J. Vasc. Endovasc. Surg., 48*(5), 565–575.

55. Yeh, C. H., Fredenburgh, J. C., & Weitz, J. I., (2012). Oral direct factor Xa inhibitors. *Circ. Res., 111*(8), 1069–1078.

56. Cabral, K. P., & Ansell, J. E., (2015). The role of factor Xa inhibitors in venous thromboembolism treatment. *Vasc. Health. Risk. Manag., 11*, 117–123.

57. Frenkel, E. P., Shen, Y. M., & Haley, B. B., (2005). The direct thrombin inhibitors: Their role and use for rational anticoagulation. *Hematol. Oncol. Clin. North. Am., 19*(1), 119–145.

58. Di Nisio, M., Middeldorp, S., & Büller, H. R., (2005). Direct thrombin inhibitors. *N. Engl. J. Med., 353*(10), 1028–1040.

59. Lee, C. J., & Ansell, J. E., (2011). Direct thrombin inhibitors. *Br. J. Clin. Pharmacol., 72*(4), 581–592.

60. Adcock, D. M., & Gosselin, R., (2015). Direct oral anticoagulants (DOACs) in the laboratory: 2015 review. *Thromb. Res., 136*(1), 7–12.

61. Franchini, M., Liumbruno, G. M., Bonfanti, C., & Lippi, G., (2016). The evolution of anticoagulant therapy. *Blood Transfus., 14*(2), 175.

62. Greinacher, A., Janssens, U., Berg, G., Böck, M., Kwasny, H., Kemkes-Matthes, B., et al., (1999). Lepirudin (recombinant hirudin) for parenteral anticoagulation in patients with heparin-induced thrombocytopenia. *Circulation, 100*(6), 587–593.

63. White, C. M., (2005). Thrombin-directed inhibitors: Pharmacology and clinical use. *Am. Heart J., 149*(1), 54–60.

64. Kelton, J. G., Arnold, D. M., & Bates, S. M., (2013). Nonheparin anticoagulants for heparin-induced thrombocytopenia. *N. Engl. J. Med., 368*(8), 737–744.

65. Greinacher, A., & Lubenow, N., (2001). Recombinant hirudin in clinical practice. *Circulation, 103*(10), 1479–1484.

66. Warkentin, T. E., Greinacher, A., & Koster, A., (2008). Bivalirudin. *Thromb. Haemost., 99*(5), 830–839.

67. Stone, G. W., Witzenbichler, B., Guagliumi, G., Peruga, J. Z., Brodie, B. R., Dudek, D., et al., (2008). Bivalirudin during primary PCI in acute myocardial infarction. *N. Engl. J. Med., 358*(21), 2218–2230.

68. Lewis, B. E., Wallis, D. E., Berkowitz, S. D., Matthai, W. H., Fareed, J., Walenga, J. M., et al., (2001). Argatroban anticoagulant therapy in patients with heparin-induced thrombocytopenia. *Circulation, 103*(14), 1838–1843.

69. ARGATROBAN - argatroban injection, solution. Smith Kline Beecham corporation http://www.fda.gov/downloads/advisorycommittees/committeesmeetingmaterials/pediatricadvisorycommittee/ucm192057.pdf (accessed 26 August, 2017).

70. Schulman, S., Wåhlander, K., Lundström, T., Clason, S. B., & Eriksson, H., (2003). Secondary prevention of venous thromboembolism with the oral direct thrombin inhibitor ximelagatran. *N. Engl. J. Med., 349*(18), 1713–1721.

71. Keisu, M., & Andersson, T. B., (2010). Drug-induced liver injury in humans: The case of ximelagatran. In: *Adverse Drug Reactions* (p. 407). Springer Berlin Heidelberg.

72. Tran, A., & Cheng-Lai, A., (2011). Dabigatran etexilate: The first oral anticoagulant available in the United States since warfarin. *Cardiol. Rev., 19*(3), 154–161.

73. Van Ryn, J., Stangier, J., Haertter, S., Liesenfeld, K. H., Wienen, W., Feuring, M., et al., (2010). Dabigatran etexilate-a novel, reversible, oral direct thrombin inhibitor: Interpretation of coagulation assays and reversal of anticoagulant activity. *Thromb. Haemost., 103*(6), 1116.

74. Dittmeier, M., Wassmuth, K. K., Schuhmann, M., Kraft, P., Kleinschnitz, C., & Fluri, F., (2016). Dabigatran etexilate reduces thrombin-induced inflammation and thrombus formation in experimental ischemic stroke. *Curr. Neurovasc. Res., 13*(3), 199–206.

75. Davie, E. W., Fujikawa, K., & Kisiel, W., (1991). The coagulation cascade: Initiation, maintenance, and regulation. *Biochem., 30*(43), 10363–10370.

76. Packham, M. A., (1994). Role of platelets in thrombosis and hemostasis. *Can. J. Physiol. Pharmaco., 72*(3), 278–284.

77. Heemskerk, J. W., Bevers, E. M., & Lindhout, T., (2002). Platelet activation and blood coagulation. *Thromb. Haemost., 88*(2), 186–194.

78. Harrington, R. A., Hodgson, P. K., & Larsen, R. L., (2003). Antiplatelet therapy. *Circulation, 108*(7), 45–47.

79. Bittl, J. A., Baber, U., Bradley, S. M., & Wijeysundera, D. N., (2016). Duration of dual antiplatelet therapy: A systematic review for the 2016 ACC/AHA guideline focused update on duration of dual antiplatelet therapy in patients with coronary artery disease: A report of the American college of cardiology/American heart association task force on clinical practice guidelines. *J. Am. Coll. Cardiol., 68*(10), 1116–1139.

80. Depta, J. P., & Bhatt, D. L., (2013). Current uses of aspirin in cardiovascular disease. *Hot Topics Cardiol., 8*, 7–21.

81. Baigent, C., Blackwell, L., Collins, R., Emberson, J., Godwin, J., Peto, R., et al., (2009). Aspirin in the primary and secondary prevention of vascular disease: Collaborative meta-analysis of individual participant data from randomised trials. *Lancet., 373*(9678), 1849–1860.

82. Smith, J. B., & Willis, A. L., (1971). Aspirin selectively inhibits prostaglandin production in human platelets. *Nature, 231*(25), 235–237.

83. Alvarez-Sabin, J., Quintana, M., Santamarina, E., & Maisterra, O., (2014). Triflusal and aspirin in the secondary prevention of atherothrombotic ischemic stroke: A very long-term follow-up. *Cerebrovasc. Dis., 37*(3), 181–187.

84. Mc Neely, W., & Goa, K. L., (1998). Triflusal. *Drugs, 55*(6), 823–833.

85. Fitz Gerald, G. A., (1987). Dipyridamole. *N. Engl. J. Med., 316*(20), 1247–1257.

86. Kimura, Y., Tani, T., Kanbe, T., & Watanabe, K., (1984). Effect of cilostazol on platelet aggregation and experimental thrombosis. *Arzneimittel for Schung., 35*(7A), 1144–1149.

87. Uchiyama, S., Shinohara, Y., Katayama, Y., Yamaguchi, T., Handa, S., Matsuoka, K., et al., (2014). Benefit of cilostazol in patients with high risk of bleeding. subanalysis of cilostazol stroke prevention study 2. *Cerebrovasc. Dis., 37*(4), 296–303.

88. Bhatt, D. L., Fox, K. A., Hacke, W., Berger, P. B., Black, H. R., Boden, W. E., et al., (2006). Clopidogrel and aspirin versus aspirin alone for the prevention of atherothrombotic events. *N. Engl. J. Med., 354*(16), 1706–1717.

89. Shalito, I., Kopyleva, O., & Serebruany, V., (2009). Novel antiplatelet agents in development: Prasugrel, ticagrelor, and cangrelor and beyond. *Am. J. Ther., 16*(5), 451–458.

90. Cattaneo, M., (2009). New $P2Y_{12}$ blockers. *J. Thromb. Haemost., 7*(1), 262–265.

91. Ueno, M., Rao, S. V., & Angiolillo, D. J., (2010). Elinogrel: Pharmacological principles, preclinical and early phase clinical testing. *Future Cardiol., 6*(4), 445–453.

92. Avanzas, P., Moris, C., & Clemmensen, P., (2014). Antiplatelet therapy. New potent $P2Y_{12}$ inhibitors. In: *Pharmacological Treatment of Acute Coronary Syndromes* (p. 31). Springer.

93. Simons, M. L., Rutsch, W., Vahanian, A., & Adgey, J., (1997). Randomised placebo-controlled trial of abciximab before and during coronary intervention in refractory unstable angina: The CAPTURE study. *The Lancet, 349*(9063), 1429.

94. Cannon, C. P., Weintraub, W. S., Demopoulos, L. A., Vicari, R., Frey, M. J., Lakkis, N., et al., (2001). Comparison of early invasive and conservative strategies in patients with unstable coronary syndromes treated with the glycoprotein IIb/IIIa inhibitor tirofiban. *N. Engl. J. Med., 344*(25), 1879–1887.

95. PURSUIT trial investigators, (1998). Inhibition of platelet glycoprotein IIb/IIIa with eptifibatide in patients with acute coronary syndromes. *N. Engl. J. Med., 339*, 436–443.

96. Pope, H., Garcia-Cortes, R., Hockett, M., Paul, L., & Bach, R., (2016). 199: Perioperative eptifibatide as bridging therapy in patients with renal impairment and coronary stents. *Crit. Care Med., 44*(12), 127.

97. Held, C., Tricoci, P., Huang, Z., Van de Werf, F., White, H. D., Armstrong, P. W., et al., (2014). Platelet thrombin-receptor antagonist, in medically managed patients with non-ST-segment elevation acute coronary syndrome: Results from the TRACER trial. *Eur. Heart J. Acute Cardiovasc. Care., 3*(3), 246–256.

98. Sikka, P., & Bindra, V. K., (2011). Emerging antithrombotic drugs: A review. *Ann. Trop. Med. Public Health, 4*, 138–142.

99. Marder, V. J., & Sherry, S., (1988). Thrombolytic therapy: Current status. *N. Engl. J. Med., 318*(23), 1512–1520.

100. Hacke, W., Kaste, M., Bluhmki, E., Brozman, M., Davalos, A., Guidetti, D., et al., (2008). Thrombolysis with alteplase 3 to 4.5 hours after acute ischemic stroke. *N. Engl. J. Med., 359*(13), 1317–1329.

101. Anderson, C. S., Robinson, T., Lindley, R. I., Arima, H., Lavados, P. M., Lee, T. H., et al., (2016). Low-dose versus standard-dose intravenous alteplase in acute ischemic stroke. *N. Engl. J. Med., 374*(24), 2313–2323.

102. El-Gengaihy, A. E., Abdelhadi, S. I., Kirmani, J. F., & Qureshi A. I., (2006). *Thrombolytics*. In: Taylor, J. B., & Triggle, D. J., (eds.), *Comprehensive Medicinal Chemistry II* (Vol. 6, p. 763), Elsevier.

103. Watkins, E., Small, J. E., & Ginat, D. T., (2015). *Tissue Plasminogen Activator (t PA)*. In: Ginat, D. T., Small, J., & Schaefer, P. W., (eds.), *Neuroimaging Pharmacopoeia* (p. 261). Springer.

104. Collen, D., & Lijnen, H. R., (2009). The tissue-type plasminogen activator story. *Arterioscler. Thromb. Vasc. Biol., 29*(8), 1151–1155.

105. Hébert, M., Lesept, F., Vivien, D., & Macrez, R., (2016). The story of an exceptional serine protease, tissue-type plasminogen activator (tPA). *Rev. Neurol., 172*(3), 186–197.

106. Gurewich, V., (2016). Thrombolysis: A critical first-line therapy with an unfulfilled potential. *Am. J. Med., 129*(6), 573–575.

107. Nordt, T. K., & Bode, C., (2003). Thrombolysis: Newer thrombolytic agents and their role in clinical medicine. *Heart., 89*(11), 1358–1362.

108. Moreadith, R. W., & Collen, D., (2003). Clinical development of PEGylated recombinant staphylokinase (PEG–Sak) for bolus thrombolytic treatment of patients with acute myocardial infarction. *Adv. Drug Deliv. Rev., 55*(10), 1337–1345.

109. Adivitiya, K. Y. P., (2016). The evolution of recombinant thrombolytics: Current status and future directions. *Bioengineered,* 1–28.

110. Wander, G. S., & Chhabra, S. T., (2013). Critical analysis of various drugs used for thrombolytic therapy in acute myocardial infarction. *Med Update, 23,* 109–116.

111. Weisel, J. W., & Litvinov, R. I., (2014). In: Saba, H. I., & Roberts, H. R., (eds.), *Mechanisms of Fibrinolysis and Basic Principles of Management, in Hemostasis and Thrombosis*. John Wiley & Sons, UK, p. 169.

112. Weitz, J. I., Eikelboom, J. W., & Samama, M. M., (2012). New antithrombotic drugs: Antithrombotic therapy and prevention of thrombosis, 9th Ed: American college of chest physicians evidence-based clinical practice guidelines. *Chest, 141*(2), 120–151.

113. Abraham, E., Reinhart, K., Opal, S., Demeyer, I., Doig, C., Rodriguez, A. L., et al., (2003). Efficacy and safety of tifacogin (recombinant tissue factor pathway inhibitor) in severe sepsis: A randomized controlled trial. *JAMA, 290*(2), 238–247.

114. Tsao, C. M., Ho, S. T., & Wu, C. C., (2005). Coagulation abnormalities in sepsis. *Acta Anaesthesiol. Taiwan., 53*(1), 16–22.

115. Abshire, T., & Kenet, G., (2004). Recombinant factor VIIa: Review of efficacy, dosing regimens and safety in patients with congenital and acquired factor VIII or IX inhibitors. *J. Thromb. Haemost., 2*(6), 899–909.

116. Howard, E. L., Becker, K. C., Rusconi, C. P., & Becker, R. C., (2007). Factor IXa inhibitors as novel anticoagulants. *Arterioscler. Thromb. Vasc. Biol., 27*(4), 722–727.

117. Povsic, T. J., Wargin, W. A., Alexander, J. H., Krasnow, J., Krolick, M., Cohen, M. G., et al., (2011). Pegnivacogin results in near complete FIX inhibition in acute coronary syndrome patients: RADAR pharmacokinetic and pharmacodynamic substudy. *Eur. Heart J., 32*(19), 2412–2419.

118. Ranieri, V. M., Thompson, B. T., Barie, P. S., Dhainaut, J. F., Douglas, I. S., Finfer, S., et al., (2012). Drotrecoginalfa (activated) in adults with septic shock. *N. Engl. J. Med., 366*(22), 2055–2064.

119. Verhamme, P., Tangelder, M., Verhaeghe, R., Ageno, W., Glazer, S., Prins, M., et al., (2011). Single intravenous administration of TB-402 for the prophylaxis of venous

thromboembolism after total knee replacement: A dose escalating, randomized, controlled trial. *J. Thromb. Haemost., 9*(4), 664–671.

120. Seale, L., Finney, S., Sawyer, R. T., & Wallis, R. B., (1997). Tridegin, anovel peptidic inhibitor of factor XIIIa from the leech, Haementeria ghilianii, enhances fibrinolysis *in vitro. Thromb. Haemost., 77*(5), 959-963.

121. Bohm, M., Bäuml, C. A., Hardes, K., Steinmetzer, T., Roeser, D., Schaub, Y., et al., (2014). Novel insights into structure and function of factor XIIIa-inhibitor tridegin. *J. Med. Chem., 57*(24), 10355–10365.

122. Baskova, I. P., & Nikonov, G. I., (1991). Destabilase, the novel epsilon-(gamma-Glu)-Lys isopeptidase with thrombolytic activity. *Blood Coagul. Fibrinolysis., 2*(1), 167–172.

123. Kurdyumov, A. S., Manuvera, V. A., Baskova, I. P., & Lazarev, V. N., (2015). A comparison of the enzymatic properties of three recombinant isoforms of thrombolytic and antibacterial protein-Destabilase-Lysozyme from medicinal leech. *BMC Biochem., 16*(1), 27.

124. Yeh, C. H., Fredenburgh, J. C., & Weitz, J. I., (2012). Oral direct factor Xa inhibitors. *Circ. Res., 111*(8), 1069–1078.

125. Kearon, C., Akl, E. A., Ornelas, J., Blaivas, A., Jimenez, D., Bounameaux, H., et al., (2016). Antithrombotic therapy for VTE disease: CHEST guideline and expert panel report. *Chest, 149*(2), 315–352.

126. Lansberg, M. G., O'Donnell, M. J., Khatri, P., Lang, E. S., Nguyen-Huynh, M. N., Schwartz, N. E., et al., (2012). Antithrombotic and thrombolytic therapy for ischemic stroke: Antithrombotic therapy and prevention of thrombosis: American college of chest physicians evidence-based clinical practice guidelines. *Chest, 141*(2), 601–636.

127. Fuster, V., Bhatt, D. L., Califf, R. M., Michelson, A. D., Sabatine, M. S., Angiolillo, D. J., et al., (2012). Guided antithrombotic therapy: Current status and future research direction. *Circulation, 126*(13), 1645–1662.

128. Alquwaizani, M., Buckley, L., Adams, C., & Fanikos, J., (2013). Anticoagulants: A review of the pharmacology, dosing, and complications. *Curr. Emerg. Hosp. Med. Rep., 1*(2), 83–97.

129. Noseworthy, P. A., Yao, X., Abraham, N. S., Sangaralingham, L. R., Mc Bane, R. D., & Shah, N. D., (2016). Direct comparison of dabigatran, Rivaroxaban, and apixaban for effectiveness and safety in nonvalvular atrial fibrillation. *Chest, 150*(6), 1302–1312.

130. Bates, S. M., Greer, I. A., Middeldorp, S., Veenstra, D. L., Prabulos, A. M., & Vandvik, P. O., (2012). VTE, thrombophilia, antithrombotic therapy, and pregnancy: Antithrombotic therapy and prevention of thrombosis: American college of chest physicians evidence-based clinical practice guidelines. *Chest, 141*(2), 691–736.

131. Monagle, P., Chan, A. K., Goldenberg, N. A., Ichord, R. N., Journeycake, J. M., Nowak-Göttl, U., et al., (2012). Antithrombotic therapy in neonates and children: Antithrombotic therapy and prevention of thrombosis: American college of chest physicians evidence-based clinical practice guidelines. *Chest., 141*(2), 737–801.

132. Baruch, L., (2013). Laboratory monitoring of anticoagulant medications: Focus on novel oral anticoagulants. *Postgrad. Med., 125*(2), 135–145.

133. Calatzis, A., Peetz, D., Haas, S., Calatzis, A., Peetz, D., Haas, S., et al., (2008). Pro-thrombinase-induced clotting time assay for determination of the anticoagulant effects of unfractionated and low-molecular-weight heparins, fondaparinux, and thrombin inhibitors. *Am. J. Clin. Pathol., 130*(3), 446–454.

134. Samama, M. M., & Guinet, C., (2011). Laboratory assessment of new anticoagulants. *Clin. Chem. Lab. Med., 49*(5), 761–772.

135. Casserly, I. P., Kereiakes, D. J., Gray, W. A., Gibson, P. H., Lauer, M. A., & Reginelli, J. P., Moliterno, D. J., (2004). Point-of-care ecarin clotting time versus activated clotting time in correlation with bivalirudin concentration. *Thromb. Res., 113*(2), 115–121.

136. Bates, S, M., & Weitz, J. I., (2005). Coagulation assays. *Circulation,* 112(4), 53–60.

137. Castellone, D. D., & Van Cott, E. M., (2010). Laboratory monitoring of new anticoagulants. *Am. J. Hematol., 85*(3), 185–187.

138. Samuelson, B. T., Cuker, A., Siegal, D. M., Crowther, M., & Garcia, D. A., (2017). Laboratory assessment of the anticoagulant activity of direct oral anticoagulants: A systematic review. *Chest, 151*(1), 127–138.

139. Pollack, C. V., (2016). Introduction to direct oral anticoagulants and rationale for specific reversal agents. *Am. J. Emerg. Med., 34*(11), 1–2.

140. Lu, G., De Guzman, F. R., Hollenbach, S. J., Karbarz, M. J., Abe, K., Lee, G., et al., (2013). A specific antidote for reversal of anticoagulation by direct and indirect inhibitors of coagulation factor Xa. *Nat. Med., 19*(4), 446–451.

141. Crowther, M., Lu, G., Conley, P., Leeds, J., Castillo, J., Levy, G., Connolly, S., & Curnutte, J., (2014). Reversal of factor Xa inhibitors-inducedanticoagulation in healthy subjects by andexanet alfa. *Crit. Care. Med., 42*(12), 1469.

142. Siegal, D. M., Curnutte, J. T., Connolly, S. J., Lu, G., Conley, P. B., Wiens, B. L., et al., (2015). Andexanet alfa for the reversal of factor Xa inhibitor activity. *N. Engl. J. Med., 373*, 2413–2424.

143. Milling, T. J., & Kaatz, S., (2016). Preclinical and clinical data for factor xa and "universal" reversal agents. *Am. J. Emerg. Med., 34*(11), 39–45.

144. Ansell, J. E., (2016). Universal, class-specific and drug-specific reversal agents for the new oral anticoagulants. *J. Thromb. Thrombolysis, 41*(2), 248–252.

145. Ansell, J. E., (2016). Reversal agents for the direct oral anticoagulants. *Hematol. Oncol. Clin. North Am., 30*(5), 1085–1098.

146. Hu, T. Y., Vaidya, V. R., & Asirvatham, S. J., (2016). Reversing anticoagulant effects of novel oral anticoagulants: Role of ciraparantag, andexanetalfa, and idarucizumab. *Vasc. Health Risk Manag., 12*, 35.

147. Aisenberg, J., (2016). The specific direct oral anticoagulant reversal agents: Their current status and future place in gastroenterology practice. *Am. J. Gastroenterol., 3*(1), 36–44.

CHAPTER 8

INSULIN THERAPY FOR DIABETES: CURRENT SCENARIO AND FUTURE PERSPECTIVES

YOGESH A. KULKARNI, MAYURESH S. GARUD, and R. S. GAUD

Shobhaben Pratapbhai Patel School of Pharmacy and Technology Management, SVKM's NMIMS, Vile Parle (West), Mumbai 400056, India, Tel.: +91-22-42332000, Fax: +91-22-26185422, E-mail: yogeshkulkarni101@yahoo.com

ABSTRACT

The discovery of insulin is one of the milestones in the history of medicine. Since the discovery in 1921, production of insulin, its formulations, and insulin delivery systems have changed from time-to-time and became more innovative. Insulin formulations fall under different categories like regular insulin, isophane insulin, lente insulin, etc. They vary in the rate of release of insulin depending on their type. Despite the controversies, insulin analogs are also getting popularity, and are being used effectively in treatment of diabetes. Stability and storage are the important issues with insulin that need to be addressed while designing the formulations as well as the delivery systems. Researchers are trying to resolve these problems and finding better ways to stabilize and administer the insulin. Use of routes other than subcutaneous for insulin administration is getting more attention of researchers. Insulin is to be administered by the non-medicinal professionals and mostly the patients itself, so the insulin delivery system needs to be patient-friendly. Insulin injections are being widely used for the administration of the insulin because of their cost-effectiveness. Insulin pens are also available, which makes the administration of insulin easy and accurate. But the repeated administration of insulin by injections or by pens makes it painful and uncomfortable for the patients. Researchers are trying various approaches and have come up with different solutions. Oral insulin is one of the approach, which is being talked since long time in the scientific community. Research laboratories are also working on automatic delivery of the insulin by Sensor-augmented pumps. In this chapter, insulin therapy has been discussed with respect to current trends and future of insulin formulations and delivery methods.

8.1 INTRODUCTION

The discovery of insulin in 1921 was a groundbreaking incident that changed the approach of diabetes treatment. Since its discovery, insulin research has seen many paradigms. Initially insulin was obtained from bovine and porcine pancreas, which were having the disadvantage of causing immunological reactions, lipodystrophy and unpredictable absorption of insulin from the subcutaneous tissue. Hence, early stage of research on insulin was focused on preparation of purified insulin. Production of insulin, its formulations and insulin delivery methods have been changed from time-to-time and became more sophisticated.

It is interesting to know about the historical background of insulin discovery (Figure 8.1). Insulin was given its name way before it was discovered. In 1890 Oskar Minkowski and Joseph von Mering from Germany observed that, if total pancreatectomy is carried out in experimental animals, it leads to the development of diabetes mellitus. This observation was then resulted in the conclusion that an unknown substance produced by the pancreas is responsible for carbohydrate metabolism [1]. Despite of less evidences about the presence pancreatic internal secretion arising from the islet cells, in 1907 J de Meyer named the secretion as "insulin." Sharpey Schafer in Britain independently suggested the same name in 1916.

In May 1921, Frederick Grant Banting, a 22-year-old orthopedic surgeon, began his research on insulin. JJR Macleod, Professor of Physiology was his advisor. He was assisted by one of the Macleod's summer students Charles Best. By the month of December, Banting and Best gathered evidences that their pancreatic extract reduced the blood glucose of diabetic dogs. JB Collip, a biochemist from the University of Alberta, joined the team. Outcomes of the experiment were the first presentation of the Toronto research and were read at the New Haven meeting of the

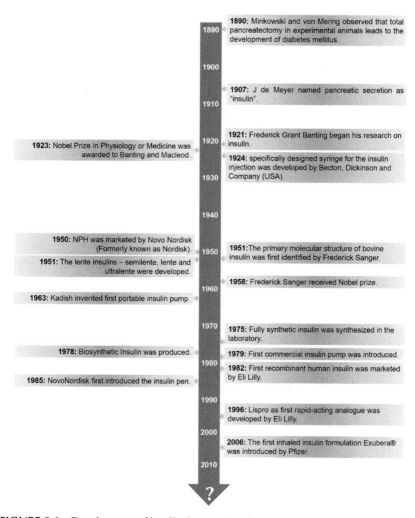

FIGURE 8.1 Development of insulin therapy at a glance.

American Physiological Association. The results were not well received because of lack of data on the side effects of the pancreatic extracts. Collip provided a scientific explanation that restoration of glycogen mobilization in the liver and its ability to clear ketones may have a role in blood glucose lowering activity of pancreatic extract [1].

In 1922, clinicians at Toronto General Hospital administered pancreatic extract to a 14-year-old, severely diabetic boy Leonard Thompson. He was injected with total 15 mL of pancreatic extract made by Banting and Best. Results were disappointing. The injection caused only slight reductions of glycemia and glycosuria. No effect was observed on ketoacidosis. Treatment was immediately discontinued. Twelve days after the first insulin administration, new series of injections was started. Thompson's glycosuria almost disappeared along with ketonuria. His blood glucose dropped to normal. It was the first time in history that there was definite evidence where scientists were able to mimic the function of endogenous insulin in diabetes. Credit for successful production of extract was given to JB Collip. He developed a method of extraction that involved changing the concentrations of slightly acidic alcohol solutions of chilled beef pancreas. The method was superior to Banting and Best's method [2].

Macleod delivered a complete summary of the work at the Washington meeting of the Association of the American Physicians on 3rd May, 1922. In 1923 Nobel Prize in Physiology or Medicine was awarded to Banting and Macleod. Banting divided his prize money equally with Best, Macleod split his with Collip [1]. The discovery of insulin brought a new sight in the management of diabetes and this way the era of insulin was started.

It took 30 years to know about the molecular structure of the insulin. The primary molecular structure of bovine insulin was first identified by Frederick Sanger in 1951 [3] He further decoded the chemistry of insulin for which he received the Nobel Prize in 1958.

8.2   STRUCTURE OF INSULIN

Similar to that of many other hormones, insulin is also a protein in nature. Insulin molecule is made up of 51 amino acids, which are arranged in two polypeptide chains A and B. Chain A is comprised of 21 amino acid residues, while chain B is comprised of 30 amino acid residues as shown in Figure 8.2. The chains A and B are bridged by disulfide linkage. Addition to these inter-chain linkage, A-chain contains an intra-chain disulfide linkages joining residue 6 and 11 (Figure 8.2). Insulin is present in the monomeric form, which has a tendency to aggregate to form dimers and hexamers [4].

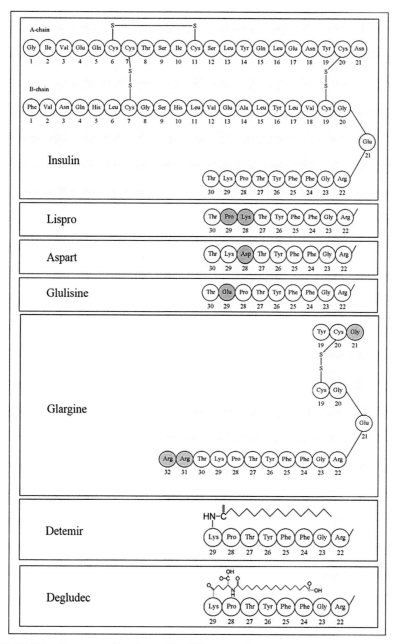

FIGURE 8.2 Structure of insulin and its different analogs.

8.3   CURRENTLY USED INSULIN PREPARATIONS

Currently various type insulin preparations are available commercially for management of diabetes. These insulin preparations fall under different categories as rapid acting, short acting, intermediate acting or long acting.

8.4   HUMAN INSULIN

8.4.1   REGULAR HUMAN INSULIN

Human insulin was the first insulin prepared by using recombinant DNA technology [5]. It has identical peptide sequence and tertiary structure to that of endogenous insulin. It is being used as a bolus insulin and is referred as short acting human insulin. It mimics the response of endogenous insulin and corrects the intermeal and premealhyperglycemia. When used in low concentration, the regular insulin solution gets rapidly absorbed in the blood stream after subcutaneous administration. But in high concentration solutions, it has the tendency to get self-aggregate into quaternary and hexamers in the presence of zinc ions [6, 7]. This causes delay in its absorption as it needs to dissociate into diamer and monomer first.

8.4.2   ISOPHANE INSULIN (NPH INSULIN)

Isophane insulin, also known as Neutral Protamine Hagedorn insulin (NPH insulin), is the only conventional insulin in clinical use. It is intermediately acting human insulin. In 1940, an attempt was made to prolong the biological action of regular insulin, which resulted in the preparation of NPH. This also reduced the frequency of insulin dosing. Protamine is added in a molar ratio of 1:6 to regular insulin to prepare NPH insulin. NPH insulin contain 0.9 molecules of protamine and two atoms of zinc per hexamer [8]. Positively charged protamine binds with the negatively charged insulin when mixed in solution of neutral pH, which is obtained by using phosphate buffer [9]. Protamine and zinc are added in NPH along with the insulin to delay insulin absorption and prolong the duration of insulin effect. Zinc is added to the insulin formulation as it plays important role in insulin biosynthesis and it is reported that zinc is involved in the formation of insulin hexamers [10]. Significant delay in the absorption of insulin from the subcutaneous tissue, resulting in an onset of action 1.5 to 4 h after injection, a pronounced peak 4 to 10 h after administration and a duration of action up to 24 h is obtained by NPH [12]. Initially the NPH insulin was prepared by using porcine insulin, subsequently it was reformulated using recombinant human insulin during 1980s. NPH has some limitations, it shows a pronounced peak action profile, where ideally it should be peakless [9]. Another limitation is that it must be uniformly suspended before administration, this leads to inter-individual and intra-individual variability in its absorption [13]. Despite of these limitations, NPH is being used clinically, alone or in a premixed insulin preparation combined with regular human insulin due to its low cost compared with the newer basal analogs [14, 15].

8.4.3   LENTE INSULIN

In lente insulin, the concentration of zinc is 10 times that added in NPH and acetate buffer are used instead of phosphate buffer. In lente insulin, there is formation of insoluble insulin-zinc complexes [16]. There are two types, one is semilente and other is ultralente. Semilente is amorphous in nature and have short duration of action. While ultralente is crystal-line with long duration of action. The lente insulin is not used commonly. They also cannot be mixed with regular insulin because of their high zinc concentration [9].

8.4.4   PREMIXED FORMULATIONS

In the premixed formulations of regular and NPH insulin, traditional short-acting, and intermediate-acting human insulin are combined in different ratios. Humulin 50/50 contain 50% of NPH and 50% of regular insulin. Another premixed formulation, Humulin 70/30 or Novolin 70/30 contain 70% NPH and 30% regular insulin [15]. These premixed insulin formulations are suitable alternatives to basal-bolus insulin therapy as there is a reduction in the number of daily injections. These formulations are to be used only in patients with type 2 diabetes [17, 18].

8.5　HUMAN INSULIN ANALOGS

For improving pharmacokinetics and pharmacodynamics of subcutaneously administered insulin, the idea of insulin analogs was introduced. Structural manipulation of the human insulin molecule was done to produce insulin analogs without affecting biological function like affinity for the insulin receptor and subsequent signaling events [19, 20]. Amino acid substitutions, inversions or additions were carried out by using protein bioengineering techniques. Till date, various insulin analogs have been prepared with modified biochemical properties. The modification in biochemical properties resulted in altered rates of self-aggregation and hence, the subcutaneous absorption of insulin into the bloodstream. Improved glycemic control and reduced hypoglycemic activity have made the insulin analog a popular option.

8.5.1　RAPID ACTING (SHORT ACTING) INSULIN ANALOGUES

Lispro, aspart, and glulisine are the three rapid-acting insulin analogs which are approved for clinical use and are being used currently. In rapid acting insulin analogs, the B insulin chain has been structurally altered to impede self-aggregation.

　　Lispro is the first rapid-acting analog developed in 1996 by Eli Lilly [15]. In lispro, there is transposition of the amino acids proline and lysine at positions B28 and B29 which leads to a conformational change that augments steric hindrance between interfaces implicated in dimerization [21] (Figure 8.2). In aspart, there is substitution of proline at B28 by a negatively charged aspartic acid residue, which lead to in repulsion of monomers [22] (Figure 8.2). While in glulisine, there is replacement of lysine at B29 for glutamine and aspartic acid at B3 for lysine (Figure 8.2). These changes resulted in increased stability and a reduced self-association tendency of the insulin [23]. Pharmacokinetic profiling of these analogs showed that they have a faster onset of action post administration (10 to 30 min), with peak of action between 0.5 to 2 h and duration of action between 3 to 5 h [15]. Less variability between injection sites has been observed for insulin analogs in comparison with that of regular insulin [18].

The pharmacodynamics of the rapid-acting insulin analogs showed enhanced clinical efficacy as compared to that of traditional insulin formulations. Patients using rapid-acting insulin analogs have also experienced an improvement in quality of life [24].

The rapid acting analogs of insulins also provide greater flexibility and convenience in the timing of administration. They can either be administered at mealtime, or immediately postprandially. Potential clinical disadvantages of rapid-acting insulin analogs include the risk of early postprandial hypoglycemia and preprandial hyperglycemia [25].

8.5.2 LONG ACTING INSULIN ANALOGS

There are three long-acting insulin analogs which are currently approved for insulin therapy. They include glargine, detemir, and degludec. Modifications in the structures of these analogs are made for prolonged absorption after subcutaneous injection and to provide a relatively peakless 24-h time-action profile.

Glargine is the first long-acting insulin analog in which modifications are made in both A and B insulin chains. At position A21, asparagine is replaced by glycine and at position B30 there is the addition of 2 arginine residues (Figure 8.2). This makes glargine less soluble at neutral and physiologic pH and stable in the acidic pH of its storage solution. After injection into subcutaneous tissue, at the neutral environment, glargine forms an amorphous precipitate from which insulin molecules are slowly released into the circulation [26]. In detemir, two modifications are made in B chain, one is deletion of amino acid threonine at position B30 and second is acylation of a 14-carbon aliphatic fatty acid (myristolytic acid) to the ε-amino group of lysine at position B29. Due to this detemir have more affinity for albumin and binds with it reversibly (Figure 8.2). This provides protracted action to detemir due to sustained release from multimeric complexes within the subcutaneous tissues post-injection as well as from albumin [27, 28].

Degludec has been recently approved ultra–long-acting basal insulin preparation. In Degludec, modification has been done in the native insulin B chain at two positions. First is the deletion of threonine at position B30 and the other is addition of a 16-carbon fatty diacid to the lysine residue at position B29 via a glutamic acid spacer (Figure 8.2). Degludec self-aggregates and forms large multihexamer complexes after subcutaneous injection, which then slowly dissociate into monomers and enters the circulation [29]. This property of degludec provides it a protracted action profile longer than 24 h.

8.5.3   PREMIXED ANALOGUE FORMULATIONS

These formulations include different ratios of rapid-acting insulin analogs. Humalog Mix 50/50 is a mixture of 50% neutral protamine lispro suspension with 50% insulin lispro, Humalog Mix 75/25 is a mixture of 75% neutral protamine lispro and 25% insulin lispro, while Novolog Mix 30 contain 70% protamine crystalline aspart and 30% insulin aspart [15]. Table 8.1 summarizes the different types of insulin and their pharmacokinetics properties.

8.6   FUTURE OF INSULIN PREPARATIONS

To achieve the maximum glycemic control, insulin replacement and supplementation strategies should focus on replication of the activity of endogenous insulin. To achieve this goal, manipulation of the pharmaceutical formulation like the addition of zinc, protamine, etc. and/or modification in insulin structure like in the insulin analogs are two main approaches studied and used by researchers. Still, the researchers have not reached to the ideal exogenous insulin preparations. Currently, various new and novel rapid-acting and long-acting insulin formulations are being developed which may provide a wide range of therapeutic options for the management of patients with diabetes in the near future.

8.6.1   INSULIN FORMULATIONS

Currently, two novel ultra-rapid-acting insulin preparations are being clinically studied for optimal postprandial glycemic control. One is insulin-PH$_2$O and the other is Linjeta. The insulin-PH$_2$O is developed by Halozyme Therapeutics (USA). It is in phase II clinical trial. It contains one of the commercially available mealtime insulin products mixed with

TABLE 8.1   Pharmacokinetic Profile of Insulin Analogs

Insulin Type	Onset of Action	Peak Action	Duration of Action
Short-Acting Insulin			
Regular	30–60 min	2–3 h	5–8 h
Intermediate-Acting			
Insulin			
NPH	2–4 h	4–10 h	10–16 h
Lente	3–4 h	4–12 h	12–18 h
Long-Acting Insulin			
Ultralente	6–10 h	10–16 h	18–24 h
Insulin Mixtures			
70/30 human mix	30–60 min	Dual	10–16 h
(70% NPH, 30% regular)			
50/50 human mix	30–60 min	Dual	10–16 h
(50% NPH, 50% regular)			
Rapid-Acting Insulin			
Analogues			
Insulin lispro	5–15 min	30–90 min	3–5 h
Insulin aspart	5–15 min	30–90 min	3–5 h
Insulin glulisine	5–15 min	30–90 min	3–5 h
Long-Acting Insulin Analogues			
Insulin glargine	2–4 h	Peak less	20–24 h
Insulin determir	2–4 h	6–14 h	16–20 h
Insulin Analogues Mixtures			
75/25 lisproanalog mix	5–15 min	Dual	10–16 h
(75% intermediate, 25% lispro)			
70/30 aspart analog mix	5–15 min	Dual	10–16 h
(70% intermediate, 30% aspart)			

recombinant human hyaluronidase (rHuPH$_2$O). There are clinical evidences that co-administration of insulin, which may be regular human insulin or insulin lispro, with rHuPH$_2$O leads to rapid absorption of insulin into the circulation followed by accelerated pharmacokinetic and glucodynamic effects. This also reduces inter-subject and intra-subject variability [30, 31]. Linjeta is developed by Biodel Inc. (USA) and is in phase III clinical trial.

It consists of regular human insulin with ethylenediaminetetra-acetic acid (EDTA) and citric acid. EDTA and citric acid act to chelate zinc ions and also prevent self-aggregation of insulin molecules after subcutaneous injection and maintain it in a monomeric state to provide rapid action [32, 33].

Two novel and potentially improved formulations of insulin glargine as basal insulin supplementation are currently under development. One is LY2963016 by Eli Lilly and the second is BIOD-Adjustable Basal by Biodel. BIOD-Smart-Basal contains insulin glargine, glucose oxidase, and peroxidase. In the presence of glucose, glucose oxidase and peroxidase react to form gluconic acid, which lowers the pH and increases the solubility of insulin glargine and promotes its release into the circulation [34]. It can be premixed with other insulin products [34, 35]. Sanofi-Aventis is also developing a new basal insulin formulation. Insulin lispro protamine now is being tested as a stand-alone basal insulin analog in patients with type 1 and 2 diabetes [36, 37].

Chemical coupling of the insulin molecule to polyethylene glycol (PEG) is another novel approach to delay insulin absorption [38]. Eli Lilly has developed a PEGylated form of insulin lispro, which is currently in phase III trials. Flamel Technologies Inc. (France) is developing a unique ultra–long-acting basal insulin product FT–105, in which insulin is non-covalently bound to a polymer consisting of a polyglutamate peptide backbone linked to vitamin E molecule within a hydrogel [39].

Glucose-responsive injectable "smart" insulin preparations are now the future of insulin treatment. These formulations release insulin in proportion to ambient subcutaneous glucose levels. Currently, these formulations are undergoing preclinical studies. SmartCells Inc. (USA) developed SmartInsulin, which is composed of insulin reversibly bound to a glucose-binding molecule. Insulin is released from the SmartInsulin polymer in the presence of glucose, which competes with insulin for the binding sites on the glucose-binding molecule [35]. Because of the novel glucose-responsive mechanism, the smart insulin formulations may result in close glycemic control and a lower risk of both hypoglycemia and hyperglycemia.

8.6.2 INSULIN ANALOGS

Last two decades have seen a remarkable change and gradual refinement in the molecular structure of insulin. Different rapid-acting and long-acting insulin analogs were prepared that can closely imitate endogenous insulin. A number of them were successfully marketed and are being used to get good glycaemic control. But there are also stories of failure. Recently, Eli Lilly have finished the development of a novel basal insulin analog, LY2605541, which was on its way to phase III trials. The reason could be related to liver toxicity [40].

8.6.3   INSULIN DELIVERY

Subcutaneous route of administration is being used for the insulin delivery using syringes and insulin pens.

8.6.4   INSULIN INJECTIONS

Two years after the discovery of insulin, in 1924, specifically designed syringe for the insulin injection was developed by Becton, Dickinson and Company (USA) [41]. In the early days, the metals and/or glass were used to make syringes. These syringes were reusable and after sterilizing by boiling them. Later in 1954 first glass disposable syringes were introduced to reduce the incidence of needle associated infections [42]. After that numerous changes have been made to insulin syringes to get it in the present form. But despite all changes, syringes has disadvantage of inducing the pain which makes them less patient-friendly. Less accuracy is another disadvantage of using syringes [43]. To overcome these disadvantages, an injection port has been made available by Medtronic, known as i-port Advance®. It combines an injection port and an inserter in one complete set that eliminates the need for multiple injections without puncturing the skin for each dose [44].

8.6.5   INSULIN PENS

Insulin pens are developed to overcome the limitations of insulin injections like inconvenience and inaccuracy. Novo Nordisk first introduced the insulin pen in 1985 [41]. Insulin pens are now reusable, more accurate and safe. Now-a-days pen needles are shorter and thinner (31–32 G ×

4–5 mm) due to which administration of insulin has become less painful resulting in improved patient satisfaction [45, 46].

Besides all these advancements and advantages, insulin pens associated with higher cost in comparison with vial and syringe.

8.7 CONTINUOUS SUBCUTANEOUS INSULIN DELIVERY

Continuous subcutaneous insulin infusion (CSII) is a progressing form of insulin delivery. It is mainly used for people with type 1 diabetes [47]. To mimic the exact effect of endogenous insulin, more physiologic delivery of insulin is necessary. There is a continuous secretion of a small amount of insulin from the beta cells of pancreas to reduce hepatic glucose output and a larger amount is secreted in the presence of food to maintain euglycemia [48]. To achieve this effect, Kadish, a doctor from Los Angeles (USA), in 1963 invented first portable insulin pump [49]. This was followed by more sophisticated modifications to provide more efficient and comfortable devices. The first commercial insulin pump was introduced in 1979 in the USA. The current generation of insulin pumps is more patient-friendly as due to their smaller size and smart features [50]. Continuous subcutaneous insulin delivery has disadvantages of higher cost, increased risk of subcutaneous infections, the inconvenience of being attached to a device, and a theoretical higher risk for diabetic ketoacidosis. Use of this technique also needs a proper patient education before its use to avoid complications [51].

8.8 SENSOR-AUGMENTED PUMP THERAPY

Research in the field of continuous glucose monitors (CGM) has made it possible to combine pump and CGM in the management of diabetes. Sensor-augmented pump (SAP) therapy is the outcome of such attempt to combine two technologies. In this therapy, CGM readings are used to adjust insulin delivery through an insulin pump. The disadvantage of SAP is that it requires patient involvement for using CGM glucose readings to adjust insulin pump delivery making it susceptible to human errors. It also requires patients to wake up to manage nocturnal hypoglycemia.

8.9   SENSOR-AUGMENTED PUMP WITH LOW GLUCOSE SUSPEND OR THRESHOLD SUSPEND PUMP

Nocturnal hypoglycemia is one of the causes of deaths related to diabetes. The making of artificial pancreas (closed-loop system) is an advanced step towards more controlled insulin delivery where the insulin delivery is interrupted once CGM glucose is at a low threshold (often 70 or 60 mg/dL) to reduce nocturnal hypoglycemia. The threshold suspends (TS) system interrupts the delivery of insulin for up to 2 h if a patient does not take action with a low glucose alarm. This feature is designed to reduce the severity and duration of hypoglycaemia [52]. In September 2016, US FDA approved Medtronic's MiniMed 670G hybrid closed looped system also described as artificial pancreas. It is the first FDA-approved device used for automatic monitoring of glucose followed by administration of appropriate basal insulin doses in type 1 diabetic patients (U.S. Food and Drug Administration).

8.10   NOVEL DELIVERY METHODS FOR INSULIN

There are many drawbacks associated with the subcutaneous route of insulin administration like pain, variable drug release, etc. It is also associated with peripheral hyperinsulinemia [53]. Hence, finding an alternate noninvasive delivery method for insulin administration has remained a point of interest. Following are the novel delivery methods which are under development.

8.10.1   INHALED INSULIN DELIVERY

Inhaled insulin delivery by the pulmonary route was the first alternative approach studied for insulin delivery other than subcutaneous route. In 1924 it was reported that insulin delivery by aerosol can reduce blood glucose [54]. Insulin delivery by the pulmonary route has the advantage of the absence of certain peptidases that breaks down insulin like in gastrointestinal tract. First-pass metabolism is also bypassed in this delivery method [55]. It is considered that transcytotic and paracellular mechanisms are responsible for insulin absorption in pulmonary route [55].

In year 2006, the first inhaled insulin formulation Exubera® was introduced by Pfizer. It was found that Exubera® have pharmacokinetics as well as the pharmacodynamics properties similar to that of insulin aspart with a faster onset of action (10–15 min) [56]. The formulation failed to become popular despite of its noninvasive route. This may be due to higher cost, the bulky delivery device and problems with pulmonary function. Hence, the product was withdrawn from the market in 2007.

Afrezza®, another promising inhaled insulin is made available by Sanofi and MannKind, which is based on Technosphere® dry powdered formulation. It was found ideal for postprandial blood glucose control because of rapid onset of action (15 min) and duration of action 2–3 h [57]. Afrezza is having side-effects of producing transient nonproductive cough and reduction in lung function side-effects [58].

8.10.2   NASAL INSULIN

Theoretically, the intranasal delivery system has many advantages. It bypasses first-pass metabolism, noninvasive and painless when compared with subcutaneous route and does not affect lung function as in inhalation route. Besides these advantages, nasal drug delivery has some shortcomings like limited permeability of a large molecule through the nasal mucosa and rapid mucociliary clearance which leads to variable absorption pattern [59]. Nasulin™, a product by CPEX pharmaceuticals and a nasal insulin by Nastech Pharmaceutical Company Inc are under investigation. The absorption enhancers like bile salt, surfactant and fatty acid derivatives are also being tested to increase mucosal permeability of insulin [60].

8.10.3   ORAL INSULIN DELIVERY

Oral route is the most exploited way of drug administration because of the ease of administration and the patient compliance. It is suggested that oral administration of insulin could closely mimic the physiological insulin delivery, which will provide more portal insulin concentration than peripheral [61]. Though oral insulin delivery could be a milestone in insulin therapy, there are many challenges in the development of oral insulin, which includes inactivation of insulin by proteolytic enzymes in the gastrointestinal tract, low permeability through the intestinal membrane due to the larger size and hydrophobicity of insulin resulting in poor bioavailability. So researchers are focusing on developing a carrier system which will protect insulin from gastrointestinal degradation as well as facilitates transport of insulin to the circulation with sufficient bioavailability [62–64].

Researchers have worked with natural and synthetic nanoparticles using chitosan, liposomes, polymeric nanovesicles, polylactides, poly ε, poly alkyl cyanoacrylate, etc. as carriers. Polymeric hydrogels are also being studied for oral insulin delivery [42].

To overcome the barrier of proteolytic enzymes, in practice, the utilization of enzyme inhibitors with protein drugs is a practical approach to increase the stability of proteins in the stomach and small intestine, so as to enhance bioavailability. Enzyme inhibitors like sodium glycocholate, aprotinin, bacitracin, soybean trypsin inhibitor, and camostatmesilate were studied by the researchers and they found that sodium glycocholate, camostatmesilate, and bacitracin can be effectively used in oral insulin formulation to provide the protection to the insulin [65]. An in-vitro study showed that duck ovomucoid, a novel class of enzyme inhibitor, can give 100% protection to insulin against trypsin and α-chymotrypsin [60]. Though it is promising to use the enzyme inhibitors to protect insulin they have the disadvantage of causing gastrointestinal enzyme deficiency in humans, and hence protein malabsorption when administered in high concentration for long duration [66].

Another approach is using absorption enhancers. It includes targeting of the lipid bilayers to enhance the absorption and permeability of protein drugs. The enhancer to be used should be non-toxic, pharmacologically, and chemically inertness, non-irritating and non-allergenic [67]. Bile salts, surfactants, calcium ion-chelating agents and fatty acids are some of the traditional absorption enhancers with prominent absorption effect [68]. Mechanism of enhancing the absorption is by modulating the structure of the cellular membrane (transcellular absorption) or by altering the tight junctions between cells (paracellular absorption) of the intestinal epithelium [69]. Zonulaoccluden toxin is another novel oral insulin absorption enhancer [70]. But, as the integrity of the cell membranes and tight junctions is changed by these enhancers, toxins and intestinal flora can get access to the systemic circulation, which further can lead to further diseases and infections [71]. Also, some of the enhancers possess intrinsic toxicity. Like bile salts, non-ionic surfactants, sodium dodecyl sulfate and lysolecithin have been shown to cause acute local damage to the intestinal wall and compromise cell viability [69].

Use of mucoadhesive polymers is another approach for oral delivery of insulin. These polymers can be used to encapsulate insulin by adhering themselves to the gastrointestinal mucus and creating the desirable insulin drug concentration gradient in the systemic circulation [72]. Chitosan is one of the ideal polymers which can be used in the oral formulation because of its non-toxic, biocompatible, biodegradable, mucoadhesive and absorption enhancing properties [73].

8.10.4   COLON SPECIFIC INSULIN DELIVERY

Oral colon delivery is one of the means of systemic therapeutic treatment. Large intestine has certain advantages due to long transit time, lower levels of peptidases which prevent destruction of peptides and also higher responsiveness to permeation enhancers [42]. So it can be a possible strategy to improve the oral bioavailability of peptide and protein drugs like insulin. Oral colon delivery systems are designed according to micro-flora-, pH-and time-dependent strategies [74].

8.10.5   BUCCAL INSULIN DELIVERY

The buccal delivery provides similar benefits as that of oral insulin. It has the advantage that it bypasses gastrointestinal degradation. It also provides a relatively larger surface area for absorption and hence results in better bioavailability [75]. Oral-lyn™ was developed by Generex Biotechnology (Canada) as a liquid formulation of short acting insulin which is administered using Generex's metered dosage aerosol applicator (RapidMist™). The phase 1 and phase 2 trials of use of Oral-lyn™ in patients with type 1 and type 2 showed promising results [76]. Now the phase 3 clinical trial is going on. Oral Recosulin® is an another formulation developed by Shreya Life Sciences Pvt. Ltd., India, which is in phase 3 clinical trials.

8.10.6   TRANSDERMAL INSULIN DELIVERY

Transdermal insulin delivery is a painless delivery method which also provides a large surface area for insulin delivery. But the penetration of insulin is restricted by the stratum corneum. There are various methods used to overcome the barrier of the stratum corneum [77]. These methods include iontophoresis [78], sonophereis or phonopheresis [79], micro-dermal ablation by removing the stratum corneum [80], electroporation [81], transfersulin [82]. Iontophoresis is the technique that uses small electric currents. Sonophereis or phonopheresis uses ultrasound waves. microdermal ablation is removal of the stratum corneum, electropora-tion uses high voltage pulses that are applied for a very short time, trans-fersulin is the insulin encapsulated in transferosome, an elastic, flexible vesicle which squeeze by itself to deliver drugs through skin pores. Freck-mannand co-workers have tested Insupatch™, a device developed as an add-on to an insulin pump, that applies local heat to the skin in order to increase the absorption of insulin [83]. Researchers have also tried recom-binant human hyaluronidase (rHuPH$_2$O) to increase insulin absorption from subcutaneous tissue [84]. Few of the researchers have also combined microneedle technology with a transdermal patch. The microneedles with 1 μm diameter and of various lengths can deliver insulin in an effective, accurate, and precise manner [85]. The transdermal delivery techniques are under development and are studied for their long-term utility, safety and usefulness in delivery of insulin.

8.10.7   INTRA-PERITONEAL OR INTRA-PORTAL INSULIN DELIVERY

Peripheral hyperinsulinemia is one of the drawbacks of intravenous and subcutaneous route of insulin delivery and is considered as non-physiolog-ical. Intra-peritoneal or intra-portal insulin delivery systems have provided new options for insulin delivery to avoid peripheral hyperinsulinemia. This route of insulin delivery has been investigated since the 1970s [86]. High portal insulin concentration can be achieved by direct delivery of insulin in the portal vein. In intra-peritoneal delivery of insulin, a pump is implanted under the subcutaneous tissue in the lower abdomen and the tip of the catheter is carefully inserted and directed towards the liver. The pump reservoir needs to be refilled at least every 3 months, depending on the individual insulin requirement [87]. Safety and efficacy of insulin delivery by intraperitoneal route has been clinically tested [88]. Still, this route is associated with limitations like it is invasive, may be associated with subcutaneous infections, cannula blockage, higher cost, portal-vein thrombosis, and peritoneal infection [89].

8.11 SUMMARY

Over a century, insulin therapy has developed in tremendous ways. It has evolved from crude animal insulin to novel insulin analogs. Pharmaco-kinetics, as well as the pharmacodynamics profile of the insulin, is also becoming superior with the changes in its forms. The insulin delivery methods are getting better with time. Though replacement of endogenous insulin action is difficult to achieve, the use of continuous glucose moni-toring system and sensor-augmented pump have made it possible to provide a close loop insulin delivery device that can closely mimic the function of pancreas. As the injectable insulin is having issues of patient compliance due to induction of pain, other routes of administration are being tested for successful delivery of insulin to increase patient acceptance and hence the strict glucose control. Various research and development laboratories are engaged in making the insulin therapy better. This continuous develop-ment will provide better options and will help to improve quality of life of diabetic patients.

KEYWORDS

- insulin analogs
- insulin delivery
- insulin pump
- oral insulin
- sensor-augmented

REFERENCES

1. Shah, R. S., Chen, M. K., Lobe, T. E., Bufo, A. J., Gross, E., & Whitington, G. L. (1997). Laparoscopic common duct exploration in a child with recurrent pancreatitis due to a primary fungus ball in the terminal common bile duct. *Journal of Laparoen-doscopic & Advanced Surgical Techniques, 7*(1), 63–67.

2. Bliss, M., Banting, F., Best, C. H., & Collip, J. (1982). Banting's, Best's, and Collip's accounts of the discovery of insulin. *Bulletin of the History of Medicine, 56*(4), 554–568.

3. Bell, G. I., Pictet, R. L., Rutter, W. J., Cordell, B., Tischer, E., & Goodman, H. M. (1980). Sequence of the human *insulin gene. Nature, 284*(5751), 26.

4. Sanger, F., & Tuppy, H., (1951). The amino-acid sequence in the phenylalanyl chain of insulin. 1. The identification of lower peptides from partial hydrolysates. *Biochem. J., 49*(4), 463–481.

5. Keen, H., Pickup, J. C., Bilous, R. W., Glynne, A., Viberti, G. C., Jarrett, R. J., et al., (1980). Human insulin produced by recombinant DNA technology: Safety and hypoglycaemic potency in healthy men. *Lancet., 2*, 398–401.

6. Binder, C., Lauritzen, T., Faber, O., & Pramming, S., (1984). Insulin pharmacokinetics. *Diabetes Care, 7*, 188–199.

7. Kang, S., Brange, J., Burch, A., Volund, A., & Owens, D. R., (1991). Subcutaneous insulin absorption explained by insulin's hysicochemical properties. Evidence from absorption studies of soluble human insulin and insulin analogues in humans. *Diabetes Care, 14*, 942–948.

8. Krayenbuhl, C. H., & Rosenberg, T., (1946). Crystalline protamine insulin. *Rep. Steno. Mem. Hosp. Nord. Insulinlab, 1*, 60–73.

9. Joshi, S. R., Parikh, R. M., & Das, A. K., (2007). Insulin-history, biochemistry, physiology and pharmacology. *J. Assoc. Physicians. India, 55*, 19–25.

10. Edmin, S. O., Dodson, G. G., & Cutfield, S. M., (1980). Role of zinc in insulin biosynthesis: Some possible zinc insulin interactions in the pancreatic beta cells. *Diabetologia., 19*, 174–182.

11. Peterson, G. E., (2006). Intermediate and long-acting insulins: A review of NPH insulin, insulin glargine and insulin detemir. *Curr. Med. Res. Opin., 22*. 2613–2619.

12. Lepore, M., Pampanelli, S., Fanelli, C., Porcellati, F., Bartocci, L., Di Vincenzo, A., et al., (2000). Pharmacokinetics and pharmacodynamics of subcutaneous injection of long-acting human insulin analog glargine, NPH insulin, and ultralente human insulin and continuous subcutaneous infusion of insulin lispro. *Diabetes, 49*, 2142–2148.

13. Jehle, P. M., Micheler, C., Jehle, D. R., Breitig, D., & Boehm, B. O., (1999). Inadequate suspension of neutral protamine Hagendorn (NPH) insulin in pens. *Lancet., 354*, 1604–1607.

14. Mooradian, A. D., Bernbaum, M., & Albert, S. G., (2006). Narrative review: A rational approach to starting insulin therapy. *Ann. Intern. Med., 145*, 125–134.

15. Borgono, C. A., & Zinman, B., (2012). Insulins: Past, present, and future. *Endocrinol. Metab. Clin. N. Am., 41*, 1–24.

16. Hallas- Moller, K., Peterson, K., & Schlitchkrull, J., (1952). Crystalline amorphous insulin-zinc compounds with prolonged action. *Science, 116*, 394–396.

17. Coscelli, C., Calabrese, G., Fedele, D., Pisu, E., Calderini, C., Bistoni, S., Lapolla, A., Mauri, M. G., Rossi, A., & Zappella, A., (1992). Use of premixed insulin among the elderly. Reduction of errors in patient preparation of mixtures. *Diabetes Care, 15*, 1628–1630.

18. Braak, E. W., Woodworth, J. R., Bianchi, R., Cerimele, B., Erkelens, D. W., Thijssen, J. H., & Kurtz, D., (1996). Injection site effects on the pharmacokinetics and glucodynamics of insulin lispro and regular insulin. *Diabetes Care, 19*, 1437–1440.

19. Owens, D. R., Zinman, B., & Bolli, G. B., (2001). Insulins today and beyond. *Lancet, 358*, 739–746.

20. Hirsch, I. B., (2005). Insulin analogues. *N. Engl. J. Med., 352*, 174–183.

21. Holleman, F., & Hoekstra, J. B., (1997). Insulin lispro. *N. Engl. J. Med., 337*, 176–183.

22. Mudaliar, S. R., Lindberg, F. A., Joyce, M., Beerdsen, P., Strange, P., Lin, A., & Henry, R. R., (1999). Insulin aspart (B28 asp-insulin): A fast-acting analog of human insulin: Absorption kinetics and action profile compared with regular human insulin in healthy nondiabetic subjects. *Diabetes Care, 22*, 1501–1506.

23. Becker, R. H., & Frick, A. D., (2008). Clinical pharmacokinetics and pharmacodynamics of insulin glulisine. *Clin. Pharmacokinet., 47*, 7–20.

24. Bott, U., Ebrahim, S., Hirschberger, S., & Skovlund, S. E., (2003). Effect of the rapid-acting insulin analogue insulin aspart on quality of life and treatment satisfaction in patients with Type 1 diabetes. *Diabet. Med., 20*, 626–634.

25. Miles, H. L., & Acerini, C. L., (2008). Insulin analog preparations and their use in children and adolescents with type 1 diabetes mellitus. *Paediatr. Drugs, 10*, 163–176.

26. Bolli, G. B., & Owens, D. R., (2000). Insulin glargine. *Lancet, 356*, 443–445.

27. Barlocco, D., (2003). Insulin detemir. Novo nordisk. *Curr. Opin. Investig. Drugs, 4*, 449–454.

28. Havelund, S., Plum, A., Ribel, U., Jonassen, I., Vølund, A., Markussen, J., et al., (2004). The mechanism of protraction of insulin detemir, a long-acting, acylated analog of human insulin. *Pharm. Res., 21*, 1498–1504.

29. Jonassen, I. B., Havelund, S., Ribel, U., Hoeg-Jensen, T., Steensgaard, D. B., Johansen, T., et al., (2010). Insulin degludec is a new generation ultra-long acting basal insulin with a unique mechanism of protraction based on multihexamer formation abstract. *Diabetes, 59*, A11.

30. Vaughn, D. E., Yocum, R. C., Muchmore, D. B., Sugarman, B. J., Vick, A. M., Bilinsky, I. P., et al., (2009). Accelerated pharmacokinetics and glucodynamics of prandial insulins injected with recombinant human hyaluronidase. *Diabetes Technol. Ther., 11*, 345–352.

31. Muchmore, D. B., & Vaughn, D. E., (2010). Review of the mechanism of action and clinical efficacy of recombinant human hyaluronidasecoadministration with current prandial insulin formulations. *J. Diabetes. Sci. Technol., 4*, 419–428.

32. Steiner, S., Hompesch, M., Pohl, R., Simms, P., Flacke, F., Mohr, T., et al., (2008). A novel insulin formulation with a more rapid onset of action. *Diabetologia, 51*, 1602–1606.

33. Heinemann, L., Nosek, L., Flacke, F., Albus, K., Krasner, A., Pichotta, P., et al., (2012). U-100, p H-Neutral formulation of VIAject((R)): Faster onset of action than insulin lispro in patients with type 1 diabetes. *Diabetes Obes. Metab., 14*(3), 222–227.

34. Simon, A. C., & De Vries, J. H., (2011). The future of basal insulin supplementation. *Diabetes Technol, Ther., 13*(1), 103–108.

35. Owens, D. R., (2011). Insulin preparations with prolonged effect. *Diabetes Technol. Ther., 13*(1), 5–14.

36. Chacra, A. R., Kipnes, M., Ilag, L. L., Sarwat, S., Giaconia, J., & Chan, J., (2010). Comparison of insulin lispro protamine suspension and insulin detemir in basal-bolus therapy in patients with Type 1 diabetes. *Diabet. Med., 27*, 563–569.

37. Fogelfeld, L., Dharmalingam, M., Robling, K., Jones, C., Swanson, D., & Jacober, S., (2010). A randomized, treat-to-target trial comparing insulin lispro protamine suspension and insulin detemir in insulin-naive patients with Type 2 diabetes. *Diabet. Med., 27*, 181–188.

38. Hinds, K. D., & Kim, S. W., (2002). Effects of PEG conjugation on insulin properties. *Adv. Drug. Deliv. Rev., 54*, 505–530.

39. Chan, Y. P., Meyrueix, R., Kravtzoff, R., Nicolas, F., & Lundstrom, K., (2007). Review on Medusa: A polymer-based sustained release technology for protein and peptide drugs. *Expert. Opin. Drug. Deliv., 4*, 441–451.

40. Hirose, T., (2016). Development of new basal insulin peglispro (LY2605541) ends in a disappointing result. *Diabetol. Int., 7*(1), 16–17.

41. Shah, R. B., Patel, M., Maahs, D. M., & Shah, V. N., (2016). Insulin delivery methods: Past, present and future. *Int. J. Pharma. Investig., 6*(1), 1–9.

42. Fry, A., (2012). Insulin delivery device technology 2012: Where are we after 90 years? *J. Diabetes. Sci. Technol., 6*, 947–953.

43. Selam. J. L., (2010). Evolution of diabetes insulin delivery devices. *J. Diabetes Sci. Technol., 4*, 505–513.

44. Burdick, P., Cooper, S., Horner, B., Cobry, E., Mc Fann, K., & Chase, H. P., (2009). Use of a subcutaneous injection port to improve glycemic control in children with type 1 diabetes. *Pediatr. Diabetes, 10*, 116–119.

45. Kreugel, G., Keers, J. C., Kerstens, M. N., & Wolffenbuttel, B. H., (2011). Randomized trial on the influence of the length of two insulin pen needles on glycemic control and patient preference in obese patients with diabetes. *Diabetes Technol. Ther., 13*, 737–741.

46. Aronson, R., Gibney, M. A., Oza, K., Berube, J., Kassler-Taub, K., & Hirsch, L., (2013). Insulin pen needles: Effects of extra-thin wall needle technology on preference, confidence, and other patient ratings. *Clin. Ther., 35*, 923–933.

47. Heinemann, L., Fleming, G. A., Petrie, J. R., Holl, R. W., Bergenstal, R. M., & Peters, A. L., (2015). Insulin pump risks and benefits: A clinical appraisal of pump safety standards, adverse event reporting and research needs. A joint statement of the European association for the study of diabetes and the American diabetes association diabetes technology working group. *Diabetologia, 58*(5), 862–870.

48. Polonsky, K. S., Given, B. D., Hirsch, L., Shapiro, E. T., Tillil, H., Beebe, C., et al., (1988). Quantitative study of insulin secretion and clearance in normal and obese subjects. *J. Clin. Invest., 81*, 435–441.

49. Kadish, A. H., (1963). A servomechanism for blood sugar control. *Biomed. Sci. Instrum., 1*, 171–176.

50. Skyler, J. S., Ponder, S, Kruger, D. F., Matheson, D., & Parkin, C. G., (2007). Is there a place for insulin pump therapy in your practice? *Clin. Diabetes, 25*, 50–56.

51. Moser, E. G., Morris, A. A., & Garg, S. K., (2012). Emerging diabetes therapies and technologies. *Diabetes Res Clin Pract., 97*, 16–26.

52. Brazg, R. L., Bailey, T. S., Garg, S., Buckingham, B. A., Slover, R. H., Klonoff, D. C. Nguyen, X., Shin, J., Welsh, J. B., & Lee, S. W., (2011). The ASPIRE study: Design and methods of an in-clinic crossover trial on the efficacy of automatic insulin pump suspension in exercise-induced hypoglycemia. *J. Diabetes Sci. Technol., 5*, 1466–1471.

53. Soares, S., & Costa, A., & Sarmento, B., (2012). Novel non-invasive methods of insulin delivery. *Expert. Opin. Drug Deliv., 9*, 1539–1558.

54. Gansslen, M., (1925). Uber inhalation von insulin. *Klin. Wochenschr., 4*, 71.

55. Heinemann, L., (2002). Alternative delivery routes: Inhaled insulin. *Diabetes Nutr. Metab., 15*, 41–422.

56. Patton, J. S., Bukar, J. G., & Eldon, M. A., (2004). Clinical pharmacokinetics and pharmacodynamics of inhaled insulin. *Clin, Pharmacokinet., 43*, 781–801.

57. Richardson, P. C., & Boss, A. H., (2007). Technosphere insulin technology. *Diabetes Technol. Ther., 9*(1), 65–72.

58. Neumiller, J. J., Campbell, R. K., & Wood, L. D., (2010). A review of inhaled technosphere insulin. *Ann. Pharmacother., 44*, 1231–1239.

59. Yaturu, S., (2013). Insulin therapies: Current and future trends at dawn. *World. J. Diabetes., 4*, 1–7.

60. Agarwal, V., Reddy, I. K., & Khan, M. A., (2000). Oral delivery of proteins: Effect of chicken and duck Ovomucoid on the Stability of Insulin in the presence of a-Chymotrypsin and Trypsin. *Pharm. Pharma. Com., 6*, 223–227.

61. Arbit, E., & Kidron, M., (2009). Oral insulin: The rationale for this approach and current developments. *J. Diabetes Sci. Technol., 3*, 562–567.

62. Ramesan, R. M., & Sharma, C. P., (2009). Challenges and advances in nanoparticle-based oral insulin delivery. *Expert. Rev. Med. Devices., 6*, 665–676.

63. Sonia, T. A., & Sharma, C. P., (2012). An overview of natural polymers for oral insulin delivery. *Drug Discov. Today, 17*, 784–792.

64. Chaturvedi, K., Ganguly, K., Nadagouda, M. N., & Aminabhavi, T. M., (2013). Polymeric hydrogels for oral insulin delivery. *J. Control. Release, 165*, 129–138.

65. Yamamoto, A., Taniguchi, T., Rikyuu, K., Tsuji, T., Fujita, T., Murakami, M., et al., (1994). Effects of various protease inhibitors on the intestinal absorption and degradation of insulin in rats. *Pharm. Res., 11*, 1496–1500.

66. Wong, C. Y., Martinez, J., & Dass, C. R., (2016). Oral delivery of insulin for treatment of diabetes: Status quo, challenges and opportunities. *J. Pharm. Pharmacol., 68*(9), 1093–1108.

67. Renukuntla, J., Vadlapudi, A. D., Patel, A., Boddu, S. H., & Mitra, A. K., (2013). Approaches for enhancing oral bioavailability of peptides and proteins. *Int. J. Pharm., 447*, 75–93.

68. Mesiha, M., Plakogiannis, F., & Vejosoth, S., (1994). Enhanced oral absorption of insulin from desolvated fatty acid-sodium glycocholate emulsions. *Int. J. Pharm., 111*, 213–216.

69. Salama, N. N., Eddington, N. D., & Fasano, A., (2006). Tight junction modulation and its relationship to drug delivery. *Adv. Drug. Deliv. Rev., 58*, 15–28.

70. Fasano, A., Fiorentini, C., Donelli, G., Uzzau, S., Kaper, J. B., Margaretten, K., et al., (1995). Zonulaoccludens toxin modulates tight junctions through protein kinase C-dependent actin reorganization, *in vitro*. *J. Clin. Invest., 96*, 710–720.

71. Whitehead, K., Karr, N., & Mitragotri, S., (2008). Safe and effective permeation enhancers for oral drug delivery. *Pharm. Res., 25*, 1782–1788.

72. Rahmani, V., Shams, K., & Rahmani, H., (2015). Nanoencapsulation of insulin using blends of biodegradable polymers and *in vitro* controlled release of insulin. *J. Chem. Eng. Pro. Technol., 6*, 1–8.

73. Chopra, S., Mahdi, S., Kaur, J., Iqbal, Z., Talegaonkar, S., & Ahmad, F. J., (2006). Advances and potential applications of chitosan derivatives as mucoadhesive biomaterials in modern drug delivery. *J. Pharm. Pharmacol., 58*, 1021–1032.

74. Maroni, A., Zema, L., Del Curto, M. D., Foppoli, A., & Gazzaniga, A., (2012). Oral colon delivery of insulin with the aid of functional adjuvants. *Adv. Drug Deliv. Rev., 64*, 540–556.

75. Heinemann, L., & Jacques, Y., (2009). Oral insulin and buccal insulin: A critical reappraisal. *J. Diabetes. Sci. Technol., 3*, 568–584.

76. Kumria, R., & Goomber, G., (2011). Emerging trends in insulin delivery: Buccal route. *J. Diabetol., 2*, 1–9.

77. Prausnitz, M. R., & Langer, R., (2008). Transdermal drug delivery. *Nat. Biotechnol., 26*, 1261–1268.

78. Kanikkannan, N., (2002). Iontophoresis-based transdermal delivery systems. *Bio Drugs., 16*, 339–347.

79. Rao, R., & Nanda, S., (2009). Sonophoresis: Recent advancements and future trends. *J. Pharm. Pharmacol., 61*, 689–705.

80. Andrews, S., Lee, J. W., Choi, S. O., & Prausnitz, M. R., (2011). Transdermal insulin delivery using microdermabrasion. *Pharm. Res., 28*, 2110–2118.

81. Charoo, N. A., Rahman, Z., Repka, M. A., & Murthy, S. N., (2010). Electroporation: An avenue for transdermal drug delivery. *Curr. Drug. Deliv., 7*, 125–136.

82. Malakar, J., Sen, S. O., Nayak, A. K., & Sen, K. K., (2012). Formulation, optimization and evaluation of transferosomal gel for transdermal insulin delivery. *Saudi Pharm. J., 20*, 355–363.

83. Freckmann, G., Pleus, S., Haug, C., Bitton, G., & Nagar, R., (2012). Increasing local blood flow by warming the application site: Beneficial effects on postprandial glycemic excursions. *J. Diabetes. Sci. Technol., 6*, 780–785.

84. Vaughn, D. E., & Muchmore, D. B., (2011). Use of recombinant human hyaluronidase to accelerate rapid insulin analogue absorption: Experience with subcutaneous injection and continuous infusion. *Endocr. Pract., 17*, 914–921.

85. Bariya, S. H., Gohel, M. C., Mehta, T. A., Sharma, O. P., (2012). Microneedles: An emerging transdermal drug delivery system. *J. Pharm. Pharmacol., 64*, 11–29.

86. Botz, C. K., Leibel, B. S., Zingg, W., Gander, R. E., & Albisser, A. M., (1976). Comparison of peripheral and portal routes of insulin infusion by a computer-controlled insulin infusion system (artificial endocrine pancreas). *Diabetes, 25*, 691–700.

87. Van Dijk P., (2016). *Intraperitoneal Insulin.* Diapedia the living textbook of diabetes. http://www.diapedia.org/type-1-diabetes-mellitus/2104588419/intraperitoneal-insulin. (Assessed 10 November.

87. Gin, H., Renard, E., Melki, V., Boivin, S., Schaepelynck-Belicar, P., Guerci, B., et al., (2003). Combined improvements in implantable pump technology and insulin stability allow safe and effective long term intraperitoneal insulin delivery in type 1 diabetic patients: The EVADIAC experience. *Diabetes Metab.*, *29*, 602–607.

88. Kumareswaran, K., Evans, M. L., & Hovorka, R., (2012). Closed-loop insulin delivery: Towards improved diabetes care. *Discov. Med.*, *13*, 159–170.

89. U.S. Food and Drug Administration, FDA approves first automated insulin delivery device for type 1 diabetes, FDA News Release (accessed March 08, 2017).

CHAPTER 9

EMERGING POTENTIAL OF *IN VITRO* DIAGNOSTIC DEVICES: APPLICATIONS AND CURRENT STATUS

SWARNALI DAS PAUL and GUNJAN JESWANI

Department of Pharmaceutics, Faculty of Pharmaceutical Sciences, Shri Shankaracharya Group of Institutions, SSTC, Bhilai, Chhattisgarh, India, Tel.: +91-9977258200, E-mail: swarnali34@gmail.com; swarnali4u@rediffmail.com

ABSTRACT

In today's era *in vitro* diagnostic devices play a crucial role in the diagnosis and treatment of any disease or disorder. They provide three-fold information on diagnostic, observing/monitoring, and compatibility purposes from human body specimens after their *in vitro* analysis or examination. In this chapter, we mainly focus on the diagnostic devices which are used for screening and testing for different diseases and their principle or applications. These diagnostic devices have the efficiency to determine the disease or condition of any human or animal body by using calibrators, reagents, software, specimen, receptacles, control materials, and similar instruments or equipments or other items. Rapid diagnostic tests are very popular IVD intended for use by individual person for instant results. Their prime advantage is that they have reduced the laboratory visit and related patient exercise. However, it should not be forgotten that these devices are an instrument so little bit of error in prediction can cause a life-threatening situation. Thus the responsibility of these diagnostic devices is very immense. In conclusion, it is expected that these devices should be used in the treatment and diagnosis to make it easier and better, but the treatment should not be blindly dependent on the result of this *in vitro* diagnostic devices for eliminating any risk of error.

9.1   INTRODUCTION

According to the definition of *In vitro* diagnostic/ medical device they are basically "a device, whether used alone or in combination, is specimen derived from the human body solely or principally to provide information for diagnostic, monitoring or compatibility purpose." They comprise of calibrators, reagents, software, specimen, receptacles, control materials, instruments or other items according to USFDA [1]. Here reagent signifies as biological, chemical, or immunological articles, proposed by the manufacturer. *In vitro* diagnostic (IVD) devices also include the genetic investigations which reveal information for healthcare decision making [2, 3]. An IVD may be either an entire test or a part of a test. Part of the test includes both non-diagnostic components, known as "general purpose reagents" (GPRs), and also the active element of the diagnostic test, called as "analyte specific reagent" (ASR). IVDs which are used in the clinical monitoring of patients also referred as a medical device. Thus, due to their critical requirements, they are subjected to regulation by the FDA.

IVDs have wide applications in the field of diagnosis, staging, screening and disease management. Today medical practices are solely dependent on the results of these diagnostic tests which affect more than 70% of healthcare resolution [4]. A data reveal in 2007 by the Centre of Disease control that 6.8 billion (approximately) clinical laboratory tests are performed annually in the USA which discloses the dependency of medical practice on this IVD's [5]. The manufacture of diagnostic kits and reagents in India is a relatively new segment of the healthcare industry. At present, India has about 25 companies manufacturing diagnostic kits [6].

FDA has described over 1700 different types of these devices under 18 therapeutic fields. The brief description is also specified for each category of devices, including intended use, class of device and data about sales requirements. Generally, IVDs can be of two types depending on their use, commercial or personal. "Rapid test devices" are used for personal or rapid diagnosis, whereas commercial devices include commercial test products and instruments, and are used for human samples laboratory analysis [7].

9.2   REGULATION AND CLASSIFICATIONS OF IVDS

The federal government has an important role to prevent the oversight of IVDs due to its potential risk and high impact on the health care. The Federal regulation of IVDs includes a number of federal agencies, like the Centers for Medicare & Medicaid Services (CMS), the Food and Drug Administration (FDA) and Therapeutic Goods Administration (TGA). All the effort of these regulatory bodies mainly focused on reliability, safety, accuracy, effectiveness and precision of IVDs [1]. However, enormous importance is also given to the quality of the testing laboratory, utilization of the final result of these tests in decision making and the fidelity of claims made by the manufacturer in marketing. Regulatory Authorities have the responsibility to act on concerned matters for a particular medical device. Essential principles of safety and performance of medical devices and labeling of medical devices should be required or requested by a regulatory authority, conformity assessment body, user or third party [8].

Classification of these *in vitro* diagnostic devices is quite difficult because a large variety and huge number of devices are in use. To classify these devices in different classes many factors have to be taken into account such as the biological effect, invasive nature, duration of use, contact with body's vital parts and the supply of energy [9, 10]. However, different regulatory bodies have classified these devices mainly according to their level of risk. In Australia, regulation of the medical devices is controlled by TGA [11]. In the USA, USFDA is the controlling authority for all the IVDs. They mainly classified the IVD in three classes: Class I, II and III. Almost 50% of IVDs are class I category, whereas 42% are class II and only 8% are class III [1, 12].

Device with a high-risk class required higher level of evaluation and monitoring. Determination of a particular class for any IVD is very difficult. Many IVD comes under more than one classification rule, highest risk classification is usually applied. Furthermore, indistinguishable devices with diverse diagnostic use are classified in different classes.

Global Harmonization Task Force, a charitable association of council from medical device Regulatory Authorities also presented the classification of the IVD Medical Device. This classification is based on factors including, proposed use and indication for use, skill needed of the proposed user, the importance of the information to the diagnosis and impact of the result to health [9]. Table 9.1 represents a general classification system for IVD Medical Devices.

TABLE 9.1   Proposed General Classification System for IVD Medical Devices

Class	Level of Risk		Device Examples
	Individual Health	Public Health	
A	Low	Low	Clinical Chemistry Analyser, prepared selective culture media
B	Moderate	Moderate and/or Low	Anti-Nuclear Antibody, Vitamin B_{12}, Pregnancy self-testing, Urine test strips
C	High	High and/or Moderate	Blood glucose self-testing, Rubella, PSA screening, HLA typing,
D	High	High	HIV Blood donor screening, HIV Blood Diagnostic

9.3   RAPID DIAGNOSTIC TESTS

Rapid diagnostic tests (RDTs) are one type of *in vitro* diagnostic procedure which can be performed under minimal resource settings and are intended for rapid test results. The global market for rapid medical diagnostic kits was $16.5 billion in 2011 and nearly $18.4 billion in 2012. This market is expected to reach $24.2 billion in 2017, at an estimated compound annual growth rate (CAGR) of 5.7% from 2012 to 2017 [13].

RDTs are already very popular for its simplicity, low cost, performance and less time taking methods. It offers many advantages, like numerous visits to diagnostic centers reduced, improved specificity of diagnosis, the practices of presumptive treatment reduced, reduced time reduces the risk of getting patient sicker. However, there are also many disadvantages of RDTs including misinterpretation of results with the patient, negative results or error in results, less sensitivity comparative of laboratory tests and lack of quality control mechanisms [14]. The major problem facing the production and implementation of RDTs is the shortage of a quality control procedure to establish their effectiveness. In a current instance, WHO stated that rapid Tuberculosis (TB) serological tests should not be used due to a lack of sensitivity and specificity. To improve the standards and commendation for use of RDTs WHO-TDR and FIND (Foundation for Innovative New Diagnostics) have lots-testing programs for RDTs for different diseases like malaria and HIV [15].

Widespread RDTs which are in use today are based on the principle of immunoassay which involves the interaction of a fixed reagent linked to specific types of detector (visible) with a patient sample. The fixed reagent may be either target antigen or antibody. Other technologies like nucleic-acid amplification is very costly due to very advanced technology involved and difficult to apply as a point of care test.

Now-a-days these rapid diagnostic tests are widely available commercially for detection of numerous diseases and conditions, including blood sugar in diabetic patients, hcg level in pregnancy, and strep throat [16]. Different FDA approved test kits are summarized in Table 9.2 according to their application. However, in the last six years (from 2010, January) more than 50 urine hCG test kit by visual color comparison tests devices have been approved by the FDA for home/prescription use. However, in those five years, only one serum HCG qualitative test kit is approved by FDA to International Newtech Development, Inc., In the same time frame, more than 370 glucose monitoring devices has been approved by FDA for home/prescription use [17].

There are total 11 records found for rapid home test kit for lutenizing hormone which determines ovulation efficiency in females in this duration. They work on the principle of visual color comparison. However, in last 6 years only one over the counter device is approved for detecting HIV antibodies. OraSure Technologies got approval for OraQuick, a Home HIV Test kit detected HIV antibodies from oral fluid in 2012. Recently in 2015, a rapid test device for Bloom syndrome (BLM) gene mutations got approval for Autosomal Recessive Carrier Screening Gene Mutation Detection System by 23andMe, Inc. [17]

9.3.1 BASIC PRINCIPLES OF RDTS

There are several different platforms commonly used to build rapid diagnostic tests. However, the diagnostic methods can be broadly classified into two: Direct methods and indirect methods. Direct methods are including microscopic, genome detection, and antigen detection. Indirect methods are included, serology screening of either IgG (immunoglobulin G) or IgM ((immunoglobulin M) type. Indirect methods are simple in use and

TABLE 9.2 Applications of Rapid Diagnostic Test Devices

Disease	Test	Description	Source
HIV	OraQuick Advance, Uni-Gold Recombigen, Clearview Stat-Pak and Complete, INSTI HIV, Determine HIV Chembio DPP	All these tests USFDA approved and are performed on whole blood from venipuncture or finger prick sample except for OraQuick Advance which also accept Oral swab	Source: http://emedi-cine.medscape.com/article/783434-overview#a5
Malaria	PfHRP2 tests, PfHRP2 and PMA test, pLDH test	Lateral flow technique is mostly used which is based on the presence of histidine-rich protein II (HRPII). These tests are largely qualitative and less helpful in endemic areas. PfHRP2 tests are a two-line test, whereas the other two tests are three lines tests.	http://www.malariasite.com/rdt/
Leishmaniasis	Crystal ® KA, DiaMed-IT LEISH, Kalazar Detect™, Signal® – KA, Onsite Leishmania Ab Rapid	Direct Agglutination Test is field gold standard. Use of a fixed antigen in this test means that a 'test of cure' for leishmaniasis is still lacking	Diagnostics Evaluation Series No.4, Visceral Leishmaniasis Rapid Diagnostic Test Performance by WHO
Oncocherciasis	Ov–16 rapid immunochro-matographic card test	Not yet available in the field, laboratory only Antibody-detection test, so unable to determine whether an infection is active	
Schistosomiasis	MAb-based urinary dipstick test	Only available for *S. haematobium*	
Lymphatic Filariasis	Circulating Filariasis Antigen (CFA) Immunochromatographic card test	Only detects *W. bancrofti* Potentially prohibitively expensive	

faster. However, direct methods are time taking and relatively complex [18, 19]. Table 9.3 describes different types of tests for their strength and weakness. The widely used methods are discussed below:

9.3.1.1 LATERAL FLOW TESTS

Lateral flow tests are the simple test which does not involve any equipment to perform the test. The test strip contains all of the reactants and detectors. In this method, the test sample is positioned into a sample well form where it migrates towards the zone of immobilized antigen or antibody. The sample reacts with the antigen or antibody within few seconds and a noticeable result appears.

9.3.1.2 FLOW-THROUGH TEST

This test requires lesser time as compared to lateral flow tests to predict the results. However, there is an additional washing and buffer step. Due to this step, portability and stability of this test reduces significantly.

9.3.1.3 AGGLUTINATION TEST

This test is also very simple. A visible clump formed as a result in this test. The clump forms due to binding of carrier particles with target analytes. Sometimes the reaction is not visible through naked eye and a microscope is needed. Moreover, the weak bonding of the carrier particles with analyte produces error(s) in the results.

9.3.1.4 TEST FOR DETECTING MULTIPLE ANTIGENS

To detect multiple antigens dipstick format RDTs have introduced. In this method a dipstick is introduced into a sample. Then the dipstick is washed and further incubated to prevent the non-specific analyte binding. These additional steps reduce their applicability in limited-resource point of care settings.

TABLE 9.3 Different Test Mechanism for Rapid Diagnostic Devices

Test principle	Strengths	Weaknesses	Applications
Lateral flow tests (immuno-chromatographic strip tests)	Rapid (5–15 min) Can be adapted for multiple sample types Easiest to use	Results are qualitative Less sensitive than an ELISA	Malaria RDTs, home pregnancy tests
Agglutination assays	Single-step Low cost per test Rapid results	Low sensitivity and ambiguous results in weak reactions Some tests require training and/or a microscope to read results Some cross-reactions can cause sensitivity problems	HIV latex agglutination assays, Leishmaniasis DAT
Flow-through	Very rapid (3–5 min)	Requires more training to perform than lateral flow Less sensitive at antigen detection compared to lateral flow	*E. coli* detection
Solid-phase ("dipstick" assays)	One strip can test for multiple parameters	Requires several intermediate steps, some training	HIV "comb" test
Microfluidic chips, immuno-sensors, "labs on a chip"	Rapid Requires no manipulation	Potentially prohibitively expensive	Largely hypothetical at this point in time

9.3.1.5   *MICROFLUIDICS OR "LABS ON A CHIP"*

Microfluidics is a rising area in the field of rapid diagnostic development. It appears as a portable cassette which contains both detectors and reactants and uses electrochemical sensors.

9.4   APPLICATION OF IVD IN MONITORING/DETECTING OF DIFFERENT DISEASES OR DISEASE CONDITIONS

In vitro diagnostic (IVD) devices have a wide variety of applications including diagnosis of diseases, screening/staging of diseases or disorders, and management of disease like help in selection and dosing of therapeutics. In this section, we attempt to compile information on IVDs which are successfully applied in major diseases or pathological conditions. Table 9.4 summarizes different approved diagnostic devices by FDA in last one year (out of more than two hundred).

Recent technological advancements have created rapid diagnostic test kit for detecting within a very short time period. However, lack of expertise and resources limit the performance of diagnostic kit. Conventional diagnostic methods are more reliable in this aspect, but time for predicting results and lack of ultra sensitivity are the limitations. To overcome the limitations of conventional methods and kits and to improve the sensitivity, reliability, applicability and consumer compliance different researchers are involved in the synthesis and application of different nanomaterials in the form of nanodiagnostic devices and also trying to develop different nanoanalytical techniques [20]. Nanodiagnostics include nanotechnology-based products like nanoparticle, nanoparticle-based immunoassays, biochips/microarrays, nanoscale visualization, nanoparticle-based nucleic acid diagnostics, nano-biosensors, nanoproteomic-based diagnostics, nano-machines, etc.

9.4.1   *APPLICATION OF IVD IN DETECTING CANCER*

Cancer is one of the deadliest diseases in the world which is spreading among the population very fast. However, the detection of cancer in its early phases can save the life of many. Researchers are facing challenges

TABLE 9.4 Approved Diagnostic Devices in Last One Year (Out of More Than Two Hundred)

S. No.	Analyte Specialty	Manufacturer	Test System	Analyte	Complexity
1	Bacteriology	Becton Dickinson and Company	BD BACTEC Peds Plus/F Culture Vials (Plastic) {mainly bacteria and yeast}	Aerobic &/or anaerobic organisms-unlimited sources	Moderate
2		Thermo Fisher Scientific	BRAHMS PCT sensitive KRYPTOR Assay	Procalcitonin (PCT)	High
3		BioFire Diagnostics	FilmArray 2.0 {For use with stool specimens collected in Cary Blair medium}; FilmArray Gastrointestinal (GI) Panel for use with the FilmArray 2.0	Gastrointestinal panel	Moderate
4	Endocrinology	Assure Tech	hCG Pregnancy Serum/Urine Combo Test (Test Strip)	HCG serum qualitative	Moderate
5		BTNX Inc.	Rapid Response hCG Pregnancy Test Cassette(Urine) (Cassette and strip)	Urine hCG by visual color comparison tests	Waived
6		—	NanoEnTek FREND System	Thyroxine free (FT4)	Moderate
7		Greenbrier International Inc.	Assured Pregnancy Test (Cassette), ASSURED THC One Step Marijuana Test Cassette)	Urine hCG by visual color comparison tests, Cannabinoids (THC)	Waived
8		Siemens Healthcare Diagnostics	Dimension Vista 1500	Testosterone	Moderate
9	General Chemistry	Carolina Liquid Chemistries	CLC720i Chemistry Analyzer	Glucose	Moderate

TABLE 9.4 (Continued)

S. No.	Analyte Specialty	Manufacturer	Test System	Analyte	Complexity
10		Abbott	Diabetes Care Inc. FreeStyle Precision Neo H Blood Glucose Monitoring System	Glucose monitoring devices (FDA cleared/home use)	Waived
11		Randox Laboratories Ltd.	Rx Daytona Plus Analyzer	Carbon dioxide total (CO_2)	Moderate
12		Sekisui Diagnostics PEI Inc	SK500 Clinical Chemistry System {Sekisui Diagnostics PEI Inc SEKURE LDL, HDLReagent}	LDL & HDL cholesterol	Moderate
13		Randox Laboratories Ltd.	Rx Daytona Plus Analyzer	Bilirubin direct	Moderate
14		ELITech Clinical Systems Selectra	ProM Analyzer	Lipase	Moderate
15		Infopia Eclipse	Blood Glucose Monitoring System	Glucose monitoring devices (FDA cleared/home use)	Waived
16	Toxicology / TDM	Germaine Laboratories Inc.	AimScreen Multi-Drug Urine Test Dip card	Amphetamines Cannabinoids (THC) Phencyclidine (PCP), Cocaine metabolites Methadone Methamphetamines Benzodiazepines Barbiturates Morphine Methylenedioxymethamphetamine (MDMA) Oxycodone Buprenorphine	Waived
17		Immunalysis Corporation	IMMTOX Clinical Chemistry Analyzer	Oxycodone	Moderate

TABLE 9.4 (Continued)

S. No.	Analyte Specialty	Manufacturer	Test System	Analyte	Complexity
18		Beckman Coulter AU680 Chemistry Analyzer	{Thermo Fisher Scientific Infinity Lithium Reagent}	Lithium	Moderate
19	Syphilis Serology	Gold Standard Diagnostics	AIX 1000 Rapid Plasma Reagin (RPR) Automated Test System (For serum only)	Treponemapallidum antibodies (includes reagin)	Moderate
20	Mycology	BIOMERIEUX INC.	VITEK 2 AST-YS {Caspofungin<=0.125 – >=8 ug/mL} (C. albicans C. glabrata C. guillier-mondii C. krusei C. parapsilosis C. tropicalis)	Yeast	High
21	Virology	Cepheid Gene Xpert Xpress System	{XpertFlu+RSV Xpress}	Respiratory viruses	Waived
22		BioFire Diagnostics with Nasopharyngeal Swabs)	FilmArray Torch {Respiratory Panel (viruses and bacteria)} (For use	Respiratory viruses	Moderate
23	Urinalysis	i-Health Inc.	AZO Test Strips Urinary Tract Infection Test	Urine qualitative dipstick nitrite, Leukocyte esterase urinary	Waived
24	General Immunology	Inova Diagnostics	HEp2 External EB Kits (IFA)	Anti-nuclear antibodies (ANA)	Moderate
25			NOVA Lite Monkey Oesophagus IFA Kit/Slides	Anti-endomysial antibodies (EMA)	Moderate
26			NOVA Lite Monkey Oesophagus IFA Kit/Slides (skin antibodies only)	Anti-skin (epidermal) antibodies	High

TABLE 9.4 *(Continued)*

S. No.	Analyte Specialty	Manufacturer	Test System	Analyte	Complexity
27			NOVA Lite Rat Liver Kidney Stomach Kit (IFA)	Anti-nuclear antibodies (ANA)	
28			NOVA Lite Rat Liver Kidney Stomach Kit (IFA)	Anti-parietal cell antibodies	
29			NOVA Lite Rat Liver Kidney Stomach Kit (IFA)	Anti-mitochondrial antibodies (AMTA)	
30			NOVA Lite Rat Liver Kidney Stomach Kit (IFA)	Anti-smooth muscle antibodies (ASMA)	
31			QUANTA Lite C1q CIC ELISA Kit	Complement C1Q	
32			QUANTA Lite HA dsDNA ELISA	Anti-DNA antibodies	
33			QUANTA Lite MPO SC ELISA	Anti-myeloperoxidase (MPO) antibodies	
34			QUANTA Lite Phosphatidylserine IgA ELISA	Anti-Phosphatidylserine Antibodies	
35			QUANTA Lite R h-fTG IgA ELISA	Anti-tissue transglutaminase (tTg)	
36		Bio-Flash Instrumentation System	{INOVA Diagnostics QUANTA Flash Jo-1}	Anti-Jo-1	Moderate
37		Fujirebio Diagnostics IncLumipulse G HE4 Immunoreaction Cartridges on LUMIPULSE G1200 Instrument}	Fujirebio LUMIPULSE G1200 SYSTEM	Cancer antigen human epididymis protein 4 (HE4)	Moderate

TABLE 9.4 *(Continued)*

S. No.	Analyte Specialty	Manufacturer	Test System	Analyte	Complexity
38	Hematology	Diagnostic Stago STAR Max	—	Factor X, Heparin, unfractionated heparin (UFH) and low molecular weight heparin (LMWH), Protein S free, Factor VII, Factor VIII, Factor IX, Factor XI, Factor XII von Willebrand factor, Antithrombin III (ATIII), Heparin unfractionated, Lupus anticoagulants, Protein S free, Alpha–2-antiplasmin, Plasminogen, Protein C, Protein S, Factor II, Factor V	High
39		DiagnosticaStago STAR Max	{STA Neoplastine CI & CI plus}	Prothrombin time (PT), Thrombin time, Activated partial thromboplastin time, (APTT), D-dimer, Thrombin time	Moderate
40		Instrumentation Laboratory (IL) Co.	ACL TOP 350 CTS {HemosIL APTT-SP}, {HemosIL Thrombin Time}, {HemosILReadiPlasTin}, {HemosIL Factor XII Deficient Plasma}	Activated partial thromboplastin time (APTT) Thrombin time Fibrinogen, Factor XII	Moderate

for developing methods of diagnosis for cancer, which can detect the type of disease in its early stage because cancer has different complex forms. Like tumor biomarkers get released in the body much time before its external indications. Thus predicting these biomarkers can ease the diagnosis of this disease. To increase the sensitivity, precision, time efficiency of these tests, researchers are incorporating nanotechnology into IVD systems. For this purpose different nanoparticles (paramagnetic nanoparticles, gold/PEG nanoparticles), nanoshells, nanotubes, quantum dots, dendrimers, micelles, etc. are currently in use. Many diagnostic companies are actively engaged in this research field for cancer biomarker [21].

Fujirebio Diagnostics, INC. is one of the best manufacturers of IVD for tumor marker assays, including pancreatic cancer, breast cancer, ovarian cancer, and other malignancies. FDA permits approval of "CYFRA 21–1″ test kit by Fujirebio's in 2011. The CYFRA 21–1 test kit that detects soluble cytokeratin 19 fragments in human serum. It has a wide scope in management of progressive disease and treatment of patients suffering from lung cancer. Another kit "KIT D816V mutational assay" helps in the selection of ASM patients who is under Gleevec® (imatinibmesylate) treatment. However, this assay method is restricted for professional use, which is to be performed under single laboratory site.

Fujirebio Diagnostics' HE4 assay is the second FDA cleared test in 2008 and is called as "The Risk of Ovarian Malignancy Algorithm" (ROMA™). It is a qualitative serum test which works by combining the results of HE4 EIA and ARCHITECT CA 125 II™. The ARCHITECT CA 125 II assay is a Chemiluminescent Microparticle Immunoassay (CMIA) for the quantitative determination of OC 125 defined antigen in human serum and plasma on the ARCHITECT i System. Premenopausal or postmenopausal woman who has an ovarian adnexal mass have a probability of tumor in surgery through the application of ROMA. Female of age more than 18 with the presence of an ovarian pelvic mass, and likelihood to have a surgery can undergo this test. ROMA must be interpreted in conjunction with an independent clinical and radiological assessment. This test is also restricted to clinical laboratories. List of other Cleared or Approved Companion Diagnostic Devices is given in Table 9.5.

TABLE 9.5 List of Cleared or Approved Companion Diagnostic Devices for Cancer Detection

S. No.	Drug Trade Name (Generic Name)	Device trade name	Device Manufacturer	Intended Use (IU)/ Indications for Use (IFU)	Type of cancer
	Imatinibmesylate	*KIT* D816V Mutation Detection by PCR for Gleevec	ARUP Laboratories, Inc.	IT is an *in vitro* diagnostic test intended for qualitative polymerase chain reaction (PCR) detection of KIT D816V mutational status from fresh bone marrow samples of patients	Aggressive systemic mastocytosis.
	Imatinibmesylate	*PDGFRB* FISH for Gleevec	ARUP Laboratories, Inc.	Intended for the qualitative detection of PDGFRB gene rearrangement from fresh bone marrow samples of patients	Myelodysplastic Syndrome/ Myeloproliferative Disease (MDS/ MPD)
	Zelboraf (vemurafenib)	COBAS 4800 BRAF V600 Mutation Test	Roche Molecular Systems, Inc.	Intended for the qualitative detection of the BRAF V600E mutation in DNA extracted from formalin-fixed paraffin-embedded human melanoma tissue. The Cobas 4800 BRAF V600 Mutation Test is a real-time PCR test on the Cobas 4800	Patients whose tumors carry the BRAF V600E mutation for treatment with vemurafenib.
	Xalkori (crizotinib)	VYSIS ALK Break Apart FISH Probe Kit	Abbott Molecular Inc.	Qualitative test to detect rearrangements involving the ALK gene via fluorescence in situ hybridization (FISH) in formalin-fixed paraffin-embedded (FFPE) non-small cell lung cancer (NSCLC) tissue specimens to aid in identifying patients eligible for treatment with Xalkori (crizotinib). This is for prescription use only.	Lung cancer

TABLE 9.5 (Continued)

S. No.	Drug Trade Name (Generic Name)	Device trade name	Device Manufacturer	Intended Use (IU)/ Indications for Use (IFU)	Type of cancer
	Xalkori (crizotinib)	VENTANA ALK (D5F3) CDx Assay	Ventana Medical Systems, Inc.	Qualitative detection of the anaplastic lymphoma kinase (ALK) protein in formalin-fixed paraffin-embedded (FFPE) non-small cell lung carcinoma (NSCLC) tissue stained with a BenchMark XT automated staining instrument.	Lung carcinoma
	Tarceva (erlotinib)	cobas EGFR Mutation Test	Roche Molecular Systems, Inc.	Qualitative detection of exon 19 deletions and exon 21 (L858R) substitution mutations of the epidermal growth factor receptor (EGFR) gene in DNA derived from formalin-fixed paraffin-embedded (FFPE) human non-small cell lung cancer (NSCLC) tumor tissue.	Lung cancer (NSCLC) tumor tissue.
	Tagrisso® (osimertinib)	cobas® EGFR Mutation Test v2	Roche Molecular Systems, Inc.	Qualitative detection of defined mutations of the epidermal growth factor receptor (EGFR) gene in DNA derived from formalin-fixed paraffin-embedded tumor tissue (FFPET) from non-small cell lung cancer (NSCLC) patients.	Non-small cell lung cancer
	Mekinist (trametenib); Tafinlar (dabrafenib)	THxID™ BRAF Kit	BioMérieux Inc.	Qualitative detection of the BRAF V600E and V600K mutations in DNA samples extracted from formalin-fixed paraffin-embedded (FFPE) human melanoma tissue.	Human melanoma tissue

TABLE 9.5 *(Continued)*

S. No.	Drug Trade Name (Generic Name)	Device trade name	Device Manufacturer	Intended Use (IU)/ Indications for Use (IFU)	Type of cancer
	Lynparza™ (olaparib)	BRACAnalysisCDx™	Myriad Genetic Laboratories, Inc.	Single nucleotide variants and small insertions and deletions (indels) are identified by polymerase chain reaction (PCR) and Sanger sequencing. Results of the test are used as an aid in identifying ovarian cancer patients with deleterious or suspected deleterious germline BRCA variants eligible for treatment with Lynparza™ (olaparib).	Ovarian cancer
	KEYTRUDA® (pembrolizumab)	PD-L1 IHC 22C3 pharmDx	Dako, North America, Inc.	Qualitative immunohistochemical assay using Monoclonal Mouse Anti-PD-L1, Clone 22C3 intended for use in the detection of PD-L1 protein in formalin-fixed paraffin-embedded (FFPE) non-small cell lung cancer (NSCLC) tissue using EnVision FLEX visualization system on Autostainer Link 48.	Non-small cell lung cancer
	Iressa (gefitinib)	therascreen® EGFR RGQ PCR Kit	Qiagen Manchester, Ltd.	The test is intended to be used to select patients with NSCLC for whom GILOTRIF® (afatinib) or IRESSA® (gefitinib), EGFR tyrosine kinase inhibitors (TKIs), is indicated.	Non-small cell lung cancer

TABLE 9.5 *(Continued)*

S. No.	Drug Trade Name (Generic Name)	Device trade name	Device Manufacturer	Intended Use (IU)/ Indications for Use (IFU)	Type of cancer
	Erbitux (cetuximab); Vectibix (panitumumab)	DAKO EGFR PharmDx Kit	Dako North America, Inc.	The EGFR pharmDx™ assay is a qualitative immunohistochemical (IHC) kit system to identify epidermal growth factor receptor (EGFR) expression in normal and neoplastic tissues routinely fixed for histological evaluation EGFR pharmDx specifically detects the EGFR (HER1) protein in EGFR-expressing cells.	Colorectal cancer

9.4.2   APPLICATION OF IVD FOR DETERMINATION OF INFECTIOUS AGENT

Detection of infectious agent is another great challenge for many diseases caused by pathogens or microorganisms. There are so many FDA approved conventional IVDs being explored for this purpose. Tests intended for the management of life-threatening infectious disease include viral load and genotyping assays for HIV, Hepatitis virus, malaria and other.

Example of BD BACTEC Peds Plus/F Culture Vials. BACTEC Peds Plus™/F culture vials (enriched Soybean-Casein Digest broth with CO_2) are for the aerobic blood cultures. Principal use is with the BACTEC fluorescent series instruments for the qualitative culture and recovery of aerobic microorganisms (mainly bacteria and yeast) from pediatrics and other blood specimens which are generally less than 3 mL in volume.

Procalcitonin (PCT) is a biomarker for the diagnosis of clinically relevant bacterial infections and sepsis. KRYPTOR®, an IVD based on a non-radiative transfer of energy by using TRACE (Time Resolved Amplified Cryptate Emission) technology for detecting PCT. The molecules of PCT (to be detected) are sandwiched between the antibodies.

Recently, electrochemical peptide-based (E-PB) sensors were also fabricated for detecting different disease causing organisms. Anti-HIV antibodies can be detected by this technique [22]. Numerous surface alteration approaches have been used to produce E-PB sensors. McQuistan *et al.* [23] stated E-PB sensors for detecting HIV in artificial human saliva using thiolated oligonucleotides. Saliva is extensively used as an analytical tool for hepatitis C virus and human immunodeficiency virus (HIV) [24].

To overcome the drawbacks of conventional IVDs several new technologies are also incorporated in this field like nanobiosensors, nanoprobes, nanorods, etc. Nano sensors can detect a single organism within 20-min [25]. Nano probes used independent hybridization reactions, multicolor nucleotides-functionalized quantum dots for detection of single-molecule [26]. This method has been employed for genetic examination of anthrax pathogen. SERS[1] technique (a spectroscopic assay) has been developed for rapid detection (60 s) of viruses by using silver nano-rods [27]. This novel technique even detects differences between viruses and viral strains.

---

[1] Surface enhanced Raman scattering.

9.4.3 APPLICATION OF IVD FOR DETECTING HORMONES

Hormonal inequity in a biological system is a specific indicator for onset of numerous diseases. Different test methods are there for detecting hormone level (estrogen, cortisol, testosterone and progesterone) in human serum or urine. However, rapid diagnostic test kits have also gained approval for this purpose. A suitable example for this is "The FREND TM Free T4 Test System," an IVD for the diagnosis of thyroid disorder. It quantitatively determines the level of free thyroxine (FT4). It is based on the principle of competitive immune assay (indirect). Researchers developed Cortisol biosensors for screening Cushing's syndrome. These IVDs are also-called immune sensors, as Ag-Ab interactions is the principle behind it. Many immune sensors also use enzyme for getting specific reaction results. Like, alkaline phosphatase (AP) enzyme [28], dithiobis (succinimidyl proprionate) were used to measure salivary cortisol concentration. Similarly salivary testosterone can be detected by an SPR biosensor in less than 13 min. Testosterone is the primary male sex hormone which also exists in women allowing the secretion of estrogen [29]. This device was developed by Mitchell and Lowe [30] by conjugating oligoethylene glycol with testosterone by forming a covalent bond and coated on the sensing surface by immobilization technique.

9.4.4 APPLICATION OF IVD IN DETECTING PREGNANCY

Positive or negative pregnancy condition is detected by identifying human chorionic gonadotropin (hCG), a placental hormone in urine or blood sample. The hCG level in non-pregnant females are less than 5.0 mIU/mL, reaching 5–50 mIU/mL three weeks after the last menstrual period, and 2,000 mIU/mL a month after conception. The rapid rise in serum levels of hCG after conception makes it an excellent marker for early confirmation and monitoring of pregnancy [31]. As per the definition by FDA, "a human hCG detecting device is a system envisioned for the early recognition of pregnancy intended to measure hCG, in plasma or urine." These devices are regulated as Class II devices (moderate risk) according to FDA and require 510 (k) clearances before marketing. A number of factors including linearity, precision, cut-off performance, accuracy, interference, and stability are reviewed for 510 (k) clearance.

Different hCG detection methods are including hCG ELISA kit, one step hCG test, and hCG Chemiluminescence Immunoassay (CLEIA) Kit, which are used to detect hCG [30]. In market mainly two kinds of pregnancy devices are available: qualitative and quantitative. Qualitative devices are used for analyzing urine samples (point-of-care or home) and serum sample (point-of-care or central lab) both. However, quantitative devices are only used for analyzing serum sample (point-of-care or central lab). Different test methods apply to these devices, including midstream, dip, droplet (cassette), etc.

Many pregnancy detecting devices are available in the market for commercial or personal use. Few of them are mentioned in Table 9.4. "On Site HCG Combo Rapid Test" is an example of one such device which is FDA approved. It follows the principle of immunoassay. Lateral flow chromatographic technique is used to perform the test and the result is concluded by visual color comparison with the cut-off level of 20 mIUhCG/mL of human serum or urine. This type of product can be categorized as a rapid test kit which only provides qualitative results.

9.4.5 APPLICATION OF IVD IN GLUCOSE MONITORING

Major achievement of IVD's has been gained by the invention of glucose monitoring devices. Glucose monitoring devices can be categorized into four types according to their principle of working. First generation devices were based on the amperometric detection of hydrogen peroxide which was expensive due to platinum electrode. The improvements were achieved with redox mediators, which are non-physiological electron acceptors. The electrons are carried to the working electrode from the enzymes easily by these mediators. Further specificity was attained by preparing reagent less biosensors based on direct transfer between the enzyme and the electrode without mediators. These are third generation glucose biosensors and preclude the disadvantage of toxic mediators.

Instead of mediators that have high toxicity, these electrodes can perform by direct electron transfer using organic conducting materials based on charge-transfer complexes. Uninterrupted *ex vivo* monitoring of blood glucose was first proposed in 1974 [32], whereas eight years later *in vivo* glucose monitoring was recognized [33]. Presently there are two types of continuous glucose monitoring systems in use. First one is a continuous subcutaneous glucose monitor and the second is a continuous blood glucose monitor. Non-invasive glucose monitoring devices are another challenge till date. The first US FDA approved transdermal glucose sensor, was manufactured by Cygnus, Inc. and is called GlucoWatch Biographer. This watch-like device was based on the transdermal extraction of interstitial fluid by reverse iontophoresis. However, many companies are currently engaged in manufacturing of these glucose monitoring devices, but 90% of the market consists of products manufactured by four major companies, including Abbott, Bayer, LifeScan, and Roche.

In 2013, Abbott introduced a new glucose monitoring technology called "Flash Glucose Monitoring." It was designed to collect continuous glucose data to permit generation of an innovative report by software called the "Ambulatory Glucose Profile." This technology was a good alternative to conventional glucose test instead of using traditional fingersticks. Table 9.6 represents information on differently recently approved glucose monitoring devices for home use by FDA.

TABLE 9.6   Recently Approved Glucose Monitoring Devices for Home Use by FDA

Document Name	Manufacturer	Test System	Effective Date
K153330	Abbott Diabetes Care Inc.,	FreeStyle Precision Neo H Blood Glucose Monitoring System	02/22/2016
K151611	LifeScan	OneTouch Ultra Plus Flex Blood Glucose Monitoring System	12/18/2015
K073492	Tyson Bioresearch, Inc.,	Gluco Dot Blood Glucose Monitoring System	12/18/2015
K142785	Prodigy Diabetes Care,	LLC PRODIGY iConnect Blood Glucose Monitoring System	12/17/2015
K151658	PhilosysGmate	Origin Blood Glucose Monitoring System	12/01/2015
K151265	SD Biosensor SD	Gluco NFC Blood Glucose Monitoring System	11/25/2015
K051285	Infopia Eclipse	Blood Glucose Monitoring System	11/17/2015
K132929	FirstVitals Health and Wellness Inc.	GlucoSec Mentor NFC Blood Glucose Monitoring System	11/09/2015
K150396	Apex BioTechnology Corp.	AutoSure Voice II Pro Blood Glucose Monitoring System	10/21/2015

9.4.6 MISCELLANEOUS

There are many more other applications of IVD. They are including tests for genetic disorder, enzyme screening tests, tests for hematology, tests for metabolism, etc. Tests intended for an inherited or acquired genetic marker, including: detecting the Philadelphia chromosome, prenatal genetic screening tests for Huntington's disease, cystic fibrosis and many more. IVDs are also helpful in monitoring biological components, including acute cardiac markers such as, Troponin T, Troponin I, and CKMB. Even congenital disorders can be detected by *in vitro* testing devices. Alpha fetoprotein (AFP) marker for fetal open neural tube defects is detected by IVD. Much software is embedded as a part of IVD. One such example is software for the analysis of results acquired in the first trimester to detecting fetal risk of trisomy. IVDs are also employed for performing coagulation testing, including factor assays, activated partial thromboplastin time (APTT) and prothrombin time testing. IVDs are also intended for therapeutic checking of the impact of incorrect use of immunosuppressive drugs such as, tacrolimus and cyclosporin, and for detecting their adverse transplantation consequence.

Intoxicology, IVDs have wide application for detecting the presence of different drug and its metabolites like cannabinoids, amphetamines, barbiturates, phencyclidine (PCP), methadone, cocaine metabolites, benzodiazepines, methamphetamines, morphine, etc. Germaine Laboratories Inc. has introduced "AimScreen Multi-Drug Urine Test Dip card" for detecting these compounds in urine.

9.5 PATENT INFORMATION

As discussed earlier *in vitro* diagnostic devices have numerous applications in different field. And till date, thousands of products got approved for marketing by FDA. However, researches are still going on with these devices to improve its sensitivity, applicability, reliability, and productivity. Many researchers have got a patent for their successful invention on these devices. Different patent information on these devices is summarized in Table 9.7 and Table 9.8.

TABLE 9.7 Patent on *In Vitro* Diagnostic Devices for Professional Use

S. No.	Title	Patent Application No.	Publication Date	Inventors
1.	Device And Method Of 3 Dimensionally Generating *In Vitro* Blood Vessels	US 20110244567 A1	6 Oct 2011	Noo Li Jeon, Ju Hun Yeon, Qing Ping Hu, Su Dong Kim, Hyun Jae Lee
2.	Fluidic Device And Perfusion System For *In Vitro* Complex Living Tissue Reconstruction	WO 2015084168 A1	11 Jun 2015	Mikhail Alexandrovich Ponomarenko
3	Process For Implementing *In Vitro* Spermatogenesis And Associated Device	EP 2886644 A1	24 Jun 2015	Marie-Hélène Perrard, Philippe Durand, Laurent David
4	Recipient Device And Method To Protect *In Vitro* Cultured Embryos And Cells Against Atmospheric Shock	WO 2009043131 A1	9 Apr 2009	Ricardo Pimentabertolla, Turco Edson Guimarães Lo, Christina Ramiresferreira, Virgilio Gustavo Da Silva,
5	Method And Apparatus For *In Vitro* Testing For Medical Devices	US 20110200976 A1	18 Aug 2011	Mari Hou, Joseph Junio
6	*In Vitro* Embryo Culture Device	WO 2002033047 A3	10 Oct 2002	Michael J Campbell, K C Fadem, Ronald J Thompson
7	*In Vitro* Metabolic Engineering On Microscale Devices	WO 2003038404 A3	4 Dec 2003	Jonathan S Dordick, Aravindsrinivasan, Jungbae Kim, David H Sherman, Douglas S Clark
8	Process For Implementing *In Vitro* Spermatogenesis And Associated Device	WO 2015092030 A1	25 Jun 2015	Marie-Hélène Perrard, Philippe Durand, Laurent David
9	*In Vitro* Device Monitoring During Minimally Invasive Ablation Therapy	US 20100331833 A1	30 Dec 2010	Michael Maschke, Gudrun Zahlmann, Sebastian Schmidt, Karstenhiltawsky
10	In-Vitro Device Support For X-Ray Based Kidney Function Test	US 20100332254 A1	30 Dec 2010	Michael Maschke, Sebastian Schmidt, Gudrunzahlmann

TABLE 9.7 *(Continued)*

S. No.	Title	Patent Application No.	Publication Date	Inventors
11	*In Vitro* Cell Culture Device Including Cartilage And Methods Of Using The Same	US 6465205 B2	15 Oct 2002	Wesley L. Hicks, Jr.
12	*In Vitro* Testing Of Endovascular Device	US 8978448 B2	17 Mar 2015	Michael V. Chobotov
13	Method And Device For The *In Vitro* Analysis For Mrna Of Genes Involved In Haematological Neoplasias	US 20120157329 A1	21 Jun 2012	Pilargiraldocastellano, Patricia Alvarez Cabeza, Miguel Pocovimieras
14	Device And Method For *In Vitro* Fertilisation.	WO 2014075153 A1	22 May 2014	Willem Ombelet
15	Device And Method For *In Vitro* Detection Of Blood	US 20020146834 A1	10 Oct 2002	Gavrielmeron, Itamarwillner
16	Process, Portable Equipment And Device For *In Vitro*, One-Step Photometric Determination Of Hemoglobin Concentration In A Diluted Blood Sample	WO 2009127024 A1	22 Oct 2009	Paulo Alberto Paes Gomes, Mauricio Marques De Oliveira, Jairribeirochagas
17	*In Vitro* Method, Use Of An Agent And Collection Device For The Inhibition Of Coagulation In Blood	EP 2772763 A1	3 Sep 2014	Emmanuel Prof. Dr. Bissé
18	Detection Device For The *In Vivo* And/Or *In Vitro* Enrichment Of Sample Material	US 20120237944 A1	20 Sep 2012	Klaus Lücke, Andreas Bollmann, Steffi Mewes, Robert Niestroj
19	Device And Method For *In Vitro* Detection Of Blood	WO 2001069212 A1	20 Sep 2001	Gavrielmeron, Itamarwillner
20	Device And Method For Creating A Vascular Graft *In Vitro*	US 6991628 B2	31 Jan 2006	Raymond P. Vito, Jack C. Griffis, Iii

TABLE 9.8 Patent on Rapid Diagnostic Test Devices

S. No.	Title	Application Date	Publication Date	Inventor
1.	Oral self-assessment device	US 20040071594 A1	2004	Sinead Malone, Michael York
2.	Glans compatible single unit semen collection and storage device,	US 20117947026 B2	24 May 2011	John C. Herr, ArabindaMandal
3.	Skin exfoliation devices and kits	WO 2013155146 A2	17 Oct 2013	Gordon Gerald Guay, David Edward Wilson
4.	Self-Inflating intragastric volume-occupying device	US 20040186502 A1	23 Sep 2004	Douglas Sampson, Michael Zanakis
5.	A single use, self-contained assay device for quantitative and qualitative measurements	WO 2006040106 A1	20 Apr 2006	Jakob EHRENSVÄRD, 4 More "
6.	Kits, apparatus and methods for magnetically coating medical devices	US 8465453 B2	18 Jun 2013	Gurpreet S. Sandhu, Robert D. Simari, Nicole P. Sandhu, Rajiv Gulati
7.	Kits, apparatus and methods for magnetically coating medical devices	WO 2005056073 A2	23 Jun 2005	Gurpreet S. Sandhu, Robert D. Simari, Nicole P. Sandhu, Rajiv Gulati
8.	Device for the early and rapid immunochromatographic detection of HIV	EP 1798556 A1	20 Jun 2007	Adnan Dr. Badwan, Murshed Abdel-Qader Murshed Mohammed, Tala SalehHamdan Dr. El-Taher
9.	Device, method and kit for *in vivo* detection of a biomarker	US 20110184293 A1	28 Jul 2011	Elisha Rabinovitz,
10.	Device, method and kit for *in vivo* detection of a biomarker	WO 2010004568 A1	14 Jan 2010	Elisha Rabinovitz,

TABLE 9.8 *(Continued)*

S. No.	Title	Application Date	Publication Date	Inventor
11.	*In vitro* protein translation micro-array device	US 20040043384 A1	4 Mar 2004	Andrew Oleinikov
12.	Device, kit and method for pulsing biological samples with an agent	US 7947450 B2	US 7947450 B2	Patrick Stordeur, Michel Goldman, Marius Tuijnder
13.	Immunoassay method, device, and test kit	US 4746631 A	24 May 1988	James A. Clagett
14.	Skin exfoliation devices and kits	US 20130274762 A1	17 Oct 2013	Gordon Gerald Guay, David Edward Wilson
15.	Diagnostic test device	US 3579306 A	18 May 1971	Margaret M Crane
16.	Immunoassay device	US 4116638 A	26 Sep 1978	Michael B. Kenoff
17.	Nano cancer barrier device(ncbd) to immobilize and inhibit the division of metastic cancer stem cells	WO 2011072482 A1	23 Jun 2011	Rutledge Ellis-Behnke, Mingtat Patrick Ling
18.	Implantable polymeric device	WO 20041110400 A2	23 Dec 2004	Rajesh A. Patel, Louis R. Bucalo, Lauren Costantini
19.	Self-contained device integrating nucleic acid extraction amplification and detection	US 5955351 A	21 Sep 1999	John C. Gerdes, Lynn D. Jankovsky, Diane L. Kozwich
20.	Self optimizing lancing device	US 7988645 B2	2 Aug 2011	Dominique Freeman, Don Alden

9.6 CONCLUSION

Diagnostics have had a tremendous impact on the management of patients with infectious diseases and are essential for outbreak detection and response, and public health surveillance. As we transition from conventional culture and antigen detection methods to newer molecular methods, the ability to provide accurate results in a clinically meaningful time frame, near the point of care, has never been greater. Yet despite these extraordinary technological advances, challenges remain. FDA approved or cleared molecular tests are available for a remarkably limited number of pathogens and generally diagnostics are notably underused. Some of the major issues that prevent broad adoption of diagnostic tests include the slow turnaround time, poor test performance characteristics, high complexity testing that cannot be easily adopted in many clinical settings, lack of understanding of the value of diagnostics, limited access to testing, and high cost. Today the diagnostics business is mainly based on technology. The goal is not just to create more tests, but to develop rapid, reliable, accurate, simple tests that will reduce time to a diagnosis and truly improve the quality of care and patient outcomes while reducing the healthcare costs.

KEYWORDS

- **CLIA**
- **diagnostic kit**
- **FDA**
- **medical devices**
- **nanodiagnostic**
- **rapid test kits**

REFERENCES

1. Regulation of Clinical Tests: In Vitro Diagnostic (IVD) Devices, Laboratory Developed Tests (LDTs), and Genetic Tests https://fas.org/sgp/crs/misc/R43438.pdf.
2. Medical device definition http://www.who.int/medical_devices/countries/regulations/mlt.pdf.
3. United Health Centre for Health Reform and Modernization, "Personalized Medicine: Trends and prospects for the new science of genetic testing and molecular diagnos-

tics," Working Paper 7, March (2012). http://www.unitedhealthgroup.com/~/media/uhg/pdf/2012/unh-working-paper-7.ashx.

4. The Lewin Group, "Laboratory medicine: A National State Report, " May (2008). p. 2.

5. Personalized Medical Devices http://www.imdrf.org/workitems/wi-pmd.asp

6. Narayan, K., (2005). Pharma/BioPharma: Promising future ahead for Indian diagnostics market. *Bio Spectrum Asia*: Pune.

7. Point-of-care diagnostic devices https://www.rivm.nl/bibliotheek/rapporten/360050025.pdf.

8. Elizabeth, M., Timothy, J. O., & Steven, I. G., (2005). "Food and drug administration regulation of *in vitro* diagnostic devices, " *Journal of Molecular Diagnostics, 7*(1), 2–7.

9. Dr. Kazunari A., (2012). Principles of medical devices classification. Endorsed by: The global harmonization task force.

10. Personalised medicine products: Evaluation of the regulatory framework https://www.rivm.nl/bibliotheek/rapporten/360211001.pdf.

11. Overview of the regulatory framework for in-vitro diagnostic medical devices https://www.tga.gov.au/overview-regulatory-framework-vitro-diagnostic-medical-devices.

12. Overview of Device Regulation https://www.fda.gov/medicaldevices/deviceregulationandguidance/overview/default.htm.

13. Global markets for rapid medical diagnostic kits. Analyst –– Peggy Lehr. Code – HLC007H.

14. Rapid Diagnostic Test http://www.who.int/malaria/areas/diagnosis/rapid-diagnostic-tests/generic_PfPan_training_manual_web.pdf.

15. Malaria Pipelines https://bvgh.org/neglected-disease-product-pipelines/malaria-pipelines/

16. Rapid Diagnostic Test (RDT) http://www.globalhealthprimer.emory.edu/targets-technologies/rapid-diagnostic-test.html.

17. *In Vitro* Diagnostics https://www.fda.gov/medicaldevices/productsandmedicalprocedures/invitrodiagnostics/default.htm.

18. Mabey, D., et al., (2004). "Diagnostics for the Developing World." *Nature Reviews Microbiology, 2.*

19. Diagnostics https://www.path.org/diagnostics/

20. Aniket, G., (2010). "Mycogenic metal nanoparticles: Progress and applications," Biotechnology Letters, 01/03/2010.

21. Cancer Biomarkers, (2013): Minimal and noninvasive early diagnosis and prognosis edited by Debmalya Barh, Angelo Carpi, Mukesh Verma, Mehmet Gunduz. CRC Press.

22. Gerasimov, J. Y., & Lai, R. Y., (2011). "Design and characterization of an electrochemical peptide-based sensor fabricated via "click" chemistry," *Chemical Communications, 47*(30), 8688–8690.

23. Mc Quistan, A., Zaitouna, A. J., Echeverria, E., & Lai, R. Y., (2014). "Use of thiolated oligonucleotides as anti-fouling diluents in electrochemical peptide-based sensors," *Chemical Communications, 50*(36), 4690–4692.

24. Radha, S. P. M., Sahba, S., Malarvili, B., & Emma, P. C., (2014). Saliva-based biosensors: Noninvasive monitoring tool for clinical diagnostics. *Bio. Med. Research International,* p. 20.

25. Liying, J., Lijie, R., Qinghua, C., & Guangzhao, C., (2010). "Glucose biosensors of anthropic saliva." *Micronanoelectronic Technology, 12.*

26. Yamaguchi, M., Kambe, S., Naitoh, K., Kamei, T., & Yoshida, H., (2002). "Gingival crevicular fluid-collecting devices for noninvasive blood glucose monitoring," In: *Proceedings of the 24th Annual Conference and the Annual Fall Meeting of the Biomedical Engineering Society EMBS/BMES Conference,* IEEE Xplore.

27. Yamaguchi, M., Kawabata, Y., Kambe S., et al., (2004). "Non-invasive monitoring of gingival crevicular fluid for estimation of blood glucose level." *Medical and Biological Engineering and Computing, 4*(3), 322–327.

28. Sun, K., Ramgir, N., & Bhansali, S., (2008). "An immune electrochemical sensor for salivary cortisol measurement." *Sensors and Actuators B. Chemical, 133*(2), 533–537.

29. Mahapatra DK, Asati V, Bharti SK. GnRH analogues as modulator of LH & FSH: Exploring clinical importance. In: Maamir S, Haghi AK, editors. Mechanical and Physicochemical Characteristics of Modified Materials. New Jersey: Apple Academic Press, 2015.

30. Mitchell, J. S., & Lowe, T. E., (2009). "Ultrasensitive detection of testosterone using conjugate linker technology in a nanoparticle-enhanced surface plasmon resonance biosensor." *Biosensors and Bioelectronics, 24*(7), 2177–2183.

31. *In Vitro* Diagnostic (IVD) Test Kits http://www.invitro-test.com/

32. Albisser, A. M., Leibel, B. S., Ewart, T. G., Davidovac, Z., Botz, C. K., Zingg, W., et al., (1974). Clinical control of diabetes by the artificial pancreas. *Diabetes, 23,* 397–404.

33. Shichiri, M., Kawamori, R., Yamasaki, Y., Hakui, N., & Abe, H., (1982). Wearable artificial endocrine pancrease with needle-type glucose sensor. *Lancet., 2,* 1129–1131.

INDEX